Ethics of the Sciences

Jan Hinrichsen · Kilian Bizer ·
Christoph Hönnige · Catrin Misselhorn ·
Thomas A. C. Reydon · Silke Schicktanz
Editors

Ethics of the Sciences

Shaping the Future of Academic Practices

palgrave
macmillan

Editors
Jan Hinrichsen
Medical Ethics and History of Medicine
University of Göttingen
Göttingen, Germany

Kilian Bizer
Chair of Economic Policy and SME Research
University of Göttingen
Göttingen, Germany

Christoph Hönnige
Institute of Political Science
Leibniz University Hannover
Hannover, Germany

Catrin Misselhorn
Department of Philosophy
University of Göttingen
Göttingen, Germany

Thomas A. C. Reydon
Institute of Philosophy
Leibniz University Hannover
Hannover, Germany

Silke Schicktanz
Medical Ethics and History of Medicine
University Medical Center Göttingen
Göttingen, Germany

ISBN 978-981-92-2820-1 ISBN 978-981-92-2821-8 (eBook)
https://doi.org/10.1007/978-981-92-2821-8

This Palgrave Macmillan imprint is published by the registered company Springer Nature Singapore Pte Ltd.
The registered company address is: 152 Beach Road, #21-01/04 Gateway East, Singapore 189721, Singapore

If disposing of this product, please recycle the paper.

Acknowledgments

This volume has been made possible through the support of numerous colleagues and several funding opportunities. Our special thanks go to Hans-Christian Kröger (Göttingen) for amazing support with the lay-out of the chapters, and to the Georg-August-University of Göttingen for funding the organization of the international conference *Ethics of the Sciences—Considering Technological, Political and societal Transformations*, which took place in Hannover in 2024 and which served as the starting point of this volume. This publication was supported by resources from the NiedersachsenOPEN Publication Fund, sponsored by *zukunft.niedersachsen*.

Contents

Notes on Contributors

Kilian Bizer holds the Chair of Economic Policy and SME Research at the University of Göttingen. He is also Director of the Institute for Small Business Economics at the University of Göttingen. He is working on sustainability and environmental issues as well as behavioral and institutional economics with a focus on policy measures and behavioral change.

Christoph Bleidorn received his Ph.D. in Biology from Freie Universität Berlin in 2004 and completed his habilitation in Zoology at Universität Leipzig in 2015. He is currently Professor of Animal Evolution and Biodiversity at Georg-August-Universität Göttingen and Scientific Director of Forum Wissen. His research focuses on biodiversity, phylogeny, and animal evolution, while his work at Forum Wissen also reflects his strong engagement in science communication.

Till Bruckermann is a professor specializing in teaching and learning research in non-formal education. His research focuses on the factors that foster individuals' learning processes and understanding of science, and the effects of participation in science and digital technologies in non-formal and informal learning contexts. His research projects address how participation can promote scientific reasoning and how inquiry-based learning can be supported. His recent publications focus on learning in citizen science projects.

Daniel Feser has been Professor of Risk and Sustainability Sciences at the Department of Social Sciences at the Hochschule Darmstadt (h_da) since 2024. He regularly conducts research projects on the intersection of innovation economics and transformative governance, with a particular focus on regulatory experiments, transition dynamics in the construction industry and regional innovation.

Simon Fink is Professor for the Political System of Germany at the University of Göttingen. He has studied political science, psychology, and law at the University of Konstanz and holds a Ph.D. from the University of Bamberg. His research interests include policy diffusion, research and technology policy, and Europeanization.

Marian Füssel is Professor for Early Modern History with special focus on the History of Science at the University of Göttingen. He studied history, philosophy, and sociology at the University of Münster. His main fields of expertise are early modern cultural history, history of knowledge, military history and history and theory. His latest monograph is *The World on Fire. A Global History of the Seven Years' War 1756-1763*, Columbia University Press (2025).

Catherine Herfeld is Professor of Philosophy and History of Economics at Leibniz University Hannover. Her research lies at the intersection of the history, methodology, and philosophy of economics. She is currently the Principal Investigator of the ERC Starting Grant project *MODEL TRANSFER: Model Transfer and Its Challenges for Science—The Case of Economics*, which combines qualitative and quantitative methods to investigate the historical trajectories and challenges of model transfer in economics.

Elisabeth Hildt is Professor of Philosophy and the Director of the Ethics Center at the Illinois Institute of Technology. She is interested in philosophical and ethical aspects of science and technology. Her research focuses on bioethics, ethics & technology, and ethical aspects of Artificial Intelligence.

Jan Hinrichsen is a cultural anthropologist of science and technology. He is currently Postdoctoral Researcher at the Department for Medical Ethics and History in Medicine, University Medical Center Göttingen, Germany. In his research he engages a praxiographic methodology and focuses on biotechnologies of immunity as well as on the ethical implications of digitalization in the health sector.

Christoph Hönnige is Professor for Comparative Politics and the Political System of Germany at the Institute of Political Science at Leibniz University Hannover. His research interests include judicial politics, legislative studies, and higher education research.

Nils Hoppe holds the Chair of Ethics and Law in the Life Sciences at Leibniz University Hannover and is Director of the Centre for Ethics and Law in the Life Sciences (CELLS). His research focuses on the intersection of health and research ethics, bioethics, and health law.

Friederike Knoke is a Research Associate at the Leibniz University Hannover, Institute for Legal Informatics (Prof. Dr. Margrit Seckelmann, Chair of Public Law and the Law of the Digital Society). From 2017–2020, she was also a Research Associate at the Leibniz Center for Science and Society at the Leibniz University Hannover. She was awarded the Preis für Wissenschaftsrecht 2023 for her Ph.D. thesis on Subjective Rights to Research Data.

Anna Kosmützky is a sociologist and Professor of Methodology in Higher Education and Science Research at the Leibniz Centre for Science and Society (LCSS) at Leibniz University Hannover, Germany. Her research expertise spans higher education studies, the sociology of science, and organization studies, with a particular emphasis on research methodology, especially in comparative research. Her work focuses on the organizational and institutional change of universities, (interdisciplinary) research collaboration, and the evolving landscape of research governance.

Daniel Kurzawe is working at the intersection of information science and humanities, focusing on digital research infrastructures. In this context, he examines the epistemology of computational knowledge production, combining approaches from philosophy of science and computer science. His work explores the impact of artificial intelligence on scientific practices, digital research methods, and digital sovereignty, with particular interest in how data-driven methods reshape knowledge systems, institutional roles, and the future of scholarly communication.

Simon Lohse is Assistant Professor of Philosophy of Science at Radboud University, focusing on pluralism in scientific policy advice and questions at the intersection of philosophy, ethics, and social studies of biomedicine.

He obtained his Ph.D. from Leibniz University Hannover, was a visiting Ph.D. researcher at the LSE (2013), and held postdoc positions in Hannover (2018–2020) and Lübeck (2021). Simon is currently working on a book project on the philosophy of scientific policy advice.

Eric M. Meslin is Adjunct Professor, Dalla Lana School of Public Health, University of Toronto; Senior Fellow, PHG Foundation, University of Cambridge; and Visiting Scholar, Centre of Genomics and Policy, McGill University. His work focuses on bioethics issues in health sciences with previous roles including President/CEO, Council of Canadian Academies; Executive Director U.S. National Bioethics Advisory Commission; Director ELSI Program, National Human Genome Research Institute; and Founding Director, Indiana University Center for Bioethics.

Catrin Misselhorn is Professor of Philosophy at the University of Göttingen since 2019. She is working on the philosophy of AI, robot, and machine ethics. Her books include *Grundfragen der Maschinenethik* (*Basic Issues in Machine Ethics*), Stuttgart 2018 (5th ed. 2022), *Künstliche Intelligenz und Empathie* (*Artificial Intelligence and Empathy*), Stuttgart 2021 (3rd ed. 2024), and *Künstliche Intelligenz—das Ende der Kunst?* (*Artificial Intelligence—the End of Art?*), Stuttgart 2023 (3rd ed. 2026).

Andreas Nehring is Professor for Science Education with a particular focus on chemistry education. His research focuses on nature of science/nature of scientific inquiry, on quality of instruction in science, on climate education, on learning analytics and AI in science education as well as on inclusivity as cross-sectional perspective within all of these fields.

Dominic Nyhuis is Professor for Quantitative Political Science Methods at the Institute of Political Science at Leibniz University Hannover. His research interests include legislative studies, party politics, as well as subnational politics.

Edoardo Peruzzi is a Postdoctoral Researcher at Leibniz University Hannover and a team member of the ERC project *MODEL TRANSFER*. His research focuses on the history and philosophy of economics. He currently investigates how model transfer contributes to scientific progress, taking the adoption of game-theoretic models in the social sciences as a case study. He also works on legal epistemology, with a focus on evidence law and expert testimony.

Thomas A. C. Reydon is Professor of Philosophy of Science and Technology in the Institute of Philosophy as well as the Centre for Ethics and Law in the Life Sciences (CELLS) at Leibniz Universität Hannover. He is also Associated Faculty in the Socially Engaged Philosophy of Science Group (SEPOS) at Michigan State University, and an elected Fellow of The Linnean Society of London. For more details about his academic work, see www.reydon.info.

Silke Schicktanz is Professor of Ethics and Cultural Studies of Biomedicine and director of the Department for Medical Ethics and History in Medicine, University Medical Center Göttingen, Germany. Her research focuses currently on ethical challenges of medical research and life sciences related to AI, big data, and risk predictions. Furthermore, she conducted various international research projects (funded by, e.g., DFG, EU, Volkswagen-Foundation, German-Israel-Foundation) on the cross-cultural analysis of bioethical discourses and how to develop further the democratization of science by public and stakeholder engagement.

Mark Schweda is a philosopher and bioethicist. He heads the Division of Ethics in Medicine at the Department of Health Services Research of the School for Medicine and Health Sciences at the University of Oldenburg (Germany). His work is concerned with ethical aspects of ageing, the life course, and human temporality, data and digital technologies in medicine and healthcare, as well as socio-cultural contexts of medicine and nursing care.

Margrit Seckelmann is Professor of Public Law and the Law of the Digital Society at Leibniz University Hannover. She completed promotion to Dr. iur. at the Goethe University Frankfurt in 2004, habilitated in 2015 at the Deutsche Universität für Verwaltungswissenschaften Speyer and was appointed an unscheduled Professor at the Deutsche Universität für Verwaltungswissenschaften Speyer in 2019.

Marc Stauch lectures law at the Leibniz University Hannover as well as at the Hochschule Bremen. He has published widely in the field of medical law and ethics.

Holmer Steinfath is Professor of Philosophy at the University of Göttingen. His research focuses on ancient philosophy, philosophical anthropology, and ethics. Currently, he investigates connections between time and good life as Principal Investigator of the Research Unit "Medicine and the Time Structure of the Good Life", funded by the German Research Council.

Philippe van Baßhuysen has a Ph.D. from the London School of Economics (2019). From 2019–2022, he was a postdoc at Leibniz University Hannover and from 2023–2024, an assistant professor at Wageningen University & Research. Since 2024, he has been professor of public health ethics at Leibniz University Hannover. He is the recipient of an ERC Starting Grant for his project "Managing Performative Science" (MAPS).

List of Figures

The Demand for Science in Politics: A Study of German Election Manifestos

Trust in Science in Modern Democracies: Determinants and Empirical Findings

Communicating Science in a Post-Truth Society

List of Tables

Ethics of Sciences in Times of Social and Technical Transformation: Introduction to the Volume

Silke Schicktanz, Christoph Hönnige, Thomas A. C. Reydon, Kilian Bizer, Catrin Misselhorn, and Jan Hinrichsen

1 Ethics of the Sciences: Shaping the Future of Academic Practices

The sciences in their broadness of disciplines, ranging from the humanities, social sciences to the natural and applied medical sciences, constitute an endeavor that is necessarily embedded in a social fabric. They are in constant interaction with societal conditions, needs, ethical and regulatory norms, and moral values. Universities and research institutions deliver findings and inventions, and educate students for society as a whole. The sciences can do this to a large extent in a self-regulated manner, but the state and the society fund research and teaching, and expect

S. Schicktanz (✉) · J. Hinrichsen
Department of Medical Ethics and History of Medicine, University Medical Center Göttingen, Georg-August-University of Göttingen, Göttingen, Germany
e-mail: sschick@gwdg.de

J. Hinrichsen
e-mail: jan.hinrichsen@med.uni-goettingen.de

C. Hönnige
Institute of Political Science, Leibniz University Hannover, Hanover, Germany
e-mail: c.hoennige@ipw.uni-hannover.de

J. Hinrichsen et al. (eds.), *Ethics of the Sciences*,
https://doi.org/10.1007/978-981-92-2821-8_1

solutions for the problems of society from the sciences. Accordingly, the sciences can transcend their boundaries with society by explicitly taking into account the needs of society. Conversely, there is often only a minimal level of understanding of scientific processes and products in society. The sciences are thus under increasing pressure to communicate their internal decision-making processes and findings in a way that non-scientists can understand.

This became most apparent during the COVID-19 pandemic. Established, but often slow, processes of knowledge generation collided with urgent invocation of expertise demanding reliable answers in insecure circumstances. This nurtured distrust in science and called the authority of scientific expertise into question. Populist politicians across the world opposed and ignored scientific advice on COVID-19 and undermined public trust in science. Mede and Schäfer (2020) termed this development "science-related populism." At the other extreme, the claim by climate activists that they are "following science" has been problematized as "scienticism" or "expertocratism" (Strohschneider, 2020), i.e., as ignoring the need for normative assessment and democratic deliberation of scientific recommendations within the process of social and political decision making. This decrease of trust and the questioning of the role of science constitute a serious problem, as the autonomy of scientific inquiry is an essential guarantee of modern and stable democracies (Ezrahi, 1990).

In this regard, the present volume, *Ethics of the Sciences: Shaping the Future of Academic Practices*, aims to illuminate a question which is vital

---

T. A. C. Reydon
Institute of Philosophy and Centre for Ethics and Law in the lIfe Sciences (CELLS), Leibniz University Hannover, Hanover, Germany
e-mail: thomas.reydon@cells.uni-hannover.de

K. Bizer
Faculty of Business and Economics, Georg-August-Universität Göttingen, Göttingen, Germany
e-mail: bizer@wiwi.uni-goettingen.de

C. Misselhorn
Philosophisches Seminar, Georg-August-Universität Göttingen, Göttingen, Germany
e-mail: catrin.misselhorn@uni-goettingen.de

to the understanding the ethical relationship between the sciences, politics, and society: How do the sciences respect the differences between the scientific and the political and societal spheres and still contribute to the needs and problems of society? In this volume, we argue that the answer lies in a nuanced, procedural and participatory "Ethics of the Sciences," as we call it—that is established first and foremost in a continuous interdisciplinary exchange—as a guiding principle for better science in society. This volume sketches the outlines of such a novel Ethics of the Sciences by means of contributions from a normative-ethical standpoint, but also from an empirical-analytical perspective. The contributors to this volume address the relation between science and society as determined largely by two normative dimensions, both of which fall within the domain of ethics, but are often discussed separately. On the one hand, there is an ethical dimension internal to the sciences regarding the question of how to define "scientific" and what "better" science could be. In this dimension, scholars are primarily concerned with "epistemic norms" for ensuring the integrity, replicability, and reliability of scientific practices. These internal epistemic norms belong to the basic core of academic practices and are often framed as "good scientific practices." However, there are various tensions in this domain. One challenge is that the increasing competition in the science system on multiple levels, not only between individuals, but also between teams, universities, and states focuses not only on publications and external funds, but also on media visibility and impact on policies (Krücken, 2021). This competition creates various wanted and unwanted additional behavioral incentives for individual scientists as well as whole institutions. It raises the overall question of how much ethical, legal, and social implications should be included in the research in different fields (Meslin et al., 1997).

Another question arises as to whether and how to balance the aim of, on the one hand, a politically neutral science and, on the other hand, academic activism, where scientists aim to deliberately influence politics and society (Nørgård & Bengtsen, 2021). This points in the direction of the external dimension: this dimension concerns the ethical context that is external to the sciences and encompasses societal views and values, for example, regarding the purpose and objectives of science. Here, the desirability and the possible benefits but also adverse consequences of new scientific findings and technologies are discussed. Results from the science process reach out to the societal, political and economic spheres. Ethical questions arise when advice is given: For example, how to deal with incomplete, uncertain information or predictive models, which, of course, include error rates that most citizens and academics from other fields cannot estimate well (Douglas, 2009). Vice versa, questions arise on the demand for research by politics and society (Boswell, 2008): funding for research or advice being asked is often bound to particular expectations regarding outcomes. This structural dependency can harm the actual freedom of research of scientists and, moreover, question the legitimacy of research or advice. Discussing aspects of demand is of the utmost importance, as it is directly connected to the trust in science. And trust is relevant, as it determines the acceptance of science by citizens (Cologna et al., 2025).

These two normative dimensions are both in themselves dynamic and closely interrelated, which is an aspect that an Ethics of the Sciences has to account for. The matter of science education and science communication lies at the intersection of the internal and external dimensions of the ethics of science. Science communication is increasingly the first essential bridge between the spheres of the internal and external normativity, as scientists are increasingly under pressure to actively communicate their findings and show the impact of their research. Science communication is increasingly becoming its own field (Gascoigne et al., 2010). The second bridge is science education at school, where teachers face ethical challenges in an increasingly science-sceptic and pluralist world (Osborne et al., 2003).

2 The Sciences in Social and Technical Transformation

The need for an interdisciplinary inquiry into the Ethics of the Sciences, its foundations, and its challenges, but also its translatability into other disciplinary contexts and beyond the academic scope, becomes ever more salient against the background of three fundamental social transformation processes that we identify as crucially challenging the sciences and the nature of their relation with society.

Firstly, the increasing significance of digital communication and the increasing availability and productivity of artificial intelligence (AI) applications are in the process of changing everyday life in an unprecedented manner. We term this development digitalization. Secondly, in light of global emergencies such as pandemics and climate change, scientific practice undergoes a shift in its temporal dimensions and trajectories that we identify as futurization: the drastically increasing political (and societal) demands to ever more predictive forms of knowledge production, such as by simulations, scenario building, and future modelling. Thirdly, in a globalized world, societies are confronted with a pluralization of world-views, values, and norms that also affect different norms of and attitudes towards academic practice, ranging from differing scientific standards to the necessary decentralization of Eurocentric epistemology, to outright skepticism about the knowledge produced by the sciences.

These three transformation processes are interlinked—they reinforce each other, resulting in what we take to be a fundamental transformation not only to the scientific system and its practices, but first and foremost to the relationship between the sciences and the societies in which they are embedded. The three transformations thus affect the internal as well as the external normativity of the sciences. In all transformation processes, questions of scientific integrity are as much problematized as those of trust, transparency, and responsibility. Therefore, in its current form, Ethics of the Sciences should be committed to the fact that future academic practice needs to be able to respond profoundly to these challenges and changes brought about by them, and thus requires sound and reliable frameworks, tools, and methodologies for ethical reflection to draw from in light of an uncertain future to come. The contributions in the present volume take up these three transformations and address them from either the perspective of the internal or external normativity.

3 An Interdisciplinary and Transdisciplinary, Comprehensive Approach to an Ethics of the Sciences

This volume emphasizes the interdisciplinary and transdisciplinary collaboration necessary to face future developments and challenges. While being aimed at fostering transdisciplinary exchange between the academic and non-academic sphere on the themes, questions, and methods of an Ethics of the Sciences, the volume is simultaneously a first result of an exchange beyond disciplinary containers, ethical domains, or methodological schools. The volume assembles expertise from a multitude of academic disciplines: applied moral philosophy, philosophy of science, bioethics (embracing medical ethics, ethics of life sciences and public health), science and technology studies, legal studies, political science, sociology, economics, cultural anthropology, science education, history, computer and data science, and biology. This allows us to enter into a collaborative approach to ethical reasoning and reflection in and on the sciences.

With this interdisciplinary group, the volume aims to achieve three things. The first of these is to reach beyond disciplinary containers in which ethical reflection usually takes place. Ethics of the Sciences rests on the view that ethical reflection needs not per se and not necessarily be confined to specific disciplinary perspectives (e.g., ethics of sociology or ethics of biomedicine). On the contrary, and in light of the all-encompassing nature of transformations to the academic system, future academic practice is dependent on interdisciplinary and transdisciplinary approaches, especially regarding the ethical reflection of the relation between society and the sciences.

Secondly, ethics is typically regarded as a mere additional layer of academic practice, as at least implicitly the concept of ELSA/ELSI (Ethical, Legal and Social Aspects/Implications) research suggests (see also the chapters of Reydon and Schicktanz, and Meslin in the first section). In contrast, Ethics of the Sciences is based on the conviction that ethics is ill-perceived as a detached domain or additional layer to knowledge production. We thus posit a comprehensive and necessarily interdisciplinary approach that presupposes scientific practice to be inherently and fundamentally imbued with ethics. Ethical issues are, we claim, at the heart of scientific practices, and thus need to be addressed, reflected, and accounted for as a foundational feature of knowledge and its production, dissemination, and implementation.

Following from this, Ethics of the Sciences reaches beyond research ethics. While research ethics is an integral part of Ethics of the Sciences, the two are not the same. Research ethics is often limited to codes of conduct and frameworks for research practice, whereas Ethics of the Sciences aims at a comprehensive perspective on the relation of the science to society and the reflection of norms internal (e.g., academic values and norms such as integrity, transparency, and reproducibility) and external (e.g., societal values and norms of responsibility, reliability, and trust) to scientific practice, as well as their intricate interplay.

4 Content and Structure of the Volume

The volume is structured into five parts corresponding to our theoretical considerations laid out above. *Part 1: Ethics and Science—The Lay of the Land* introduces the concept of the ethics of science and lays out the conceptual orientation of the entire volume. *Part 2: The Internal Ethics of Science* focuses on the internal epistemic norms, which are designed to ensure the quality, integrity, replicability, and reliability of scientific procedures and outcomes. The chapters in this part also address digitalization and pluralization. *Part 3: Science in the Societal and Political Sphere* leaves the internal perspective and focuses on the external perspective of the relationship between science, politics, and society, especially the demand for science in the context of pluralization and futurization, but also its acceptance. *Part 4: Ethics in Science Communication and Education* focuses on the intersection between the internal and external perspectives as becoming prominent in the current discussion over digitalization. The effects of science depend on trust and acceptance of processes and findings in society, and thus on communication of science and society, and the integration of science in the educational system. *Part 5: Social and Technological Transformations* outlines all three transformations as examples: The impact of AI will be fundamental and all-encompassing; the climate crisis is not an issue of a distant future; and aging in Western science raises questions of justice and injustice in the delivery of medical assistance. In what follows, we provide a more detailed overview of each of these five parts.

Part 1: Ethics and Science—The Lay of the Land provides an overview of the lay of the land for the topic of ethics of science in three chapters. It outlines the theoretical concept of the volume and provides an historical overview of the development of the ethical, legal, and social implications (ELSI) funding program that firstly covered the case of genomics but now has spread into so many fields of sciences. The chapters in this part also discuss the implications of multiple competition in the science system for the ethics of the sciences.

The opening chapter by Thomas Reydon and Silke Schicktanz explores the broad area of ethical reflection on the production of knowledge in the sciences as well as on scientific practice. This chapter presents the theoretical framework of the volume. It explores the different dimensions of interaction between ethics as an academic discipline and the various sciences. Existing approaches use different prepositions to indicate the various ways in which ethics and science can be related: as ethics of science, ethics and science, ethics in science, and ethics for science. The chapter provides an overview of these different modes of interaction. It addresses the overarching question of where ethical reflection should be located in relation to scientific practice. The authors examine the internal and external normativity of the sciences—i.e., standards of good scientific practice and academic integrity, and ethical assessment of the impact of science on society and the interplay of societal values and scientific practice, respectively. They discuss approaches to resolving the relationship between ethics and science in a variety of ways, including applied bioethics education for life scientists, formal research ethics guidelines, the role of institutional review boards (IRB) and research ethics committees, and the role of moral philosophers in interdisciplinary research projects. Each approach represents a different understanding of the interaction between ethics and the sciences, and each has its advantages and disadvantages. Based on this assessment, the authors argue for an integrative program that can overcome the limitations set by conceptual differences and differences in styles of argumentation between moral philosophers and researchers in other academic domains. This integrative program involves a continuous exchange between stakeholders on the epistemic and normative dimensions of the topics in focus in their research.

The chapter by Eric Meslin charts the origins, architecture, impact, and futures of the ELSI philosophy in genomics. He shows ELSI was "designed in" to the Human Genome Project, grounded in anticipatory governance, public engagement, and policy option development, and later codified in U.S. law. Its signature innovation was a fixed budget set-aside (ultimately 5%) for ELSI research within NIH's genome program, with peer-reviewed grants. ELSI priorities co-evolved with scientific milestones and expanded post-sequence under "Genomics and Society." The program institutionalized multidisciplinary methods across ethics, law, social science, and economics. Meslin documents global diffusion (ELSA in Europe, GE3LS in Canada, and initiatives in Asia, Africa, and Latin America) and policy influence, from genetic non-discrimination and human-subject rules to the United Nations Educational, Scientific and Cultural Organization (UNESCO)'s genome declaration, GA4GH data-sharing standards, and ELSI services embedded in research infrastructures. Facing technological and political disruption, Meslin sketches two scenarios: a politically driven "end of ELSI" via funding cuts and anti-expertise, and a more plausible diversification-through-convergence in which ELSI principles guide AI, biodigital convergence, and biodiversity initiatives. He concludes that ELSI's designed-in model—anticipatory, well-funded, and multidisciplinary—offers a durable template for aligning innovation with societal values.

In her chapter, Anna Kosmützky focuses on the relationship between the ethics of sciences and multiple competition in the scientific arena. Her chapter is conceptual in nature. Competition has long been understood as a key mechanism in the self-organization of science. Traditionally, she argues, research has emphasized a reputation-driven competition within scientific communities, but contemporary higher education and research systems are increasingly shaped by multiple and overlapping competitions—for publications, third-party funding, positions, impact, visibility, and institutional prestige. The chapter conceptualizes multiple competition as a structural condition of academic work and analyzes how it transforms the environments in which epistemic values are pursued, responsibilities are allocated, and evaluative standards are negotiated. The multiple competition for publications, third-party funding, positions, impact, visibility, and institutional prestige, of course, puts pressure on scientists, but also universities and the state financing the universities, and thus creates behavioral incentives. The chapter develops an interdisciplinary dialogue between the sociology of science and the Ethics of the Sciences. It shows how multiple competitions create normative tensions, diffuse responsibility across institutional layers, and generate forms of normative pluralism that challenge established ethical frameworks. Kosmützky argues for a reflexive and procedural Ethics of the Sciences that relates to the organizational and systemic conditions under which knowledge generation takes place. She does not consider ethical standards as an external oversight mechanism. In bringing both perspectives together, she demonstrates how analyses of multiple competition and ethical reflection can mutually enrich one another.

Part 2: The Internal Ethics of Science covers the internal dimension of the Ethics of the Sciences in four chapters. It discusses current aspects of digitalization and pluralization in topics such as academic activism, digital sovereignty, data protection in research data management, and the transfer of theoretical models in interdisciplinary research.

The chapter by Marian Füssel examines academic activism as an ethical challenge through a historical sociology of knowledge, arguing that historicizing the shifting meanings of "freedom" clarifies today's tensions between university autonomy and activist claims. Füssel distinguishes activism aimed at society-wide change from intra-institutional reform and contrasts the roles of "intellectuals" and "activists," highlighting how campus "free spaces" simultaneously enable protest and safeguard academic inquiry. Focusing on Germany's *longue durée*, he sketches key phases of student activism: liberal-national mobilization (1800–1866); rightward shifts and antisemitism (1867–1917); Weimar crisis; Nazi coordination and repression; postwar ambivalence leading to the 1968 transnational wave and its fragmentation; and recent diversification shaped by identity politics and digital contention. Activism's tactical, time-limited practices (occupations, demonstrations, pamphleteering) are set against the university's structurally "passivist," defensive reforms; space (campus vs. city) and time (rapid student turnover vs. professorial continuity) shape dynamics, learning, and escalation. Counter-movements (e.g., Bund/Netzwerk Freiheit der Wissenschaft) illustrate a dialectic in which opposing sides invoke "freedom" to police speaking positions. The chapter concludes that an ethics of science should analyze what activism does—its practices, forums, and evolving semantics—rather than prescribe abstractions, using historicization to make implicit normative commitments explicit and align academic responsibilities with democratic pluralism.

In his chapter, Daniel Kurzawe examines challenges and opportunities through Large Language Models in knowledge production. Large Language Models (LLMs) are reshaping scientific practice by embedding digital tools across all phases of scientific knowledge production and thus shifting cognitive labor to automated systems. Kurzawe shows the challenges that AI poses in terms of how academic institutions must shape and reshape their practices of knowledge generation in the future. He argues that cloud-based infrastructures not only archive and supply but also transform data (especially through blurred boundaries between sources, metadata, and analytical products). LLMs synthesize heterogeneous data via probabilistic token relations in high-dimensional spaces, enabling fluent outputs while displacing provenance, transparency, and interpretability. They decouple scientific information from centralized knowledge institutions and increase scientific dependence on opaque, commercially controlled platforms. This raises ethical and epistemic concerns around bias, accountability, hallucinations, and energy consumption. The discussion focuses on the direct influence the service providers have on the visibility and findability of information and, furthermore, illustrates how the use of LLMs pose problems on the question of authorship and ownership. In response, the author positions libraries and research infrastructures as key counterbalances that can ensure traceability, curatorial framing, and high-quality training data. The chapter concludes with a call to embed guiding frameworks (CARE and FAIR) which help build transparent, accountable, open infrastructures to realize AI's benefits while preserving epistemic agency.

The chapter by Friederike Knoke and Margrit Seckelmann discusses ethical aspects of science in relation to data use and data protection, in particular focusing on challenges that arise from existing regulations that apply to data management in research projects. Knoke and Seckelmann explore the conflicts that may arise between the interests of researchers who work with personal data in their projects and the (sometimes very restrictive) data protection regulations that are in force in the country or countries in which a project is carried out. They show that the interplay of different regulations that apply to a research project, in particular in cases in which multiple local regulations apply, as well as the vagueness of legal terms and concepts, often lies at the basis of researchers' feeling that meeting the applicable regulations poses an enormously complex task. This feeling is enhanced by a perceived clash between certain scientific values, such as openness and transparency, and the restrictions that are imposed by applicable regulations. Knoke and Seckelmann suggest that further concretization and interpretation of regulations, an increased interest among the actors involved in building data infrastructures for the interests of researchers and research subjects, and in particular expert advice to researchers by legal experts are key to meeting the challenges that researchers face in this context.

The chapter of Catherine Herfeld and Edoardo Peruzzi discusses from a conceptual-analytic perspective the issue of model transfer as a form of interdisciplinary research. Science is often highly specialized and thus suffers from siloization of the different fields, both in terms of career paths and organizational units. Interdisciplinary research can overcome these silos, but scientists need to find a common theoretical and empirical base. One of the key aspects of this is the transfer of models across domains, raising questions as to how such transfers relate to successful interdisciplinary research. Herfeld and Peruzzi examine processes of model transfer, particularly the movement of mathematical models between fields, as instances of research across disciplines that occur alongside increasing specialization. Model transfer processes are manifold in that models from some source domain undergo different kinds of modification processes in order to become applicable in a new domain. The core of their paper is the development of a taxonomy of the modification processes that enable a transfer of models across fields. They identify model imperialism, model migration, model translation, and model integration as different modes of this transfer. The chapter also discusses the normative problems arising from this transfer, which may hamper the applicability and may lead to problematic conclusions from the model. This may stem from misapplication or mis-reformulation of assumptions and idealizations as well as a lack of transparency and continuity.

Part 3: Science in the Societal and Political Sphere leaves the internal perspective on looks at the relationship between the political sphere and the science system. It covers the aspects of futurization, pluralization and digitalization in four chapters and deals with the problems of policy advice, the demand for science in politics, the role of science in democracies and the use of regulatory experiments.

In their chapter, Simon Lohse and Philippe van Baßhuysen focus on predictive policy advice and point out three major problems of such futurization. They argue that scientific policy advice can support rationality, trustworthiness, legitimacy, and societal responsiveness to policies, and they note that this might hold true especially in multi-faceted crisis. Focusing on predictive policy analysis and on external scientific advice only, they look at three problems: (1) For a given predictive problem, it is difficult to ascertain which discipline or disciplines should be asked for advice. (2) How to manage the implicit value influences in predictive policy advice. (3) The prediction may also influence what is actually happening. After describing these problems briefly, the authors concentrate on interdependencies between them. The interdependency between selecting disciplines and the influence of implicit value influences is clear: the more disciplines are consulted on, e.g. COVID-19, the more implicit values must be made transparent to help trustworthiness, legitimacy, etc. Also, the link between performativity and what disciplines are asked for advice is obvious: If you ask only researchers of epidemiology on COVID-19, they focus on containing the virus; social scientists will instead point out unintended consequences. A more pluralistic perspective might be helpful in bringing about a deeper societal understanding, but might reduce responsiveness to policy.

Christoph Hönnige and Dominic Nyhuis find that policy models do not sufficiently include science or scientists or higher education in policy making, and discuss attempts to close this gap by looking at trust in the science system in modern democracies. They view science as an instrument to de-personalize and objectify power and they compare science in this respect to bureaucracy or the judiciary system. Science can only rely on making the better argument and showing empirical consequences. But politicians do not have to follow science policy advice. Hence, trust in science in the general public is crucial to measure the general influence sciences may have. The authors criticize the available political science surveys (Eurobarometer, the Academic Freedom Index, and the World Value Survey) for not providing consistent and continuous data on trust in science; instead, these deal with the issue rather sporadically or are inconsistent with data on faith in science. In conclusion, the authors suggest that policy models should explicitly include the science system, regular survey data should cover trust in science to observe changes over time and across countries, and to identify factors which de- and increase trust in the science system.

Simon Fink's chapter looks at the demand for science in politics. With that aim in mind, he looks at German election manifestos. He finds that there is no increase in demand for science in party manifestos. This holds true both at the aggregate level and at the party level. An increase cannot even be observed in specific policy fields. But apparently almost all parties agree that science can be a driver of innovation and economic growth. Parties also make use of science if they attempt to depoliticize issues. Only the very right-wing political party Alternative für Deutschland (AFD) frequently refers to science negatively. Fink does not view party positions on science as being stable over time, but rather as depending on external factors and short-term trade-offs to gain voters' approval. In other words, if science comes in handy, party manifestos will rely on it. In this sense, there is no greater dependence on science within the political arena, but rather a constant process of renegotiation of political agendas within parties. The competition of parties for voters adds more volatility to the use that political parties make of science. Apparently, our society and our political parties are not subject to a general development toward a more science-based interpretation of the world, but make use of what serves their purposes for the time being.

In their chapter, Kilian Bizer and Daniel Feser discuss regulatory experimentation as a mechanism for learning in two arenas: the academic arena, which aims for epistemic learning, and the political arena, which is interested in political learning—i.e. if a policy option is better than another in terms of addressing the electorate or specific parts of it. Regulatory experiments took place in Finland with basic income and in the Netherlands with electricity markets. These adopted an experimental rule or set of rules and observed changes in behavior which could be compared to a control group. Making use of randomized control treatments can potentially increase quality of legislation. The authors argue with the help of an additional case study in the county of Nienburg, Germany that the local level of legislation can also experiment successfully with regulation. The county of Nienburg along with most of its municipalities created a fund for inner development of villages and towns. They point out that even though scientists aim for internal validity and methodological rigor while policy makers look at political risk and external validity as well as feasibility, it is possible to succeed in serving both interests, even if a local wicked problem is addressed. Regulatory experiments can help to further develop evidence-based policy.

Part 4: Ethics in Science Communication and Education connects the internal and external perspective by focusing on ethics of science in science communication and education. It covers the topics of the integration of AI in education, the communication in a post-truth society, and the role of pluralism in science in education about science.

AI and particularly LLMs show a high degree of potential for supporting and transforming teaching and learning. The chapter by Andreas Nehring and Elisabeth Hildt discusses ethical considerations associated with the integration of AI in educational contexts, focusing on five key areas. Firstly, the issue of bias and fairness in AI algorithms is addressed. AI systems are only as unbiased as the data they are trained on, and biased datasets can lead to discriminatory outcomes. Secondly, as education-related disparities are very well documented in many countries around the world, ethically driven efforts in education aim at improving learning for all students and disadvantaged groups in particular. Numerous approaches to using AI in education address this and claim to reduce these disparities. However, it is questionable whether this can be achieved in all cases, or whether differences in access to and use of AI will in turn create new disparities. Thirdly, the issue of data privacy in AI educational tools is addressed, highlighting how the collection and analysis of student data, while beneficial for personalized learning, create concerns regarding consent, data security, and the potential misuse of personal information. Fourthly, the impact of AI on student autonomy and learning processes in education is discussed. While AI can provide tailored learning paths, there is a risk of over-reliance on technology, potentially diminishing critical thinking. Fifthly, many AI systems operate as "black boxes," offering little transparency into how inputs are transformed into outputs. This opacity hinders stakeholders' ability to understand, audit, or challenge the decisions made by these systems, posing significant obstacles to achieving accountability and transparency. In conclusion, the chapter underscores the significance of ethical guidelines and policies that address these concerns, advocating for a collaborative approach involving educators, technologists, policy makers, and students to harness the benefits of AI in education responsibly and inclusively.

Christoph Bleidorn's chapter engages with the ethical aspects of science communication. It addresses the role of science communication in what can be called a "post-truth society": the kind of societies that large parts of the world's population are currently living in and in which emotions, personal beliefs, and appeals to identity often outweigh objective facts and evidence in the shaping of public opinion and in decision-making contexts. Bleidorn argues that in such societies, the traditional knowledge deficit model of science communication, according to which the aim of science communication should be to provide audiences with scientific knowledge that they are lacking but need as a basis for their decisions, fails. Instead of such a view of science communication as aimed at knowledge transfer, Bleidorn suggests that effective science communication in post-truth societies should be participatory and should aim at engaging the public in the scientific process, thus fostering trust in science. This entails a need for increased attention for ethical aspects of science communication, such as the promotion of equity and inclusion with respect to access to scientific knowledge as well as to knowledge production processes, the empowerment of citizens with respect to decision-making about scientific issues, and so on.

In their chapter, Andreas Nehring, Thomas Reydon, and Till Bruckermann explore how pluralism about the nature of science (NOS) is related to and generates ethical challenges for science education and science communication. Pluralism about NOS is examined from a history and philosophy of science (HPS) perspective, focusing on the demarcation problem and the diversity of scientific practices. The authors present and discuss pluralist NOS approaches within science education, analyze the ethical implications, and conclude with guiding principles for future work. Drawing on HPS perspectives, they show a plurality of conceptions of NOS underlying the epistemic practices, methods, and values across the sciences. By integrating philosophical and educational perspectives, the chapter analyzes approaches and their implications for teaching, curriculum design, and communication, emphasizing that curricular and conceptual choices are inherently value-laden. Conceptual pluralism regarding NOS, while epistemically adequate, entails ethical questions that have too rarely been discussed. By connecting pluralism, epistemic diversity, and ethical responsibility, the chapter demonstrates that fostering informed and reflexive trust requires transparency about the normative assumptions embedded in NOS frameworks. The authors conclude that interdisciplinary collaboration between HPS and science education/communication is essential to developing ethically responsible pluralism in terms of how science is taught, communicated, and understood.

Part 5: Social and Technological Transformations discusses the major social and technical transformations with a focus on futurization and digitalization in the field of the Ethics of the Sciences. It focuses on three major examples in four chapters: machine ethics, climate change, and demographic changes by an aging population.

In her chapter, Catrin Misselhorn discusses the new dimensions added to ethics by agential AI. She scrutinizes whether machines can and should be moral agents—i.e., whether machines can have the capacity to make moral decisions and whether they should do so. Analogous to AI, this field is called "artificial morality." Just as AI is concerned with replicating or simulating human cognitive abilities, artificial morality deals with modeling or simulating human moral decision making and behavior. This is relevant in areas like care, autonomous weapon systems, and autonomous driving. Machine ethics is the discipline concerned with the theoretical and ethical reflection of artificial morality. The most fundamental question of machine ethics is whether machines can be moral agents at all. Only if this is the case can the project of artificial morality get off the ground. The chapter distinguishes various types of moral agents and argues that machines can be explicit moral agents. They can take moral decisions in a functional sense insofar as they are capable of moral information-processing. However, in contrast to humans, they are not full moral agents, since full moral agency requires consciousness and intentionality which machines are still lacking. However, machine ethics deals not just with the question whether artificial moral agents are possible and how morality can be implemented in artificial systems; it also considers whether from a moral standpoint, it is desirable to delegate moral decisions to machines. The chapter provides an overview about this debate and discusses the major moral arguments for and against moral machines. It argues for a balanced position, stressing that the moral evaluation of artificial moral agents should take into account the diversity of ethical approaches and contexts of application.

The chapter by Holmer Steinfath argues that the persistent gap between what we know about climate change and how we act is rooted in a clash of temporalities. He contrasts everyday, individualistic time—shaped by present bias, near-future discounting, short planning horizons, and fragmented attention—with the long, cumulative, and partially irreversible timescales of anthropogenic climate change. Drawing on distinctions between A- and B-series time, he shows how agency oriented toward proximate futures obscures the backloaded impacts of past emissions, the millennial persistence of CO_2, and tipping-point dynamics that can trigger self-reinforcing, irreversible harms. He situates this temporal misalignment within political and economic short-termism, which amplifies moral corruption and diffuses responsibility across generations and borders. Rather than proposing new duties, the chapter explores how reconfiguring our self-understanding in time could motivate change: shifting from strictly individual lifetimes to transgenerational horizons (while avoiding narrow communitarianism), linking climate action to broader solidarities against exploitation and colonial legacies, and cultivating attachments to natural rhythms and planetary time that can sustain meaning for secular publics. Steinfath concludes that climate ethics needs compelling images of a good life in time to align shared norms with lived motivations and to make urgent, long-horizon action politically and personally intelligible.

The chapter by Mark Schweda focuses on the issue of aging in many industrialized countries, considering both ethical and socio-political perspectives. Once considered pure chance, aging in good health is now often seen as a long-term project by state organizations and individuals. In light of scientific, technological, and societal developments, aging and old age are becoming target points of individual and societal predictions, plans, and programs. Schweda argues that this reframing of later life in terms of a calculable and plannable individual or societal future project has far-ranging ethical implications. He analyzes these implications from an ethical perspective: He outlines the preconditions and steps in the historical emergence of the project perspective and highlights the relevance of the modern sciences of aging. His main focus is on the ethical implications of this development, focusing on individual perspectives for a good life in old age as well as on responsibilities for aging well. He argues that the ethical discussion of the implications of aging is not only an individual responsibility but has to analyze the complex interaction of the various individual, communal, social, economic, political, and scientific actors, factors and structures influencing health and old age.

In their chapter, Nils Hoppe and Marc Stauch also deal with the matter of aging in society. They address the increasingly unavoidable problem of scarcity in publicly funded healthcare, a challenge that was brought into sharp relief by the COVID-19 pandemic. This creates a need for fair, publicly defensible rationing mechanisms. Against this background, they examine age-based rationing and develop a justice-based version of the "fair innings" idea: society should prioritize enabling citizens to reach a politically agreed "fair age" in reasonable health, after which entitlements to further publicly funded curative treatment are weakened, while leaving room for privately financed curative care beyond the cut-off point. The proposed "equitable fair innings" approach adopts the standpoint of a just public resource distributor and identifies two purposes of public health spending: firstly, to secure or restore health to support normal functioning; and, secondly, when this is no longer achievable, to palliate suffering and protect dignity. Based on the principles of distributive justice, the fairest division is to allocate resources so that, as far as possible, everyone reaches the fair age; spending scarce curative resources beyond that point is unjust because it imposes opportunity costs on those who have not yet reached their entitlement. The authors anticipate objections concerning efficiency, "ageism," socio-economic inequality, quality-of-life comparisons, and burdens on the doctor–patient relationship. They emphasize that reducing suffering and providing dignity-preserving care remains unconditional and is not affected by the proposed approach.

References

Boswell, C. (2008). The political functions of expert knowledge: Knowledge and legitimation in European Union immigration policy. *Journal of European Public Policy, 15*(4), 471–488. https://doi.org/10.1080/13501760801996634

Cologna, V., Mede, N. G., Berger, S., et al. (2025). Trust in scientists and their role in society across 68 countries. *Nature Human Behavior, 9*, 713–730. https://doi.org/10.1038/s41562-024-02090-5

Douglas, H. E. (2009). *Science, policy, and the value-free ideal.* Pittsburgh University Press.

Ezrahi, Y. (1990). *The descent of Icarus.* Harvard University Press.

Gascoigne, T., Cheng, D., Claessens, M., Metcalfe, J., Schiele, B., & Shi, S. (2010). Is science communication its own field? *Journal of Science Communication, 09*(03), C4. https://doi.org/10.22323/2.09030304.

Krücken, G. (2021). Multiple competitions in higher education. A conceptual approach. *Innovation, 23*(2), 163–181. https://doi.org/10.1080/14479338.2019.1684652

Mede, N. G. & Schäfer, M. S. (2020). Science-related populism. Public understanding of science *29*, 473–491. https://doi.org/10.1177/0963662520924259

Meslin, E. M., Thomson, E. J., & Boyer, J. T. (1997). The ethical, legal, and social implications research program at the national human genome research institute. *Kennedy Institute of Ethics Journal, 7*(3), 291–298. https://doi.org/10.1353/ken.1997.0025

Nørgård, R. T., & Bengtsen, S. S. E. (2021). The activist university and university activism—An editorial. *Policy Futures in Education, 19*(5), 507–512. https://doi.org/10.1177/14782103211102658

Osborne, J., Collins, S., Ratcliffe, M., Millar, R., & Duschl, R. (2003). What "ideas-about-science" should be taught in school science? A Delphi study of the expert community. *Journal of Research in Science Teaching, 40*, 692–720. https://doi.org/10.1002/tea.10105

Strohschneider, P. (2020). Zumutungen: Wissenschaft in Zeiten von Populismus, Moralisierung und Szientokratie. kursbuch.edition.

Part 1: Ethics and Science—The Lay of the Land

Ethics and the Sciences: Dimensions of Interaction and Interdisciplinarity

Thomas A. C. Reydon and Silke Schicktanz

1 Three Ways of Relating Ethics to the Sciences

Ethics of science[1] in an institutionalized and professional format has established itself as an academic area of scholarship only in the second

[1] While "ethics of science" is the most widespread term, we will use "ethics of science" and "ethics of the sciences" interchangeably. We have a preference for the latter term, as our concern is not with science in the strict sense (i.e., the natural and life sciences), but rather with the whole spectrum of fields, areas of work, subjects, and disciplines that are part of institutionalized knowledge production in contemporary academia. The term as we use it thus reaches beyond the natural sciences, life sciences, and health sciences to include the humanities, the social, political, and economic sciences, the formal sciences, the engineering disciplines, legal studies, and so on. The term as we use it corresponds to the Dutch term "*wetenschap(pen)*" or the German term "*Wissenschaft(en)*," the latter having been proposed by Rabinow (2011) to express the commonality across academic disciplines.

T. A. C. Reydon (✉)
Institute of Philosophy and Centre for Ethics and Law in the Life Sciences (CELLS), Leibniz University Hannover, Hanover, Germany
e-mail: thomas.reydon@cells.uni-hannover.de; reydon@ww.uni-hannover.de

S. Schicktanz
Department of Medical Ethics and History of Medicine, University Medical Center Göttingen, Georg-August-University Göttingen, Göttingen, Germany
e-mail: sschick@gwdg.de

J. Hinrichsen et al. (eds.), *Ethics of the Sciences*,
https://doi.org/10.1007/978-981-92-2821-8_2

half of the twentieth century.[2] An important factor that contributed to its establishment as a recognizable area was growing concerns regarding the moral responsibility of researchers for intended as well as unintended adverse consequences of their research and its applications. Two early, but still widely discussed, examples that laid the groundwork for the later development of the field are the case of chemist Fritz Haber and his active contribution to the development of poisonous gases for use in the trench warfare of the First World War, and of the development of the atomic bomb in the Manhattan Project in the Second World War (after which several physicists expressed feelings of responsibility for bringing such weapons into the world) (see, for instance, Reed, 2014 or Miller, 2018). The increasing visibility of adverse consequences of results of scientific research in the decades after the Second World War added momentum to these concerns. Consider for example Rachel Carson's *Silent Spring* (Carson, 1962), a widely read account in the 1960s and 1970s of the adverse effects of the application of pesticides on human health, on non-human species, and on nature in general, which led to a broad concern for the environmental effects of the products of scientific research.

Adding to this, misconduct by scientists in their research practices has become increasingly visible in the twentieth century too, even though scientific misconduct is far from a new phenomenon. Cases in which scientists exhibited morally reprehensible behavior while carrying out their research presumably have occurred since the origin of science as a social institution and as a profession (for brief discussions of some notorious cases from the late eighteenth century to today, see European Commission, 2010, pp. 15–18). Perhaps the most egregious case encompasses the Nazi experiments on concentration camp prisoners in the Second World War (Weindling, 2008), but morally reprehensible practices in science had already been publicly criticized well before that time. For example, the use of animals in experiments was widely criticized as inhuman in the nineteenth century (Maehle, 1993), as was the procurement of body parts from cemeteries or morgues for anatomical research (Fangerau, 2014). In the second half of the twentieth century and the first decades of the twenty-first century, numerous cases of major and minor scientific misconduct have been discussed in the public media and addressed by investigative committees (for some infamous cases, see

[2] For more on this history, see Briggle and Mitcham (2012, pp. 1–13).

Resnik, 1998, pp. 1–12), and surveys suggest that so-called "questionable research practices" are widespread (Fanelli, 2009).

Notwithstanding the establishment of ethics of science as a recognizable academic domain with its own textbooks, journals, and professorships, the term "ethics of science" does not denote a unified field of research. In the literature the term often features as the overarching name for the broad area of ethical reflection on science and its governance that encompasses a variety of fields that are concerned with different normative aspects of scientific practice and the outcomes of scientific research. These include various subdomains of applied ethics, as well as ELSI/ELSA research (see Meslin, this volume), responsible research and innovation (RRI) (see van Basshuysen & Lohse, this volume), research ethics and good scientific practice, and technology assessment (TA). In addition, many other disciplines that cannot be subsumed under the header of "ethics of science" have also long been concerned with normative aspects of the sciences—these include science and technology studies (STS), history and philosophy of science, medical humanities, sociology of medicine, science education (see Nehring & Hildt, this volume, and Nehring, Reydon & Bruckermann, this volume), science communication (see Bleidorn, this volume), legal studies, political science (see Fink, this volume, and Hönnige & Nyhuis, this volume), psychology (Rest & Narvez, 1994), and sociology (Zussman, 2000). Even though there are varying degrees of overlap between these different academic communities, they remain largely separate: theories, viewpoints, principles, and methods do not travel smoothly between them. In addition, there often are considerable differences between these fields with respect to the ways in which scholars view the relation between ethical reflection and the various sciences. As a result, ethical reflection on the sciences remains fragmented to a considerable degree.[3]

In the face of such fragmentation, our aim in this chapter (which we understand as a programmatic contribution to the meta-discussion on the ethics of science) is to assess how ethical reflection is best located in relation to scientific practice and to develop a perspective towards a more integrative approach to ethics of the sciences. Contemporary ethics of science, understood in the broad sense described above, encompasses different views regarding the way in which ethical reflection and the

[3] For more discussion, see Lohse et al. (2020).

various sciences should interact. For instance, while some think of ethics of science as a field of work that looks at science from the outside and understand ethics as part of the forward-looking public governance of the sciences, others hold the view that ethical reflection must be integrated into the sciences themselves and see it rather as reactive to new scientific practices and innovations. We examine different views of how ethical reflection and the sciences relate to each other and develop a perspective for an integrative approach to ethics of the sciences that incorporates both these reactive and prospective parts of reflection into a dynamic process of reflection on the sciences.

We will discuss three stances regarding the relation of ethical reflection to science, which we call "ethics *and* science," "ethics *in* science," and "ethics *for* science" (with "ethics *of* science" as the overarching name for all ethical reflection on practices of academic research and scholarship, as discussed above). These propositional relations are not strictly delimited or well-defined approaches, and there is no established usage of these terms in the literature.[4] At first glance, it seems that terms like "ethics of science," "ethics and science," "ethics in science," and "ethics for science" can be used interchangeably.[5] For example, the first three all are titles of widely used introductory textbooks that provide a broad coverage of the field (Briggle & Mitcham, 2012; D'Angelo, 2019; Resnik, 1998). However, in this chapter we use the terms differently, with "ethics and science," "ethics in science," and "ethics for science" as markers for three different *normative attitudes* regarding how ethical reflection can and should relate to the sciences—that is, three different general ways in which ethics as an area of expertise can be positioned in relation to another field of expertise (such as life science, health science, computer science, social science, etc.). We argue that these three normative attitudes can be found in the broad area of ethical reflection on science as it exists today and that they manifest themselves in various ways: in differences in the topics

[4] For a similar approach to analyzing the relation between philosophy of science and the sciences, see the section "Philosophy of science at, in, and with..." in Veigl and Currie (2025).

[5] Our analysis is based on terminology that is used in English-language literature. Clearly, the analysis would look different when carried out in another language. In German and Dutch, for instance, the common terms "*Wissenschaftsethik*" and "*wetenschapsethiek*" obscure the differences in perspective that we want to highlight with the of/and/in/for tetrachotomy.

that researchers address and questions they ask, in methodological differences, and in differences in the aims that researchers set themselves. At base, they embody divergent, but in our view also complementary, views of how ethical reflection on science is best done.

The term "ethics *and* science" suggests that ethics on the one hand and the sciences (in the broad sense of "*Wissenschaften*") on the other hand are best understood as separate areas of work that may interact or collaborate, while each following their own independent trajectories.[6] Newly created specializations in applied ethics, such as "bioethics," "nanoethics," "genethics," or "AI ethics," are examples of this understanding. As subdisciplines of moral philosophy, they attempt to bring various ethical theories, such as utilitarianism, virtue ethics, or Kantian ethics, to bear on moral issues that stem from new developments in specific areas of science and technology. "Genethics," for instance, can be defined somewhat broadly as "the study of the ethical issues that arise out of the science of genetics and the uses of genetic technologies" (Lewens, 2004, p. 326). Less broadly and more precisely, "genethics" refers to the reflection on ethical issues that can arise from genomic medicine and applications of genetic engineering in clinical and public health contexts.[7] Similarly, "nanoethics" is understood as the discipline that examines "ethical and social questions arising from nanotechnology" (Lin & Allhoff, 2007, p. 8). These definitions suggest that ethical reflection principally follows developments in science, takes place in an area that stands beside the science it reflects on, and provides normative assessments of these developments after the fact.[8]

The term "ethics *in* science," to the contrary, suggests that ethics should be embedded within the sciences and play an active role in the practices of the respective field, involving a social and epistemological internalization within the respective field of science.[9] And the phrase

[6] Of course, on this broad understanding of science ethics and ethics of science are themselves also sciences. Indeed, in Dutch and German they count as *wetenschappen/ Wissenschaften*.

[7] See for instance the *Genethics Forum* (http://genethicsforum.ning.com and https:// www.ethox.ox.ac.uk/ethics-support/genethics-club).

[8] In German this is sometimes expressed by calling such fields "*Begleitwissenschaften*"—literally, "accompanying sciences."

[9] This view parallels a similar movement in the philosophy of science of promoting the integrating of work in the philosophy of science in scientific research itself, thus doing

"ethics *for* science" suggests still another way of relating ethics and the sciences, namely the idea of philosophy as the handmaiden of science, whose task it is to be "an under-labourer in clearing the ground a little, and removing some of the rubbish that lies in the way to knowledge," as Locke aptly put it in his *Essay Concerning Human Understanding*.

In what follows, we will examine these three stances in more detail. We will begin with a brief sketch of the coarse-grained structure of ethics *of* science as a broad field of ethical reflection, using the general distinction between the *internal* and *external* normativity of the sciences. This discussion highlights two different kinds of questions that any form of ethical reflection in relation to science should have in view. We then examine the three normative understandings mentioned above, ethics *and* science, ethics *in* science, and ethics *for* science. We will do so by highlighting the kinds of issues that they can address and the kinds that they tend to miss. On the basis of this assessment, we sketch a perspective for an integrative program that can overcome the limitations set by conceptual differences and differences in styles of argumentation between moral philosophers and researchers in other academic domains.

2 Ethics of Science: The Internal and External Normativity of Science

A fundamental distinction that is commonly made in ethical reflection on science is between the *internal* and the *external* normativity of the sciences (Reydon, 2013, pp. 60–97; cf. Pennock, 2019, Chapter 2). Ethics of science as a broad field of scholarship is understood as including both. Internal normativity encompasses those values and standards of conduct that pertain to the various aspects of the knowledge production process and the professional role of "being a scientist" (Committee on Science, Engineering, and Public Policy, 2009). The internal normativity of the sciences is usually discussed under the headers of "responsible conduct of research," "good scientific practice," or "scientific integrity." External normativity, in contrast, pertains to normative assessments of the impact of science on society, i.e., to the question how potential and actual applications of scientific knowledge are to be valued, as well as the broader

philosophy *in* science rather than philosophy *of* science (Laplane et al., 2019; Pradeu, 2025; Pradeu et al., 2024; Sober, 2022).

question how the sciences should relate to the values and interests that are present in the societies within which they are embedded.

2.1 *The Normativity Internal to Science*

Reflection on the internal normativity of the sciences aims at fostering appropriate conduct of academics as actors in knowledge producing institutions. In this context, we understand ethical reflection as a learning process that integrates reactive as well as prospective actions in a dynamic reflective process. Often, conceptions of appropriate conduct are codified in regulations and guidelines for safeguarding good scientific practice and avoiding scientific misconduct that are issued by research institutes and universities as well as national and international funding agencies, academies, and other institutions (examples include: ALLEA, 2023; Deutsche Forschungsgemeinschaft, 2022; KNAW et al., 2018; Max-Planck-Gesellschaft, 2022). Such guidelines typically include specifications of what kinds of actions qualify as scientific misconduct, suggestions as to how cases of scientific misconduct should be treated, and specifications of the kinds of values good scientists should adhere to. The *European Code of Conduct for Research Integrity*, for example, mentions reliability, honesty, respect, and accountability as core values that researchers should embody (ALLEA, 2023, p. 5). The guidelines and regulations that are developed as part of efforts at codifying the internal normativity of the sciences can be understood as a form of governance that is not only aimed at regulating everyday work in the sciences, but also at fostering reflection on existing practices and norms, the possible revision of existing practices and norms, as well as the prospective establishing of new rules to govern scientific work.

Attempts to specify the core values of science can be traced back at least to sociologist Robert Merton's work in the mid-twentieth century on the social structure of science. In his famous paper, "The normative structure of science" (Merton, 1973; originally published with the title "Science and technology in a democratic order": Merton, 1942), Merton inquired into the "ethos" or the "mores" of science, which he held to be "that affectively toned complex of values and norms which is held to be binding on the man of science" (Merton, 1973, pp. 268–269). Merton famously identified four norms—or rather, normative complexes—as constituting the ethos of science: universalism (the norm that knowledge claims should be assessed independently of "the personal and social attributes of their protagonist"; Merton, 1973, p. 270), communism or communalism (the norm that scientific knowledge is not the property of particular individuals or institutions, but "assigned to the community (…), a common heritage"; Merton, 1973, p. 273), disinterestedness (the norm that research should not be conducted to pursue particular interests of individuals or institutions, but only to advance human knowledge), and organized skepticism (the norm that scientific claims should not be believed overly hastily, but that judgment should be suspended until a claim has been sufficiently scrutinized).

Two things are noteworthy about the ethos of science as specified by Merton, as well as about contemporary formulations of the core values of science. First, notwithstanding decades of work on this issue, no exhaustive set of core norms and/or values of science has emerged. Most guidelines mention four or five such norms or values, often also called "principles," but according to a recent study on guidelines adopted in different European countries there seems to be only weak convergence among the various guidelines on which norms are mentioned (Desmond & Dierickx, 2021). Interestingly, even though Merton's four norms are widely cited in textbooks on research ethics, they hardly appear in contemporary guidelines and codes of conduct, which tend to mention different norms (honesty, reliability, fairness, impartiality, responsibility, accountability, respect, and others). Second, the core norms and values that are listed have less to do with morality than with goal-orientation: they are norms and values that do not describe and proscribe what morally appropriate conduct is, but rather that express what kind of behavior promotes the aims of science, i.e., the production of reliable knowledge (Reydon, 2016, 2020).

The internal normativity of the sciences thus primarily is *goal-directed normativity—it* is normativity that pertains to the promotion of behavior that is beneficial to the process of knowledge production and the discouraging of behavior that harms that process. Accordingly, the internal normativity of science can be conceived of as addressing normative questions that ensure from the assumption of a particular role, namely the role of the scientist, and at promoting *professional conduct* (Reydon, 2015, 2016, 2020). Accordingly, authors have discussed these issues in terms of the *role responsibility* of scientists (Douglas, 2003), i.e., the responsibilities that someone assumes when taking up the role of scientist, or the scientific virtues (Pennock, 2019), i.e., those virtues and character traits that an exemplary scientist would exhibit, or the professional attitude that all "good" scientists should embody (Reydon, 2020). This is not to say that moral norms and values do not play a role in the internal normativity of the sciences. The *European Code of Conduct for Research Integrity*, for instance, mentions "[r]espect for colleagues, research participants, research subjects, society, ecosystems, cultural heritage, and the environment" (ALLEA, 2023, p. 5) as one of four fundamental principles of research integrity. Clearly, this is a complex of moral values that is less connected to the aims of scientific research. And norms of morally good behavior indeed *must* always be part of guidelines and codes of conduct, as regrettably it remains necessary to remind actors in scientific institutions how people should treat each other properly. The point is, though, that such moral norms (treat people with respect, do not commit sexual harassment, do not abuse the power you have over co-workers, and so on) are not specific to how scientists should behave—they are general moral norms that apply to all of human life and are not in any way specific to the role of scientists *as scientists*. Besides such general moral issues, the internal normativity of science is concerned with the question what makes for a "good scientist" in a professional sense, that is, with the question how individuals should act as parts of the institution of science. This question is addressed with an eye on the aims of that institution, however, which is what connects internal to external normativity.

2.2 External Normativity

The external normativity of science is concerned with science as an institution within society. Ethical reflection that falls in the fold of external normativity seeks to clarify the responsibilities of individual scientists as well as the institution of science with respect to serving the various interests of the members of society as well as humanity in general, and not causing harm to people, non-human animals, the environment, or societies.

With respect to the avoidance of causing harm, the actual and potential consequences of scientific research and of applications of scientific knowledge are in focus. An important question in this context is whether and if so, to what extent scientists are responsible for (intended or unintended) adverse consequences that ensue from their research. In many cases, scientists have expressed feelings of responsibility and even guilt for technological applications that were only possible because of their research. The case of the atomic bomb, already mentioned above, is well-known: not only researchers who were immediately involved in the development of the first bomb expressed such feelings, but also prominent scientists who had done theoretical work that provided fundamental knowledge that was needed for the Manhattan Project expressed feelings of responsibility (Reydon, 2013, pp. 80–82). But *feeling* responsible for adverse consequences is not the same as *being* responsible. Some scientists have emphasized a strict distinction between science and technology, according to which science provides knowledge that is free from normative aspects and technology turns this knowledge into applications that can cause benefits or harms (e.g., Bridgman, 1947; Wolpert, 2005). On such a view, scientists who "only" produced knowledge would not be responsible for adverse consequences and responsibility would lie with the engineers who turned this knowledge into practical applications. But such a strict separation between science and technology does not seem feasible and the question of responsibility of scientists for the consequences of their work remains an open issue.

Another major issue in the domain of external normativity is what the aims of science should be. While one view may be that science should simply aim to produce knowledge about the world, such a view is both historically and practically hard to uphold. Historically, science never was a straightforward quest for true knowledge. From the start, science was conceived of as an enterprise *in the service of humankind*. For example, Francis Bacon, Lord Chancellor of England in the early seventeenth century and one of the fathers of modern science, held a view of science as ultimately aimed at the improvement of the living conditions of the people (Reydon, 2013, pp. 82–84). In the Preface of the *Instauratio Magna*, a work of which only parts were published but that in many ways can be seen as one of the starting points of modern science, Bacon admonishes readers to "consider what are the true ends of knowledge, and that they seek it not either for pleasure of the mind, or for contention, or for superiority to others, or for profit, or fame, or power, or any of these inferior things; but for the benefit and use of life" (Bacon, 2017, p. 17).

This raises the question what exactly the aim of doing science in the service of humankind encompasses and how it is best achieved. For Bacon, the picture was straightforward: science should uncover all the laws of nature that, once known, could be applied in practice to build machines, to devise processes, and to make products that could improve the lives of people and promote the common good. But unfortunately things are more complicated. As resources for conducting research are very limited—research staff, funding, and time are not infinitely available, after all—, choices have to be made regarding the allocation of resources to the problems that can be investigated (Kitcher, 2001, 2004, 2011). This means that decisions must be made in which the interests and values of researchers, stakeholders, and society more broadly must be taken into account, and it has to be determined how decisions will be made and how the various values and interests in society should be weighed. Individual scientists always play some role in such decision processes, such that to some extent these issues transgress the conceptual boundary between external and internal normativity.

2.3 *Internal and External Normativity Are Inseparably Connected*

One thing that follows from the preceding discussion is that the internal and external normativity of science are intimately connected. While the internal normativity of science follows from the aims of science (as a normativity that specifies actions that are desirable because they promote the aims of science), the external normativity of science is concerned with what those aims should be. But the question what the aims of science should be is also a question for individual scientists themselves to answer: the question how science can best serve the interests of humankind is as much a question for individual scientists (as specialists for the various topics that are being investigated) and individual members of society outside the sciences (as stakeholders and people whose lives are generally affected by applications of scientific knowledge) to address. While for analytic purposes a distinction between internal and external normativity of science is useful, the preceding discussion should have made it clear that such a distinction should not be taken as a strict separation between two areas of ethical reflection. Any ethical reflection related to science should cover both internal and external normativity equally and should do so in an integrative manner. Accordingly, any ethics *in* science cannot be confined to internal normativity, and any ethics *for* science should not only have the interests of the scientific community in view but also the interests of society and humankind as a whole. This is a first indication that an integrative approach to ethics of the sciences is needed. With this in hand, we now turn to a deeper exploration of ethics *and*, *in* and *for* science.

3 Ethics *and*, *in*, and *for* Science: Proximity to and Distance from the Sciences

Social studies of science and academia provide helpful insights for discussing the underlying assumptions of the three different stances regarding the ways in which ethics and other disciplines can come together in practice. In this section, we will draw on the so-called "spatial turn" in the social and cultural sciences (Warf & Arias, 2008) and use it for our own reflections on three typological cases of an institutionalized relationship between ethics and science. Reflecting on the spatial arrangements (e.g., proximity and distance) of academic disciplines and subjects is here understood as a specific methodological perspective that allows us to pursue the question of the "localization" of ethics *and*, *in*, and *for* science in more detail. In this context, "localization" refers to both the structural anchoring in the university landscape in the concrete, material sense and the interdisciplinary profile in an abstract, immaterial sense.[10] Note, again, that while in the literature the labels of ethics *and*, *in*, and *for* science are often used interchangeably, we use them for analytical and expository purposes as indicators of distinct stances regarding how ethics and science can relate to each other.

3.1 Ethics and Science

The title of the book, *Ethics and Science: An Introduction*, by Briggle and Mitcham (2012) constitutes an example of how ethics and science are conceived of as two different disciplinary spheres coming together. Briggle and Mitcham's book provides an overview of a variety of approaches in ethical reflection and of the various ways in which ethical and political approaches can interact with the sciences. The book is part of a series of books that relate ethics as an academic field to other domains (other titles include *Ethics and Business* and *Ethics and the Media*) and the authors frame their topic by highlighting several developments in the

[10] The general idea relies on Georg Simmel's approach of socialization and the role, material, and immaterial forms of localization play for this process; it has been adopted for academia (Simmel, 1923; see also Schicktanz, 2011).

sciences that invite responses from ethicists.[11] While such an approach positions ethics as an independent field next to the various sciences, there are different ways to think about such a positioning. Most importantly, researchers can position their work at a distance from the field of science that they reflect on, or much closer to that field.

Nowotny and her colleagues (2001) have convincingly demonstrated that spatial categories of proximity and distance between the various domains of research and the various domains of society should play a central role in our understanding of modern science. While science used to be conceived of as universal (in the sense of producing knowledge that is valid and applicable everywhere), today the locality of knowledge is emphasized. Especially transdisciplinary and interdisciplinary research is characterized by very heterogeneous knowledge and methods. Nowotny and colleagues refer to this form as "Mode 2" science and distinguish it from the traditional form (which they call "Mode 1" science; the terms were coined in Gibbons et al. (1994)).

Mode 1 science is characterized as being university-based, hierarchical, knowledge-oriented, disciplinary, and having little interest in the social impacts of research. Mode 2 science, on the other hand, aims at providing technical, practical, or societally desired solutions. For this, cooperation between different disciplines and between academia and non-academic domains is needed. A certain proximity between such collaborators is necessary to achieve mutual loyalty and understanding, but at the same time a certain degree of distance is needed to allow for critical distancing and productive skepticism. The interplay between proximity and distance is particularly central to the success and social acceptance of such academic cooperations.

For example, spatial proximity (e.g., a shared laboratory, interdisciplinary centers of research, or placing a institute of ethics of medicine within a medical clinic) enables the development of trust between the different participants, which in turn allows distance in terms of content (e.g., in the choice of methods or research questions). This new form of trust enables a transdisciplinary science based on the division of labor, which dispenses with the idea of unity in terms of content, but at the same time is socially dependent on cooperation between actors with

[11] These developments include the increasing capabilities of science to produce results that exceed human control, the increasing occurrence of scientific misconduct, and the increasing commercialization of science (Briggle & Mitcham, 2012, pp. 5–10).

different approaches. The categorization of collaborations in terms of proximity and distance proposed by Nowotny and collaborators helps to better understand the close relationship between ethics and the sciences. According to Nowotny and collaborators, Mode 2 science is transdisciplinary, problem-oriented, and often organized through cooperation between academia, politics, and social groups (e.g., NGOs, governments, or industry partners). Furthermore, it is characterized by an enormous heterogeneity of both investigative methods and content, as well as a lack of a canonical set of theories.

Ethics and engineering, we suggest, is an example of ethics and science done along the lines of Mode 2 as described above. The field is characterized by broad heterogeneity of approaches to technology assessment or "ethics and technology" (Van de Poel & Royakkers, 2023), the absence of widely shared ethical views, an orientation on concrete problems, and cooperation between academics and non-academic stakeholders. But even though the field is strongly focused on risk assessment (as for instance the assessment of potential failure is crucial in engineering), interaction with the field of ethics often remains at an anecdotal level without systematic training of researchers in ethical reflection (Guntzburger et al., 2017; Jasanoff, 2016). This is not the case throughout the field, however. Recently, for instance, in artificial intelligence (AI) and computer science the discussion how "values" should be integrated in engineering has been revived (see also Misselhorn, this volume). As Friedemann and Kahn (2007) distinguish, there are three main ideas about how values and ethical principles are related to the development of new technologies. In the *embodied* approach, values are built into technology by designers. In the *exogenous* approach, values are determined and imposed by users after a technology has been implemented. *Interactional* approaches focus on the interaction between designers and users and include approaches such as value-in-design and participatory design. Interestingly, all three approaches can be found in current design practices in AI technology (Shneiderman, 2022).

3.2 Ethics in Science

The term "ethics *in* science" is often used to indicate that ethical reflection should be situated within a specific field of work and is seen as an integral part of that field. This is often the case in academic fields that have a strong practical component: consider, for example, "ethics in nursing" where ethical reflection draws on practical experiences or focuses on ethical choices that must be made in everyday practices (Benjamin & Curtis, 1992). Such situating of ethical reflection within specific fields of practice can be done in two different ways: the first is by creating what can be called a new "pivot point" within an existing system; the second is by creating new intermediate spaces between existing areas of research.

The first model—creating "pivot points"—has been particularly successful in the field of medicine. Over the past decades many chairs and even departments specializing in ethics have been established within medical faculties and university hospitals (in Germany often complemented with medical history as a second area). Such ethical units serve as "pivot points" in the sense that they can bring together the reflective aspects of research, teaching, and patient care that arise in the other units in their environment, and can feed ethical knowledge back into the practices of these other units (e.g., as clinical ethics committee or ethics counseling). They take up issues from the other units and reflect on their ethical and social aspects, and offer results to the units in their home institution, while marking their own field of expertise. The institute within which an ethics unit is embedded benefits from the integration of ethical reflection in teaching, research, and health care. Here, professionalization and institutionalization go hand in hand. However, for their functioning such chairs or departments depend materially on the preexisting infrastructure of the faculty or larger institution in which they are established.

The second model encompasses the finding and filling of available intermediate spaces between existing areas of research by means of establishing new areas of work. This is probably easiest to do where the conventional boundaries between different fields or between research and application are already blurred (e.g., between medicine and biology, or between physics and applied engineering). A prominent example of this model is the establishment of institutes for technology assessment (TA) since the late 1960s (Boavida & Moniz, 2016). Many of these exist independently of universities or other academic institutions, even though they are academic research institutes that combine expertise from ethics, law, social sciences, engineering, biology and physics. They are closely connected to public and political interests, e.g., by carrying out research that has been commissioned by the government or other public actors or institutions. In contrast to the first model, in which an established academic field is embedded in an existing environment, here a new area of work is created that reflects on ethical and social aspects of technology in part from an "inside" perspective, i.e., as part of scientific research or engineering endeavors.[12]

In both models, so our understanding, the primary goal is to establish a new institution (a chair, department, or independent institute) with material conditions (workplaces, a library, etc.), specific work tasks (e.g., a curriculum and a research agenda) and resources that can be distributed (e.g., positions, supervision of early-career researchers). We also assume that one should not underestimate that, despite all the materiality that such institutions can encompass, they can be dissolved within short periods of time, precisely because their status within an established environment or in the interstitial space between fields may remain controversial. Examples are the closing of the US Office of Technology Assessment (OTA) in 1995 during the Reagan administration or the closing in 2003 of a Technology Assessment institution that existed since 1992 and was widely recognized in Germany (the *Akademie für Technikfolgenabschätzung in Baden-Württemberg*). To function well, new units and institutions have to be accepted by the existing units and institutions in their environment.

While such new institutions are "invented" in order to solve specific ethical and societal problems, at the same time they must create a new canon of approaches, methods, and reference works on the topics they address to be visible on their own.

3.3 A Critical Perspective: Ethics for Science

As noted in the introduction, the term "ethics for science" strongly indicates a role for ethics as a field standing in the service of the sciences. While the term itself is rarely used by ethicists themselves, the term appears in criticisms by social scientists and political scientists who see ethics as being abused as an instrument for legitimizing projects in the modern life sciences. Such legitimization occurs, for instance, by screening research projects off from social criticism or by replacing risk discourses in society (which can be efficient means for assessing

[12] Grunwald (2019, p. 1) defines technology assessment as neither a scientific discipline nor a field of practice, but rather a "specific relationship between practice and theory" that can manifest itself in various contexts, including in science and engineering.

research projects and technologies) by specialized ethical discourse. Francis Fukuyama, who was a member of the President's Council on Bioethics in the USA, certainly promoted this assessment with his popular book, *Our Posthuman Future*, when he wrote that the community of bioethicists had emerged in tandem with the biotechnology industry and should be seen as "sophisticated (and sophistic) justifiers of whatever it is the scientific community wants to do" (Fukuyama, 2002, p. 204).

The idea of ethics in the service of science is connected to the question how independently, self-determinedly and self-consciously a field can act in relation to the sciences on which it is supposed to reflect. A dependency of ethical reflection on the subject matter of a particular discipline, e.g., a particular area of life science, medicine, or engineering, can on the one hand be rooted in the way in which applied ethicists of science have a multi- or interdisciplinary background. Being primarily trained as a biologist, a medical doctor or an engineer and then secondarily trained as an ethicist may impact one's own cognitive framing: what one sees as a problem, how one interprets a given problem, what rationale one develops behind a decision, or what one perceives as morally important. On the other hand, material and economic dependencies can also play a role when ethical reflection is carried out as part of scientific research projects, even though many applied ethical projects can draw on financial resources that have been created specifically for them (such as ELSI/ELSA programs that have their own funding, see Meslin, this volume).

However, one can understand the occurrence of such dependencies as a challenge related to the social-spatial categories of proximity and distance. Ethical reflection should—and always strive to—rest on an understanding of the specific, practical problems and needs of the respective scientific field. For this, proximity to the field in question is indispensable, i.e., a precise examination, understanding, and knowledge of the situation in the field is necessary. On the other hand, a lack of distance to the field under consideration can make it difficult to change perspectives. The risk is then that a certain pragmatic logic imposes itself and becomes embedded in both the description of the problem and the proposed solution. The alternative, to have too large a distance from the field under consideration, carries the risk of being too far removed from the field's actual research practices and therefore remaining too abstract and providing implausible or impractical proposals.

The critique of an ethics "for" science, mentioned above, is related to the former risk. It can result in an instrumentalization of ethics for the purpose of developing and implementing new technologies, involving a possible loss of its intellectual authority and its position as part of public governance of technology. So-called "ethics washing" by ethics boards that search for the most fitting expert statements to justify prioritizing technological progress (or market placement) over social concerns and the protection of vulnerable populations severely harms the image of ethics as a field of scholarship (Van Maanen, 2022; Morley et al., 2021). Socio-political controversies over the implementation of new technologies, such as stem cell research (Jasanoff, 2005; Thompson, 2013) or the development of artificial intelligence applications, illustrate the problem that arises when ethics of the sciences is reduced to an ethics *for* science, i.e., a service discipline for technology that loses its meaning as a reflective, deliberative, and responsible field of scholarship.

4 A Perspective for an Integrative Approach to Overcome Existing Fragmentation

Using the framework of the proximity and distance between academic disciplines in the preceding section we have attempted to identify some of the problems that the stances of "ethics *and* science," "ethics *in* science," and "ethics *for* science" can be faced with. The discussion showed that the fragmentation that exists within the broad domain of ethics of science can be analyzed in terms of two dimensions, namely the dimension of internal vs. external normativity and the dimension of proximity to vs. distance from the sciences.

A focus on proximity or distance can pull ethical reflection in opposite directions, as can a focus on internal or external normativity. By taking these two dimensions as the two axes of a two-dimensional space, the various disciplines that in some way or another are concerned with ethical reflection on the sciences (see Sect. 1) can be positioned onto these two axes to achieve more clarity about the nature of the work that is done by these various communities. Researchers interested in ethics of the sciences can locate their own approach in relation to other approaches, which may help them see more clearly which aspects of ethical reflection might remain unaddressed in their approach and to identify needs for further research in relation to the topic they are working on.

Ethical reflection on the sciences only functions well, we suggest, if a good balance is achieved between proximity and distance, as well as between a focus on internal and external normativity. Ethical reflection should be sufficiently close to the actual practices of the sciences in order to be well attuned to the issues that give rise to ethical concerns, but at the same time ethical reflection should be sufficiently distanced from scientific practice to be able to achieve an unbiased assessment of these issues and avoid becoming a service field that is tasked primarily with legitimizing research projects and applications. Similarly, ethical reflection should be aimed at improving the everyday practices that are internal to the sciences, as well as at providing clarity on adverse consequences that may occur externally to the sciences. Since the internal and external normativity of science is inseparably connected, as we have discussed, here too a good balance between a focus on internal and external normativity is needed. The challenge thus is to determine what a good balance between the two poles of each of these two dimensions would be, and how such a balance can be achieved.

All three stances discussed above fall short in this respect. "Ethics *and* science" denotes a perspective that leaves both the questions of proximity and distance, and of internal and external normativity largely open. Ethical reflection is seen as an endeavor that is separate from the sciences, and it is left to individual researchers to decide how closely they want to engage with the discipline on which they reflect and whether they want to focus on internal or external aspects of that discipline. The stances of "ethics *in* science" and "ethics *for* science," in contrast, provide a clear view of the relation between ethical reflection and the sciences by choosing proximity over distance (while leaving the internal–external question open). While this has advantages with respect to the fine-tuning of ethical reflection to the content and practice of the fields that scholars reflect on, it also incurs the risk of overly strong dependencies of ethics on science and the accompanying loss of a sufficiently unbiased viewpoint from which ethical reflection can take place. This is particularly clear in the case of "ethics *for* science," where ethical reflection is explicitly conceived of as occurring in the service of the scientific projects with which it is connected. In both "ethics *in* science" and "ethics *for* science" the need for proximity tends to be overemphasized and the need for distance underappreciated.

One way to meet this challenge, we suggest, is to take an integrative approach that connects the various disciplines mentioned in Sect. 1, each having its own focus and approach, into an encompassing ethics of the sciences in which results that have been achieved from different perspectives are connected into a comprehensive framework. Developing such an integrative approach in sufficient detail will be a considerable undertaking that goes far beyond the scope of the present chapter. Here, we want to conclude by briefly discussing three kinds of integration—conceptual, socio-spatial, and competence-oriented integration—that can be used to develop such an integrative approach. We take this as a starting point for future work.

Conceptual integration is the comparison and possible integration of different ideas, theories, and concepts that underlie areas of practice. A relevant question in this context would, for example, be whether central concepts and approaches in clinical practices which originate in the health sciences share basic meanings with ethical theories of care. If potential conceptual gaps on one or both sides are identified, the question arises how this requires reflection by productively asking what is missing in each original conception. Such a process of conceptual matching and integration could help to advance both sides and even to develop applied ethical approaches for the particular scientific context.

Socio-spatial integration involves the close collaboration of ethicists and experts from the field under consideration in academic social settings. It may also involve the creation of real places (e.g., centers, publication platforms, conferences, and so on) where researchers can meet, as well as cooperative formats of work (for example, through jointly organized lecture series or training programs). Socio-spatial integration can even involve the creation of shared workspaces for researchers from the various disciplines, which would allow for regular personal encounters. However, socio-spatial integration often does not only involve the creation of social and/or spatial areas of overlap, but also the creation of new boundaries to highlight the innovative aspects of such integration as well as to demarcate relevant differences between the integrated domain of work and the separate "ancestral" disciplines. An example of such boundary setting is the establishment of institutionalized ethics committees for policy advice in controversial areas, such as stem cell research, applications of AI, dual use research, or specific biotechnologies (e.g., Wainwright et al., 2006). The widespread successful (albeit laborious) implementation of ethics committees in health research as well as the establishment of separate ethics units in medical schools can be taken as a model for the way in which ethics in other sciences could be realized. When following this model, however, it is necessary to reflect on the challenges and limits of such a form of socio-spatial integration. A particular task in this regard is to reflect on the implications that social-spatial relation of ethics to science has with

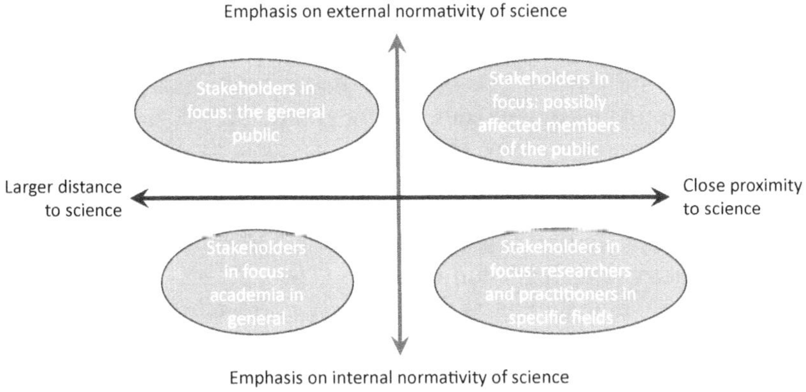

Fig. 1 The horizontal axis represents the social-spatial relation of ethics to science. The vertical axis represents placement of emphasis (more on external or on internal normativity)

regard to the stakeholders taken into account, depending on whether the emphasis is more on external or on internal normativity (see Fig. 1).

When the emphasis lies more on the internal normativity of science, the closer one is to scientific practice, the more specific researchers and practitioners will be the focus. Otherwise, one generalizes by addressing academia in general. When the emphasis is placed on the external normativity of science, close proximity requires addressing affected groups in society, whereas when considering science from a distance, it is the general public that is addressed.

The creation of new functions and the assumption of new tasks are essential for the establishment of a field of ethics in the life sciences, which sometimes results in a formalistic, even bureaucratic understanding of ethics (as happened with "ethics" as part of the work of research ethics committees). Spatial development is usually followed by permanent social demarcation processes, as it creates new sub-divisions and new struggles over resources, which also is an important factor in attempts at socio-spatial integration.

Finally, in *competence-oriented integration* the focus lies on the integration of different types of scientific and scholarly competencies. Ethical competency can be understood as a field-specific competency to analyze problems and to apply ethical theories or concepts to these problems.

Scientific competency, in contrast, can be understood as having sufficient knowledge of the state of the art in the science on which ethical reflection is supposed to take place, being able to assess the general potential and limits of scientific methodologies, having thorough background knowledge of the most important facts about the topic in question, and so on. The integration of both kinds of competencies requires researchers to explicitly integrate normative claims with empirical, factual knowledge.

Such efforts can be supported by new methodological approaches that have been developed during the last two decades, as found for example in revised meta-ethical discussions (e.g., Marchetti & Marchetti, 2016) and in work following the so-called empirical turn in applied ethics (e.g., Ives & Dunn, 2017; Mertz et al., 2014; Molewijk et al., 2004; Schicktanz, 2009). In both these approaches, the gap between the factual domain and the normative—the gap between the "is" and the "ought"—is not considered an unbridgeable divide. Instead, empirical research is used in a stepwise manner to identify normative conflicts or to provide a plausible basis for justifying social norms. Furthermore, empirical premises are stripped of their implicit, often hidden normativity (think in this respect of "thick concepts" such as "health," "welfare," "sustainability," or "diversity"), which allows us to discuss the justification of using related empirical data in political or social decision-making. A competence-oriented approach requires, we suggest, multidisciplinary training that does not prioritize ethical expertise over scientific expertise or vice versa. This overcomes the problem of fragmentation by incorporating methods of reflection over subject knowledge expertise. In so doing, the professional distinction between "the" ethicist and "the" scientist is at least partially abandoned as individuals to a certain degree become experts in both domains. This process involves collaborative reflection that allows all involved parties to develop their competencies beyond their original expertise. Joint training programs at different stages of continuous professional development, rather than as a single exercise, can support the implementation of such competence-oriented integration.

We do not see these three ways of integrating ethical reflection and science as exclusive strategies, but rather as complementary approaches that can—and should—be pursued together. While the analysis we presented in this chapter certainly is far from a final answer to the question how ethical reflection and the sciences could and should be joined, we hope that the discussion presented in this chapter of various ways in which ethics and the sciences can relate to each other, and the various aspects in which a better integration can be conceived can serve as a guide to possibilities as well as potential pitfalls on the way to a future ethics of the sciences.

Competing Interests The authors have no conflicts of interest to declare that are relevant to the content of this chapter.

References

ALLEA. (2023). *The European code of conduct for research integrity: Revised edition 2023*. ALLEA/All European Academies.

Bacon, F. (2017). *New Atlantis and the great instauration* (2nd ed.). Jerry Weinberger (Ed.). Wiley-Blackwell.

Benjamin, M., & Curtis, J. (1992). *Ethics in nursing*. Oxford University Press.

Boavida, N., & Moniz, A. B. (2016). Technology assessment in Non-PTA countries: An overview of recent developments in Europe. In A. Moniz & K. Okuwada (Eds.), *Technology assessment in Japan and Europe* (pp. 75–88). KIT Scientific Publishing.

Bridgman, P. W. (1947). Scientists and social responsibility. *The Scientific Monthly, 65*, 148–154.

Briggle, A., & Mitcham, C. (2012). *Ethics and science: An introduction*. Cambridge University Press.

Carson, R. (1962). *Silent spring*. Houghton Mifflin.

Committee on Science, Engineering, and Public Policy. (2009). *On being a Scientist: A guide to responsible conduct in research*. National Academies Press.

D'Angelo, J. G. (2019). *Ethics in science: Ethical misconduct in scientific research* (2nd ed.). CRC Press.

Desmond, H., & Dierickx, K. (2021). Research integrity codes of conduct in Europe: Understanding the divergences. *Bioethics, 35*, 414–428.

Deutsche Forschungsgemeinschaft. (2022). *Guidelines for safeguarding good research practice – Code of conduct (Revised Version 1.1)*. Deutsche Forschungsgemeinschaft.

Douglas, H. E. (2003). The moral responsibilities of scientists (Tensions between autonomy and responsibility). *American Philosophical Quarterly, 40*, 59–68.

European Commission. (2010). *European textbook on ethics in research*. Publications Office of the European Union.

Fanelli, D. (2009). How many scientists fabricate and falsify research? A systematic review and meta-analysis of survey data. *PLoS ONE, 4*, Article e5738.

Fangerau, H. (2014). Geschichte der Forschung am Menschen. In C. Lenk, G. Duttge, & H. Fangerau (Eds.), *Handbuch Ethik und Recht der Forschung am Menschen* (pp. 169–176). Springer.

Friedmann, B., & Kahn, P. H. (2007). Human values, ethics, and design. In J. A. Jacko & A. Sears (Eds.), *The human-computer interaction handbook* (pp. 1267–1292). CRC Press.

Fukuyama, F. (2002). *Our posthuman future: Consequences of the biotechnology revolution*. Farrar Straus & Giroux.

Gibbons, M., Limoges, C., Nowotny, H., Schwartzman, S., Scott, P. & Trow, M. (1994). *The new production of knowledge: The dynamics of science and research in contemporary societies*. Sage.

Guntzburger, Y., Pauchant, T. C., & Tanguy, P. A. (2017). Ethical risk management education in engineering: A systematic review. *Science and Engineering Ethics, 23*, 323–350.

Grunwald, A. (2019). *Technology assessment in practice and theory*. Routledge.

Ives, J., & Dunn, M. (2017). *Empirical bioethics: Theoretical and practical perspectives*. Oxford University Press.

Jasanoff, S. (2005). *Designs on nature: Science and democracy in Europe and the United States*. Princeton University Press.

Jasanoff, S. (2016). *The ethics of invention: Technology and the human future*. W.W. Norton & Company.

Kitcher, P. (2001). *Science, truth, and democracy*. Oxford University Press.

Kitcher, P. (2004). Responsible biology. *BioScience, 54*, 331–336.

Kitcher, P. (2011). *Science in a democratic society*. Prometheus Books.

KNAW, et al. (2018). *Netherlands code of conduct for research integrity*. https://www.nwo.nl/en/netherlands-code-conduct-research-integrity

Laplane, L., Mantovani, P., Adolphs, R., Chang, H., Mantovani, A., McFall-Ngai, M., Rovelli, C., Sober, E., & Pradeu, T. (2019). Why science needs philosophy. *Proceedings of the National Academy of Sciences, 116*, 3948–3952.

Lewens, T. (2004). What is genethics? *Journal of Medical Ethics, 30*, 326–328.

Lin, P., & Allhoff, F. (2007). Nanoscience and nanoethics: Defining the disciplines. In F. Allhoff, P. Lin, J. Moor, & J. Weckert (Eds.), *Nanoethics: The ethical and social implications of nanotechnology* (pp. 3–16). Wiley.

Lohse, S., Wasmer, M. S., & Reydon, T. A. C. (2020). Integrating philosophy of science into research on ethical, legal and social issues in the life sciences. *Perspectives on Science, 28*, 700–736.

Maehle, A.-H. (1993). The ethical discourse on animal experimentation, 1650–1900. In A. Wear, J. Geyer-Kordesch, & R. French (Eds.), *Doctors and ethics: The earlier historical setting of professional ethics* (pp. 203–251). Rodopi.

Marchetti, G., & Marchetti, S. (Eds.) (2016). *Facts and values: The ethics and metaphysics of normativity*. Routledge.

Max-Planck-Gesellschaft. (2022). *Leitplancken: Hinweise für verantwortliches Handeln/Guidelines for Responsible Conduct*. Max-Planck-Gesellschaft zur Förderung der Wissenschaften.

Merton, R. K. (1942). Science and technology in a democratic order. *Journal of Legal and Political Sociology, 1*, 115–126.

Merton, R. K. (1973). The normative structure of science. In R. K. Merton (Ed.), *The sociology of science, theoretical and empirical investigations* (pp. 267–278). University of Chicago Press.

Mertz, M., Inthorn, J., Renz, G., Rothenberger, L.G., Salloch, S., Schildmann, J., Wöhlke, S., & Schicktanz, S. (2014). Research across the disciplines: A road map for quality criteria in empirical ethics research. *BMC Medical Ethics, 15*, 1–14.

Miller, S. (2018). *Dual use science and technology*. Springer.

Molewijk, B., Stiggelbout, A. M., Otten, W., Dupuis, H. M., & Kievit, J. (2004). Empirical data and moral theory. A plea for integrated empirical ethics. *Medicine, Health Care and Philosophy, 7*, 55–69.

Morley, J., Elhalal, A., Garcia, F., Kinsey, L., Mökander, J., & Floridi, L. (2021). Ethics as a service: A pragmatic operationalisation of AI ethics. *Minds and Machines, 31*, 239–256.

Nowotny, H., Scott, P., & Gibbons, M. (2001). *Rethinking science: Knowledge and the public in an age of uncertainty*. Polity Press.

Pennock, R. T. (2019). *An instinct for truth: Curiosity and the moral character of science*. MIT Press.

Pradeu, T. (2025). "Philosophy in science", a distinctive approach combining philosophy and science. In S. J. Veigl & A. Currie (Eds.), *Methods in the philosophy of science: A user's guide* (pp. 271–287). MIT Press.

Pradeu, T., Lemoine, M., Khelfaoui, M., & Gingras, Y. (2024). Philosophy in science: Can philosophers of science permeate through science and produce scientific knowledge? *British Journal for the Philosophy of Science, 75*, 375–416.

Rabinow, P. (2011): *The accompaniment: Assembling the contemporary*. University of Chicago Press.

Reed, B. C. (2014). *The history and science of the Manhattan project*. Springer.

Rest, J. R., & Narvez, D. (Ed.) (1994). *Moral development in the professions: Psychology and applied ethics*. Lawrence Erlbaum.

Resnik, D. B. (1998). *The ethics of science: An introduction*. Routledge.

Reydon, T. A. C. (2013). *Wissenschaftsethik: Eine Einführung*. Eugen Ulmer.

Reydon, T. A. C. (2015). Plagiate als Professionalisierungsproblem. In C. Lahusen & C. Markschies (Eds.), *Zitat, Paraphrase, Plagiat: Wissenschaft zwischen guter Praxis und Fehlverhalten* (pp. 293–304). Campus.

Reydon, T. A. C. (2016). Die Wissenschaftsethik von Übermorgen: Zwischen Werten und Verantwortung. In M. Maring (Ed.), *Zur Zukunft der Bereichsethiken – Herausforderungen durch die Ökonomisierung der Welt* (pp. 303–316). KIT Scientific Publishing.

Reydon, T. A. C. (2020). What attitude should scientists have? Good academic practice as a precondition for the production of knowledge. In K. R.

McCain & K. Kampourakis (Eds.), *What is scientific knowledge?* (pp. 18–32). Routledge.

Schicktanz, S. (2009). Zum Stellenwert von Betroffenheit, Öffentlichkeit und Deliberation im empirical turn der Medizinethik. *Ethik in der Medizin, 21*, 223–234.

Schicktanz, S. (2011). Ethik in den Lebenswissenschaften – Überlegungen zur Frage der Verortung. In L. Kovács, J. Clausen, & T. Potthast (Eds.), *Darwin und die Bioethik* (pp. 335–351). Verlag Karl Alber.

Simmel, G. (1923). *Soziologie: Untersuchungen über die Formen der Vergesellschaftung, 3.* Duncker & Humblot.

Shneiderman, B. (2022). *Human-centered AI*. Oxford University Press.

Sober, E. (2022). Philosophy in science: Some personal reflections. *Philosophy of Science, 89*, 899–907.

Thompson, C. (2013). *Good science: The ethical choreography of stem cell research*. MIT Press.

Van de Poel, I., & Royakkers, L. (2023). *Ethics, technology, and engineering: An introduction*. Wiley-Blackwell.

Van Maanen, G. (2022). AI ethics, ethics washing, and the need to politicize data ethics. *Digital Society, 1*, 9.

Veigl, S. J., & Currie, A. (Eds.) (2025). *Methods in the philosophy of science: A user's guide*. MIT Press.

Warf, B., & Arias, S. (Eds.) (2008). *The spatial turn: Interdisciplinary perspectives*. Routledge.

Wainwright, S. P., Williams, C., Michael, M., Farsides, B., & Cribb, A. (2006). Ethical boundary-work in the embryonic stem cell laboratory. *Sociology of Health & Illness, 28*, 732–748.

Weindling, P. J. (2008). The Nazi medical experiments. In E. Emanuel, C. Grady, R. A. Crouch, R. Lie, F. G. Miller, & D. Wendler (Eds.), *The Oxford textbook of clinical research ethics* (pp. 18–30). Oxford University Press.

Wolpert, L. (2005). Is science dangerous? *Philosophical Transactions of the Royal Society B, 360*, 1253–1258.

Zussman, R. (2000). The contributions of sociology to medical ethics. *Hastings Center Report, 30*, 7–11.

Future Prospects for the ELSI Philosophy in a Time of Disruption

Eric M. Meslin

1 Introduction

The only time I met James Watson, co-discoverer of the structure of DNA and founding director of the US-led Human Genome Project (HGP), was during a coffee break at a Cold Spring Harbor Laboratory conference in 1997. I thought I had hit the mother lode: this was the opportunity to hear from the horse's mouth how the Ethical, Legal, and Social Implications (ELSI) program came to be and its prospects for the future. "Was it a political strategy, a personal commitment to ethical issues, both?" I remember asking Watson, giving him some latitude to explain his version of events. Oddly, he dismissed my question altogether—never answering—and continued down the hall with his coffee cup.

E. M. Meslin (✉)
Dalla Lana School of Public Health, University of Toronto, Toronto, Canada
e-mail: emeslin1020@gmail.com

Centre of Genomics and Policy, McGill University, Montreal, Canada

PHG Foundation, Cambridge, UK

J. Hinrichsen et al. (eds.), *Ethics of the Sciences*,
https://doi.org/10.1007/978-981-92-2821-8_3

Undaunted, this experience only increased my interest in the program I played a small role in shepherding in its early years.[1] As the world would come to appreciate over the next three decades, the seeds planted by Watson and cultivated by many others including his successor directors Francis Collins and Eric Green, researchers who benefited from the program's support, and a global community of diverse end-users, grew into a flourishing bush extending well beyond the walls of the US-centered grants program on the NIH campus in Bethesda, Maryland.

Other countries adopted their own versions of the ELSI program such as the Ethical, Legal, and Social Aspects (ELSA) initiatives found in many European countries, and the Genomics, Ethical, Environmental, Economic, Legal, and Social Issues (GE^3LS) program in Canada. New areas of research were explored, new methods developed, and new knowledge disseminated. Along the way ELSI would become the accepted term adopted for studying important moral and societal issues in genomics.[2]

This is the context for this chapter. It begins with a brief account of the origins of the US ELSI program, outlines some of its unique features, and discusses its wider impact and influence before considering future prospects for the ELSI "philosophy" at an especially challenging time when political and economic disruptions have affected the normal functioning of research, innovation, investment, and scientific progress.

2 A Brief History of ELSI's Origins and Evolution

Watson's silence in that hallway was disappointing but not debilitating since the ELSI origin story was already being curated by influential advisory committees from the US National Research Council (NRC, 1988), US Congress' Office of Technology Assessment (OTA, 1988),

[1] I was appointed Branch Chief at the National Human Genome Research Center in July 1996. Shortly after, NCHGR was renamed the National Human Genome Research Institute, and along with it came re-naming the Branch as a "Program," as did my title to Bioethics Program Director. I remained in that role until October 1998 when I became Executive Director of the US National Bioethics Advisory Commission.

[2] It is noteworthy that "ELSI" has transformed from an acronym used by the US government attached to a specific program to take on the identity of a regular word (like scuba, radar, NASA, or laser) in other parts of the world. It has yet to attain the status of an *anacronym*: an acronym where "knowledge of the words... has faded from common recall" (Shorter Oxford Dictionary, 2007).

US Department of Health and Human Services and Department of Energy (US DHHS/DOE, 1990); and from scholarly perspectives by those with deep inside knowledge (Juengst, 1994; Langfelder & Juengst, 1993; Cook-Deegan, 1994; Hanna, 1995). Most acknowledge Harold Schmeck's October 4, 1988 *New York Times* story reporting Watson's overall plan for the HGP as the first public mention of what would become the ELSI program:

> Watson discussed the human as well as the technical side of the project. A substantial amount of the project's future budget, *perhaps as much as 3 percent*, ought to be devoted to ethical issues, he believes. (Schmeck, 1988, p. C16; italics added)

Less than six months later, Watson's utterance was translated into practice with the release of a Program Announcement (PA) issued by the NIH Office of Genome Research (precursor to the National Center for Human Genome Research and later the National Institute for Human Genome Research) entitled "Ethical and Legal Studies Relating to the Program to Map and Sequence the Human Genome" (NIH, 1989). In 48 words, the Introduction to the PA describes what would become one of the most innovative research initiatives in bioethics history:

> The National Institutes of Health, under the auspices of the Human Genome Project, is interested in receiving applications for research grants or conference grants addressing the ethical, social, and legal issues that may arise from the application of knowledge gained as a result of the Human Genome initiative. (NIH Guide, 1989, p. 9)[3]

The establishment of the ELSI program was a significant milestone both in the history of genomics and science, and for bioethics generally. To understand how significant, the next section unpacks what was unique about what was established. In other words, what made ELSI...ELSI? Four features are noteworthy.

[3] It is noteworthy that many of the words and word order were still variably used in this Program Announcement. For example, the title of the announcement only lists "ethical and legal issues"; the Introduction refers to ethical, *social,* and legal issues (italics added); while the Research Scope refers to "ethical, legal, and social issues." Even the science itself is referred to as the Human Genome "Project," the Human Genome "initiative," and the Human Genome "Program."

First, it was *"designed-in" to the Human Genome Project from the start*. It has become common in emerging areas of technology to adopt the strategy of embedding principles or values directly into the governance structure from the start. This has been referred to as "by design," as evidenced by "privacy by design" (Cavoukian, 2009), "sustainability by design" (UNEP, 2011), "security by design" (Souppaya et al., 2022), and "ethics by design for AI" (Brey & Dainow, 2024). The idea is that by building ethical and social considerations in from the start, they will be (and be seen to be) necessary features of the policy architecture throughout its lifecycle, rather than as add-ons, impediments to overcome, or cosmetic afterthoughts.

Among the innovative features of the designed-in approach was the intention to build capacity to *anticipate and address* issues, an idea first proposed by the "NIH and DOE Joint Working Group on Ethical, Legal, and Social Issues" (shortened to ELSI Working Group). The full set of aims was as follows:

1. Anticipate and address the implications for individuals and society of mapping and sequencing the human genome;
2. Examine the ethical, legal, and social consequences of mapping and sequencing the human genome;
3. Stimulate public discussion of these issues; and
4. Develop policy options to assure that genetic information is used for the benefit of individuals and society (Fink, 1990).

Juengst (2021) has observed that ELSI's emphasis on anticipation aligned with the policy philosophy of *anticipatory governance* where the focus of analysis and advice begins upstream in the policy lifecycle, engages many voices, and looks to the future; as contrasted with approaches that are reactive, do not involve the public, and focus downstream on the impact of policy (Guston, 2014). Anticipatory governance can be traced to many sources including the futurist Alvin Toffler who first coined the phrase *anticipatory democracy* which proposed that decision-making in the face of overwhelming information overload requires novel, innovative tools to involve the public in decision-making (Toffler, 1975). Indeed, before there was an ELSI program, anticipatory analysis was being implemented in technology assessment, as evidenced in the US Congress' Office of Technology Assessment which adopted techniques of foresighting and horizon-scanning into its assessments (Bimber, 1996; Herdman, 1989). This led Juengst to conclude "the ELSI program's vision is the 1990's offspring of ...these intellectual movements" (Juengst, 2021, p. 3363).

Further confirmation of the intention for ELSI to "design ELSI in" to the HGP is found in the passage of the *National Institutes of Health Revitalization Act of 1993* which created the National Center for Human Genome Research (NCHGR). After listing the general purpose of NCHGR, the legislation identified six specific purposes, the last of which was "reviewing and funding proposals to address the ethical and legal issues associated with the genome project (including legal issues regarding patents)" (National Institutes of Health, 1993, (Sec 6)). Looking back on the language today, this subsection reads as a quaint description of the ELSI program. Yet in both symbolic and instrumental ways it set the stage for a new model of public science: where consideration of bioethical issues was designed-in as a necessary component of the broader science initiative.

Second, a *percentage of funding was committed to support ELSI research.* In US research politics, funding plays an outsized role in validating the importance given to any priority. This "power of the purse" is exercised in the appropriations that Congress approves every year. The ELSI program was unprecedented in the way it was funded and administered. Indeed, it has been called the "ELSI hypothesis—the idea that the best way to produce informed science policy for the HGP would be a formal budget set-aside within it" (Dolan et al., 2022, p. 1). The ELSI program had a specific mandate to administer an extramural grants program, like other NIH programs. This meant that the full administrative and policy infrastructure of the NIH, including peer review, grants administration, and a national advisory committee overseeing the process was also built-in to the funding model (Hanna, 1995).

Yet as important as embedding ELSI in the HGP and the adoption of an extramural funding program was, it was the funding model itself that was especially novel: a percentage of overall funding for the HGP itself would be set aside for funding ELSI projects. Until this time, funding for research on ethical issues in science, to the extent that academic scholars in the humanities and social sciences were funded at all, tended to be limited to fellowships, scholarships, and small awards for individuals. To be funded by an extramural program of the US National Institutes of Health, the largest biomedical research body in the world, was almost unheard-of for ethics research. To be funded using a percentage model was a brilliant economic approach, especially given the historical trend of increasing budgets at the NIH.

Table 1 provides a high-level view of how this funding grew by showing for each fiscal year the total extramural funding spent by the NIH on the HGP, the amount spent on ELSI, and the percentage that the ELSI amount reflected. Here we see that Watson's initial idea that "as much as 3%" be set aside was almost fully met in the 1990 budget that funded the 1989 Program Announcement, with grant awards totalling $1,571,036, or 2.6% of the overall extramural budget of $27.4 M spent on the HGP. Keen readers will note that the ELSI percentage varied for the first four years of the program, until passage of the *NIH Revitalization Act* that finally established 5% as the benchmark amount for ELSI funding. The key subsection of that law states:

> (c) (1) Except as provided in paragraph (2), of the amount appropriated to carry out subsection (a) for a fiscal year, the Director of the Center shall make available *not less than 5 percent* for carrying out paragraph (6) of such subsection. (italics added)[4]

Annual increases continued almost unabated through the first quarter of the twenty-first century, as shown in the table: by 2021 the NHGRI extramural budget had grown to $437 M, of which $23 M (5.26%) was spent on the ELSI program.

Third, *ELSI priorities co-evolved with the HGP's science goals.* From the outset, the ELSI program was seen as a fellow traveler with the HGP science goals and therefore as an essential feature of the HGP project itself. For example, in its 1990 five-year plan, the DHHS-DOE advisory committee devoted an entire chapter to "Ethical, Legal and Social Considerations." Among the goals to be reached by 1995 were to "Identify and define the major issues and develop initial policy options to address them" (US DHHS-DOE, 1990, p. 21). Juengst observed that

[4] Sometimes underappreciated is the significant role that the US Department of Energy (DOE) played in the ELSI history including their own significant financial commitment to projects which capped its ELSI set-aside at 3% (Patrinos & Drell, 1997). For example, between 1990 and 1996, DOE's ELSI program spent between—$1M and $2.2M each year, with DOE's total ELSI support by 1999 estimated at about $20M (ELSI Research Planning and Evaluation Group, 2000).

Table 1 NIH ELSI expenditures 1990–1996, 2018–2021

Fiscal Year	*Extramural NIH HGP ($ millions)*	*ELSI Expenditures ($ millions)*	*ELSI % of HGP NIH (%)*
1990	$27.4	$1.57	2.6
1991	$59.7	$3.93	6.2
1992	$70.7	$5.14	5.1
1993	$85.5	$5.33	6.2
1994	$93.7	$5.43	5.4
1995	$106.1	$5.30	5.0
1996	$117.2	$6.20	5.2
2018	$391.0	$21.0	5.37
2019	$405.0	$21.0	5.19
2020	$430.0	$22.0	5.12
2021	$437.0	$23.0	5.26

Adapted from Meslin et al. (1997), Dolan et al. (2022), NHGRI (2025)
Figures are rounded and exclude DOE amounts

the co-evolution of the ELSI goals along with the science goals of the HGP was intentional:

> Since the availability of this map will dramatically expedite the subsequent search for particular genes, 1995 has been set as the ELSI Program's deadline for offering policy approaches to the highest priority problems that the management of new genetic information is likely to pose. (Juengst, 1994, p. 122)

While the emphasis on anticipation was no longer highlighted as a prominent feature in the "post-sequence" period (i.e., after the complete draft was announced in 2003), co-evolution of the science and ELSI goals continued until the present day as evidenced by the five-year plans prepared in 1998 (Collins et al., 1998), 2003 (Collins et al., 2003), and in the more general strategic visions published in 2011 (Green et al., 2011) and 2020 (Green et al., 2020). It is noteworthy that following 2003, the ELSI program evolved along with the strategic thinking, adopting the term "Genomics and Society" as the overarching structure, a term that was to be taken up by many countries and initiatives. This also gave rise to a robust discussion about whether this post-sequence period called for a post-ELSI research agenda (Balmer et al., 2016; Myskja et al., 2014).

Fourth, there was an *intentional embedding of multidisciplinary expertise throughout the ELSI research portfolio.* The ELSI program may have been the first to use an acronym to connote the many disciplines needed to anticipate and address issues, but multidisciplinarity has always been a feature of the broader field of bioethics research (Mathews et al., 2016), teaching (Allana et al., 2024), and policy analysis (Meslin, 2025).[5] From its beginning when philosophy and theology were its principal disciplines (Reich, 1995; Pellegrino, 1999; Sass, 2008) bioethics regularly included new disciplines inside its informal tent including law, public health, psychology, sociology, philosophy of science, history, and economics and an equally diverse collection of empirical, conceptual, descriptive, normative, analytic, and summative methods (Sugarman & Sulmasy, 2010). Over the decades, further developments in transdisciplinary, interdisciplinary, and translational methods have been added (Austin et al., 2008; Choi & Pak, 2006; Jones & Hens, 2023). Whether ELSI has a set of methods separate from those in bioethics is the subject of continued discussion (Ogbogu & Ahmed, 2022; Walker & Morrisey, 2014).

What made the ELSI program unique, however, was the expectation that multidisciplinary approaches would be embedded into the science goals of the HGP itself. As Juengst has asserted:

> For the ELSI WG, the anticipation of consequences was counted alongside genetic mapping and the development of sequencing technologies as part and parcel of the Human Genome Project. At the same time, however, this effort could not be limited to genome scientists. To gain foresight on the possible consequences of the genomic revolution the effort's forecasts required the scope that could only be achieved by combining as wide a range of perspectives as possible. (Juengst, 2021, p. 3371)

This multidisciplinary commitment has had pros and cons. Suffice it to say, as with most government acronyms ELSI's creation was an important efficiency tactic: letting single letters stand for clusters of topics without having to stipulate each in detail provided a convenient heuristic for notifying the public that three broad domains of concern—ethical, legal, social—were being emphasized. For example (E)thical is presumed to include moral, value, and other normative matters; (L)egal must

[5] See also Reydon and Schicktanz's chapter in this volume "Ethics and the Sciences: Dimensions of Interaction and Interdisciplinarity".

include jurisprudential, regulatory, legislative, judicial, and other governance considerations; (S)ocial captures all civic, political, cultural, and related topics and so on. As discussed below, however, the "acronymization" would soon present its own set of challenges, from good natured ribbing about the use of "E,L,S,I" itself (Powledge, 2003)[6] to more serious concerns about allowing letters to stand for larger domains of knowledge blurring important disciplinary approaches, meanings and values which one critic referred to it as "ELSIfication" (López & Lunau, 2012).

Having described the early history, key characteristics, and unique features of ELSI, let us turn to its impact and influence.

3 Impact and Influence of the ELSI Philosophy

It is unlikely the early HGP architects could have predicted the global uptake and experimentation with the ELSI philosophy. Indeed, one need only look at the widespread use of the acronym to see evidence of what the social historian Yuval Harari calls the "contagion model" of information dissemination (Harari, 2024). But focusing academic and governmental attention on ethical concerns about the use of genetic science had a long history pre-dating ELSI and the HGP. Two well-studied areas were the unethical uses of genetics in the name of political ideology, including the horrific eugenics, genocide, and ethnic cleansing policies in many countries both before and after WWII (Annas, 2010; Lombardo, 2022), and the more normative investigations of new reproductive and genetic technologies that emerged in the latter half of the twentieth century (Walters, 1987).

It is common in academic circles, but ultimately unhelpful, to equate the number of publications with impact. That said, several efforts have been made to count the number of ELSI-related publications in the peer-reviewed literature from the early days of the HGP until the present (Callier et al., 2016; Smit et al., 2024; Walker & Morrisey, 2012) which largely draw from databases such as PubMed, Scopus, and Web

[6] In her assessment of NHGRI's 2003–2008 5-year plan, Powledge remarked: "Ethical, legal, and social issues have been saddled with the unfortunate acronym ELSI, providing an inevitable source of jokes about NHGRI's ELSI program serving as a cash cow for bioethics" (Powledge, 2003). The humor in the comment is based on the word-play of "Elsie the Cow" a widely recognized mascot for a leading dairy company in the US.

of Science. Estimates from these studies are in the order of many thousands of papers, but likely a significant undercount, as it would exclude non-peer-reviewed papers such as books and book chapters, the gray literature, technical reports and commissioned work, blogs, and the growth in non-English publications. Indeed, the output of the ELSI program's productive history of funded research, conferences, workshops, and training programs has grown sufficiently to warrant its curation in the growth of databases including at NHGRI (National Human Genome Research Institute, 2025), the ELSIHub within the Center for ELSI Research and Analysis (ELSIHub, 2025), and Georgetown University's National Center for Bioethics Literature (Georgetown University Library, 2025).

Awareness of the need to evaluate US ELSI program's impact was also built-in from the start (Burke et al., 2015; ELSI Research Program Evaluation Group, 2000; NIH-DOE Task Force, 1993; US DHHS-DOE, 1990; Juengst, 1994; Langfelder & Juengst, 1993; Meslin et al., 1997; McEwen et al., 2014). Equally well known in policy circles is the difficulty in assessing components of the policy process and its outcomes. Below are two categories of ELSI impact and influence over the years: its uptake and adaptation in different forms by others, and its role in informing policy.

Uptake and Adaptation. If mimicry is the sincerest form of flattery, the US ELSI program is entitled to the compliment, as evidenced by the diversity of global programs and initiatives over the past three decades. ELSI is found either by name or design in research findings, programs, conferences, curricula, and policy documents in dozens of countries, on every continent, including outer space (Feys et al., 2024). In the early years, ELSI was adapted by several formal funding programs using similar letters, such as the Medical, Ethical, Legal, and Social Implications (MELSI) program created in 1993 by the Canadian Genome Analysis and Technology initiative (López & Robertson, 2007), or the Ethical, Legal, and Social Aspects (ELSA) program adopted in 1994 by the European Union in its 4th EU Framework Program (Hilgartner et al., 2016). ELSAGEN was adopted in 2009 as the name of a multinational initiative established by Austria, Finland, and Germany (Chadwick & Zwart, 2013; Zwart & Nelis, 2009). From 2005 to 2010 an initiative involving many European countries, referred to as ERA-SAGE (European Research Area on Societal Aspects of Genomics) used the ELSA model (Nydal et al., 2016).

Among the more innovative of these initiatives is Genome Canada's GE^{3}LS program. Now a quarter century old GE^{3}LS was unique in three ways: first, by adding "economic" and "environmental" it acknowledged the importance of these disciplines as distinct from the monolithic "ethical" considerations. Second, the GE^{3}LS program applied to other sectors beyond human health including forestry, agriculture, aquaculture, and natural resources. Third, Genome Canada required as a condition of funding awards that GE^{3}LS research (and researchers) be fully "integrated" into all science proposals submitted for funding (Kosseim & Chapman, 2011, p. 205).

Other countries provided funding for ethical, legal, and social issues though not always under the formal banner of an acronym. For example, South Korea funded an ethical, legal, and social implications program in 2001 (Yoon et al., 2010); the Dutch government set aside 4% of the country's €189 M 2001 budget to study ethical, social, economic, psychological, and legal aspects of genomics; the Netherlands included a 5% set-aside in 2004 to study "the societal components of genomics research" (Zwart et al., 2014); and in 2008 the Research Council of Norway funded ELSA aspects of biotechnology, nanotechnology, and cognitive science (Research Council of Norway, 2014). France's current *Genomics Plan 2025* explicitly includes "Ethical considerations at every step of the France Genomic Medicine Plan 2025" (Alliance Nationale, 2025). To this list one can add many other countries who have recognized the value of ELSI major projects, such as the Brazilian Longitudinal Study of Aging, called ELSI-Brazil (Lima-Costa et al., 2018), the many initiatives in Southeast Asia (Mikami et al., 2021; Yoshizawa et al., 2014), China (Chen et al., 2022), and the African continent (Akinyemi et al., 2020; Sommerfeld & Oduola, 2007).

The ELSI philosophy has also been taken up by philanthropies using different strategies, including The Bill and Melinda Gates Foundation's Ethical, Social, and Cultural (ESC) program (Lavery et al., 2013; Singer et al., 2007); The John Templeton Foundation's support of a Global Observatory for Gene Editing (Jasanoff & Hurlbut, 2018); and the Indianapolis-based Lilly Endowment whose $105 M gift establishing the Indiana Genomics Initiative at the Indiana University School

Medicine, included $5 M (5%) to establish the Indiana University Center for Bioethics (Indiana University, 2001).[7]

Informing Policy. From the outset, the US ELSI program was intended to serve as a policy advice resource. Among the first actions undertaken after Watson's 1988 press conference, was the establishment of the ELSI Working Group to help coordinate the work by the NIH and the DOE (Yesley, 1991; McCain, 2002; Wolfe 2002). It provided advice that would eventually result in the 1989 Program Announcement, and then later on specific policy issues. It was authorized to provide policy options to government to ensure that genetic testing not be misused, with advice about testing for Cystic Fibrosis among the first cases of influence (Cook-Deegan, 1994). For more than three decades thereafter, the ELSI hypothesis was tested on a regular basis. The same databases that curate academic publications include hundreds of papers focused on policy matters though an accurate count would be impossible, in part because policy itself is a complex concept, referring to dozens of different instruments including regulations, legislation, guidelines, standards of practice, declarations and treaties, and because policy development by its nature is messy for both epistemic and practical reasons (Meslin, 2022). Rare, therefore, is the situation when a publication arising from ELSI research has findings so compelling and unavoidably clear that decision-makers have no choice but to adopt a policy based on them. But ELSI's DNA can be found in policies, regulations, guidelines across the globe.

In the US, ELSI research informed the development of the *Genetic Information Non-Discrimination Act of 2008* legislation that was informed by early ground-breaking research on genetic privacy by ELSI-funded researchers such as Annas (1995). It also helped set the stage for revisions to the US federal regulations governing the protection of human participants in research to account for genomic studies (Hudson & Collins, 2015), and continues to the present day (Chapman et al., 2025).

While the US effort enjoys a long history of influence and impact, the ELSI philosophy's most significant may have been its uptake and use globally by international organizations to convert advice into policy. International organizations and professional bodies established ELSI committees and offices to provide capacity for advising, developing policy, and updating their advice to members.

[7] I was Founding Director of the Indiana University Center for Bioethics from 2001 to 2016, benefiting from this generous $5 M endowment in the early years.

Describing the sweeping breadth of these institutions and programs would require a separate chapter, but consider two illustrative examples: the first is the historic adoption by the United Nations of the *Universal Declaration on the Human Genome and Human Rights* (United Nations, 1997).[8] Stimulated by the HGP and other developments including the announcement of Dolly the cloned sheep, UNESCO was tasked with developing an international declaration to guide policy in the context of bioethics and human rights. It became the first legal and ethical framework at the global level to discuss such issues (Lenoir, 1999), from which followed other UN declarations including one on bioethics itself (Andorno, 2007).

The second example is the establishment of Global Alliance for Genomics in Health (GA4GH), an organization of more than 1000 collaborators in 90 countries with a focus on data sharing, which has facilitated the development of dozens of guidelines, standards, and best practices (Rehm et al., 2021). A key feature of the GA4GH structure is its Regulatory and Ethics Workstream, further illustrating the value of designing-in ethics from the start. This is illustrative of the ELSI's philosophy's policy influence which now extends to networks and innovative consortia (Kaye et al., 2012), embedding ELSI expertise within institutional governance (Karabin et al., 2024), and the growth of ELSI-specific services for scientists such as the ELSI service desk at BBMRI-ERIC (2025), and the ELSI Office at the ELIXIR-Europe bioinformatics platform (ELIXIR, 2022).

ELSI's history was not without challenges. Accompanying the enthusiasm for the well-intended recognition of the ELSI philosophy was an equally cautious view about its relevance and value (Marshall, 1996; Roberts, 1993), about ways in which the ELSI goals appeared to diverge from and potentially conflict with the goals of the HGP (Andrews, 1999), and that its preoccupation with identifying concerns and challenges on the one hand and hype on the other may have restricted its legitimacy and progress, stifling innovation (Caulfield et al., 2013; Evans et al., 2011). These considerations are also part of ELSI's legacy to be acknowledged as we contemplate its future.

[8] What follows is based in part on Meslin and Lavery (2025), which also discusses the role I played in the Declaration's drafting and final approval while representing the HGP ELSI Program on behalf of the US government.

4 Future Prospects in a Time of Disruption

It is always difficult to predict the future during a time of rapid technological change. The speed of innovation itself makes predictions tenuous at best (Friedman, 2016), and because technology by its nature can be disruptive, in both a positive and negative sense (Perleth et al., 2022). Genomics is an example of both forms of disruption, from the early days of recombinant DNA research when the public was fearful of escaping genetically engineered molecules, to present-day innovations in genome sequencing, stem cell science, gain-of-function studies, and CRISPR. With the arrival of robotics, quantum computing, 3D printing, and the suite of applications of artificial intelligence (Sajeer, 2023), any foresighting for the future role of ELSI will be that much more challenging.

As this chapter is being written in 2026 when the world is experiencing significant political, economic, and social disruption, several realities are adding complexity to any serious conversation about the future prospects for the ELSI philosophy. Among the most visible is the widespread damage done by US President Trump's attack on science, the rule of law, public health, and social policy. During his first 100 days US government actions have shut down or hollowed out key regulatory and oversight programs for medical research, food safety, and public health; withdrawn the US or threatened its departure from key global organizations such as WHO and NATO; fired or hastened the resignations of respected experts from leadership roles in the US research ecosystem; tried to limit the capacity of universities to function as places of academic freedom; and significantly cut or eliminated broad areas of health and science research reflected in the President's FY 2026 federal budget request that included a 40% (~$17.5B) cut to the US National Institutes of Health, plus cuts of ~$4.7B from the National Science Foundation, $3.5B from the Centers for Disease Control and Prevention, and many other areas of research (Vought, 2025).

The combination of emerging technology disruption and political disruption does not fully capture the context for imagining the prospects for ELSI. Indeed, three other disruptive influences are at work: the erosion of public trust in science and expertise generally (Priest et al., 2003); the ominous increase in the use of misinformation and disinformation (Council of Canadian Academies, 2023); and the polarization of civic discourse (Pew Research Center, 2023). Taken together, we find ourselves at a crossroads where trust and truth are considered suspect values, to say nothing of longstanding ethical principles and moral virtues that are essential for social cohesion. It is against this backdrop that two futures are presented, beginning with the most ominous.

The End of ELSI. Fifteen years ago, the respected philosopher and distinguished founding parent of bioethics Ruth Macklin provided a dystopian analytic of what she called "the death of bioethics":

> In a steady progression over the years, the field became increasingly fragmented and bureaucratized. Disagreement and dissension were rife, and this once flourishing, multidisciplinary field began to splinter in multiple ways. Prominent journals folded, one by one, and were replaced with specialized publications dealing with genethics, reproethics, nanoethics, and necroethics. Mainstream bioethics organizations also collapsed, giving way to new associations along disciplinary and sub-disciplinary lines. (Macklin, 2010)

I used Macklin's quote when I presented the arguments in this chapter to a public audience in April 2025, rewriting her phrase as "The End of ELSI," since it—and a similar one, "The Death of"—are catchphrases in popular book titles like Fukuyama's *The End of History* (1992), Horgan's *The End of Science* (1996), de Chosal's *The End of Democracy* (2017), Nichols' *The Death of Expertise* (2017) and Kakutani's *The Death of Truth* (2018). The idea that ELSI would end seemed dramatic enough for a public lecture, but given the widespread disruption the world is experiencing, we may need to take this possibility more seriously.

On the one hand, this future is not without justification. After all, the ELSI program was designed-in from the start to anticipate and address issues that would arise with the mapping and sequencing of the human genome. Once that goal was met in 2003, ELSI's original work was done, its purpose having been fulfilled, even though five-year plans and strategies were adopted for this post-sequence period. This view sees government support for innovative programs as the equivalent of financial backing of technology start-ups: with risk being taken on by investors in the early years but who expect the program to eventually be self-sustaining. Moreover, even with its demonstrated impact and influence one could imagine a future in which society does not believe public funds should be used to support ELSI research, but that such support should come from private or philanthropic organizations.

On the other hand, ELSI could end for more ominous reasons: that its very existence is inconsistent with prevailing views of the government in power. In a world where increased censorship and restriction of human rights are on the rise, the public support of a philosophy that challenges this trajectory is unlikely to enjoy support. Given funding restrictions for a wide variety of humanities and social sciences research adopted by the US government in 2025, it is easy to see how the 5% set-aside committed to ELSI research could end, and with it the harm of lost opportunity (Datz, 2025).

Either way, the prospect of learning lessons (Fisher, 2005; Henderson et al., 2012; Walker & Morrissey, 2012) ensures that, despite the current plausibility of this dire future, it is not assured. While there is a strong likelihood in the short run of scaling back or eliminating the US ELSI program and one cannot presume that this will be the case everywhere. A sign of such optimism may be found in the recent publication from the national genome programs of Australia, Canada, Denmark, Qatar, Singapore, UK, and US which identified "embedding ELSI considerations into human genomics and genomic medicine research" as one of six "converging priorities" they jointly share (Howley et al., 2025). Nor should one presume that decisions about growing, maintaining, or constraining ELSI *programs* are identical with similar actions about the idea, the philosophy of ELSI. Combined with the embedding of ELSI features in major research consortia and organizations, the expansion of an active and engaged ELSI research community, the imminent death of the idea is far less likely. Perhaps the threat of reduced funding will force a type of reckoning, similar to what international global health and development is facing as the US pulls back from its international commitments: time for a new business model.

Diversification Through Convergence. Another plausible future follows from Macklin's foresighting about specialization in bioethics. Her concern was that bioethics was in a death spiral arising from a fragmentation of the field, with specialized language, journals, and an emphasis more on bureaucracy than innovation. One can see evidence of the specialization claim in genomics as its tent enlarged to include many "omics" (e.g., pharmacogenomics, metabolomics, lipidomics), the use of new sequencing technologies, and biobanking and big data initiatives. ELSI's tent expanded accordingly: early priorities such as privacy, discrimination, and return of genetic information extended into regenerative medicine, direct-to-consumer marketing, public health, data sharing, and enhancement.

We should note that increasing specialization can also be a sign of a healthy ecosystem—the policy equivalent of natural selection and genetic diversity. But as more technologies (genomic and otherwise) are brought online the continuation of the specialization trajectory may heighten risks for ELSI, especially given recent cautions about whether bioethics stretched itself too thinly responding to the COVID-19 pandemic (Vonderschmidt et al., 2023) and whether it has sufficient expertise to advise on artificial intelligence (Čartolovni et al., 2022; McLennan et al., 2020).

Specialization should not be confused with diversification, which may occur either by splitting up or combining, where the creation of new connections and collaborations produce specialized knowledge. A compelling example can be found in the work on the technology convergence including recent treatises on a triple convergence of genetics, biotechnology, and artificial intelligence called "superconvergence" (Metzl, 2024), or the merging of digital and biological technologies into an area called biodigital convergence (IEC, 2024; Policy Horizons Canada, 2019). This diversification-by-convergence presents an exciting case study in how multidisciplinary expertise can be brought to bear on problems larger than any single type of technology.

Another form diversification may take is around foundational theory used to ground future work. Physicists have long coveted a unified field theory to explain their world, and it is no less compelling a search to construct a theory that helps explain ethical, legal, and social challenges in science (Meslin & Blasimme, 2013). This is easier said than done, as highlighted in discussions about whether ELSI is a standalone discipline or a sub-field of bioethics (Myskja et al., 2014). Nor is this new, as found in early discussions about any theory differences between bioethics, philosophy of medicine, and medical humanities (Engelhardt, 1990), and more recently between bioethics and science and technology studies (Müller et al., 2023).

A final example of diversification draws on the anticipatory governance model discussed above, where ELSI's anticipatory philosophy is now employed in the ever-widening science of genomics and its "omics" spin-offs such as vaccinomics (Ozdemir et al., 2011) and metabolomics (Fuse et al., 2019). It is also evident in Genome Canada's GE^{3}LS program whose sector approach to genomics extends beyond human health. We also find ELSI considerations being applied in areas both adjacent to and well outside of genomics including nanotechnology (Guston, 2014; Nakagawa et al., 2010), technology assessment (Ali-Kahn et al., 2015; Krahn et al., 2020), security studies (Jerković & Siedschlag, 2021), and other emerging technologies (OECD, 2024). The OECD framework proposes an approach that would be familiar to the ELSI community under the heading "Embed values throughout the innovation process":

> Innovation actors should strive to integrate values throughout the innovation process, not merely as an aspirational initial statement. This process includes, but is not limited to, four phases: (1) agenda-setting and research and development (R&D) design, (2) R&D, (3) early commercialisation, and (4) diffusion and deployment... This requires innovation actors to seek to align technological development with values in particular institutional contexts at particular loci in that process. (OECD, 2024, p. 17)

But perhaps ELSI's most intriguing, forward-looking, future-proofing trajectory will result from the global attention to the planet itself. Already there are projects associated with ELSI and biodiversity where scholars with specialized knowledge in science, environment, ethics, and governance have begun to collaborate (Doezema, 2023; Sherkow et al., 2022). This expertise will fill a growing need in the private sector, among NGOs, and at the highest levels of international governance. There is a symmetry to this perspective, coming full circle to van Rensellaer Potter's view more than five decades ago that bioethics focuses on matters much broader than biomedical issues, and should be seen as a "bridge" between the present and the future, nature and culture, science and values, and humankind and nature (Potter, 1971; ten Have, 2012). Having observed the roll-out of the UN Convention on Biodiversity (1993) and follow-on instruments including the *Cartagena Protocol on Biosafety* (2000) and the *Nagoya Protocol on Access and Benefit-Sharing* (2014), one can expect greater engagement between bioethics, ELSI, and biodiversity in the years to come.

5 Concluding Thoughts

Reflecting on my hallway encounter with Watson three decades later, I may have been wrong to expect a definitive "answer" to my question about ELSI's origin. I was prepared for the answer to be filtered through his recollections and biases because I was familiar with the unitary origin story myth in any accounting of history, including bioethics (Gaines & Juengst, 2008), but I may not have been sufficiently open to the idea that he may not have had a definitive answer himself (nothing wrong with that). In retrospect, his silence did me a favor: it planted a seed of inquiry that has served me well: to understand developments in science and policy as more flexible than rigid, especially when people are involved. But what is not in doubt is that ELSI unleashed new ways of looking at science which brought ethics inside the policy tent.

Potter understood the object of bioethics study as a large tent from the smallest molecule to the largest ecosystem—akin to the spectrum from DNA to global health systems (Pang, 2002). “All of these problems” wrote Potter, “call for actions that are based on values and biological facts” (Potter, 1971, p. viii). ELSI’s emergence at the outset of the HGP was a response to such a call for action. It was an intentional feature built into the HGP by enlightened scientists, legislators, humanists, and a public willing to trust their intentions. Its legacy includes the publications, policies, and people that have participated in its activities, and countless more who have benefited from it. There will be a future for ELSI, and it will be fascinating to watch it unfold.

Acknowledgements A version of this paper was presented as “Future Prospects for ELSI/GE^{3}LS in a Time of Disruption” at the Health Services, Policy, and GE^{3}LS Workshop convened by the Canadian Institutes of Health Research, Institute on Genetics, Ottawa, Canada, April 2, 2025. I thank the reviewers of this chapter for their input and suggestions.

Competing Interests The author has no conflicts of interest to declare that are relevant to the content of this chapter.

References

Akinyemi, R. O., Jenkins, C., Nichols, M., Singh, A., Wahab, K., Akpalu, A., Sarfo, F. S., Owolabi, L. F., Obiako, R., Akinyemi, J., Ojebuyi, B., Adigun, M., Musbahu, R., Bello, A., Titiloye, M., Calys-Tagoe, B., Ogunronbi, M., Uvere, E., Laryea, R., ... Arulogun, O. (2020). Unraveling the ethical, legal, and social implications of neurobiobanking and stroke genomic research in Africa: A study protocol of the African neurobiobank for precision stroke medicine ELSI project. *International Journal of Qualitative Methods*, *19*.

Ali-Khan, S. E., Black, L., Palmour, N., Hallett, M. T., & Avard, D. (2015). Socio-ethical issues in personalized medicine: A systematic review of English language health technology assessments of gene expression profiling tests for breast cancer prognosis. *International Journal of Technology Assessment in Health Care, 31*(1–2), 36–50.

Allana, A. A., Ali, S. K., & Ghias, K. (2024). Bioethics curriculum for undergraduate medical students: An evaluation study utilizing mixed methods approach. *BMC Medical Education, 24*(1), 385.

Alliance Nationale pour les sciences de la vie et la santé. (2025). *Genomic medicine France 2025.* Retrieved May 2, 2025, from https://sante.gouv.fr/IMG/pdf/genomic_medicine_france_2025.pdf

Andorno, R. (2007). Global bioethics at UNESCO: In defence of the universal declaration on bioethics and human rights. *Journal of Medical Ethics, 33*(3), 150–154.

Andrews, L. B. (1999). *The clone age: Adventures in the new world of reproductive technology.* Henry Holt and Co.

Annas, G. J. (2010). *Worst case bioethics: Death, disaster and public health.* Oxford University Press. http://ndl.ethernet.edu.et/bitstream/123456789/58107/1/19.pdf

Annas, G. J., Glantz, L. H., & Roche, P. (1995). The genetic privacy act and commentary. Boston University School of Public Health; Boston, MA [8 July 2008]. Retrieved May 5, 2025, from http://www.ornl.gov/sci/techresources/Human_Genome/resource/privacy/privacy1.html

Austin, W., Park, C., & Goble, E. (2008). From interdisciplinary to transdisciplinary research: A case study. *Qualitative Health Research, 18*(4), 557–564.

Balmer, A. S., Calvert, J., Marris, C., Molyneux-Hodgson, S., Frow, E., Kearnes, M., Bulpin, K., Schyfter, P., Mackenzie, A., & Martin, P. (2016). Five rules of thumb for post-ELSI interdisciplinary collaborations. *Journal of Responsible Innovation, 3*(1), 73–80.

BBMRI-ERIC. (2025). BBMRI-ERIC ELSI services and research. Retrieved July 1, 2025, from https://www.bbmri-eric.eu/services/elsi/

Bimber, B. A. (1996). *The politics of expertise in congress: The rise and fall of the office of technology assessment.* State University of New York Press.

Brey, P., & Dainow, B. (2024). Ethics by design for artificial intelligence. *AI and Ethics, 4*(4), 1265–1277.

Burke, W., Appelbaum, P., Dame, L., Marshall, P., Press, N., Pyeritz, R., Sharp, R., & Juengst, E. (2015). The translational potential of research on the ethical, legal, and social implications of genomics. *Genetics in Medicine, 17*(1), 12–20.

Callier, S. L., Abudu, R., Mehlman, M. J., Singer, M. E., Neuhauser, D., Caga-Anan, C., & Wiesner, G. L. (2016). Ethical, legal, and social implications of personalized genomic medicine research: Current literature and suggestions for the future. *Bioethics, 30*(9), 698–705.

Čartolovni, A., Tomičić, A., & Lazić Mosler, E. (2022). Ethical, legal, and social considerations of AI-based medical decision-support tools: A scoping review. *International Journal of Medical Informatics, 161*(104738), 104738.

Caulfield, T., Chandrasekharan, S., Joly, Y., & Cook-Deegan, R. (2013). Harm, hype and evidence: ELSI research and policy guidance. *Genome Medicine, 5*(3), 21.

Cavoukian, A. (2009). Privacy by design: The 7 foundational principles. Information and Privacy Commissioner of Ontario.

Chadwick, R., & Zwart, H. (2013). From ELSA to responsible research and Promisomics. *Life Sciences, Society and Policy, 9*(1).

Chapman, C. R., Quinn, G. P., Natri, H. M., Berrios, C., Dwyer, P., Owens, K., Heraty, S., & Caplan, A. L. (2025). Consideration and disclosure of group risks in genomics and other data-centric research: Does the common rule need revision? *The American Journal of Bioethics: AJOB, 25*(2), 47–60.

Chen, Z., Jiang, W., Xu, J., Jiang, L., & Wang, G. (2022). Ethical, legal, and social implications of genomics in China: A scoping review and implications for precision public health. *China CDC Weekly, 4*(32), 706–710.

Choi, B. C., & Pak, A. W. (2006). Multidisciplinarity, interdisciplinarity and transdisciplinarity in health research, services, education and policy: 1. Definitions, objectives, and evidence of effectiveness. *Clinical Investigative Medicine, 29*(6), 351–364.

Collins, F. S., Patrinos, A., Jordan, E., Chakravarti A., Gesteland, R., Walters, L., & the members of the DOE and NIH planning groups. (1998). New goals for the U.S. Human Genome Project: 1998–2003. *Science (New York, N.Y.), 282*(5389), 682–689.

Collins, F. S., Green, E. D., Guttmacher, A. E., & Guyer, M. S. (2003). A vision for the future of genomics research. *Nature, 422*(6934), 835–847.

Cook-Deegan, R. (1994). *The gene wars: Science, politics and the human genome*. WW Norton.

Council of Canadian Academies. (2023). *Fault lines: Expert panel on the socioeconomic impacts of science and health misinformation*. CCA. Retrieved May 1, 2025, from https://cca-reports.ca/wp-content/uploads/2023/02/Report-Fault-Lines-digital.pdf

Datz, T., What's lost when federal funding for research is abruptly terminated. Retrieved July 2, 2025, from https://hsph.harvard.edu/news/whats-lost-when-federal-funding-for-research-is-abruptly-terminated/

de Chosal, C. B. (2017) *The end of democracy*. Tumblar House.

Doezema, T. (2023). The promise of ELSI: Coproducing the future of life on earth. *Science as Culture, 32*(4), 461–485.

Dolan, D. D., Lee, S. S.-J., & Cho, M. K. (2022). Three decades of ethical, legal, and social implications research: Looking back to chart a path forward. *Cell Genomics*, *2*(7).

ELIXIR. (2022). Ethical, Legal and Social Implications (ELSI) services officer. Retrieved July 1, 2025, https://elixir-europe.org/about-us/vacancies/ethical-legal-and-social-implications-elsi-services-officer

ELSI hub. (2025). *ELSI publications*. Retrieved July 1, 2025, from https://elsihub.org/search/publications

ELSI Research Planning and Evaluation Group (ERPEG). (2000). *A review and analysis of the Ethical, legal, and Social Implications (ELSI). Research programs of the national institutes of health and the department of energy*. Retrieved April 28, 2025, from https://www.genome.gov/sites/default/files/genome-old/pages/Research/DER/ELSI/erpeg_report.pdf

Engelhardt, H. T., Jr. (1990). The birth of the medical humanities and the rebirth of the philosophy of medicine: The vision of Edmund D. Pellegrino. *The Journal of Medicine and Philosophy*, *15*(3), 237–241.

Evans, J. P., Meslin, E. M., Marteau, T. M., & Caulfield, T. (2011) Genomics. Deflating the genomic bubble. *Science*, *331*, 861–862.

Feys, R., Uygun, K., Von Reiterdank, I. F., Wolf, S. M., & Isasi, R. (2024). Biopreservation beyond the biosphere: Exploring the ethical, legal & social implications of suspended animation in space. *The Journal of Law Medicine & Ethics*, *52*(3), 648–665.

Fink, L. (1990). Human genome project progress: Backgrounder; NCHGR bioethics program addresses social and ethical implications of knowledge about human genes. *National Center for Genome Research*. Retrieved April 22, 2025, from https://repository.digital.georgetown.edu/handle/10822/556962

Fisher, E. (2005). Lessons learned from the ethical, legal and social implications (ELSI). *Technology and Society*, *27*(3), 321–328. https://www.sciencedirect.com/science/article/abs/pii/S0160791X05000278

Friedman, T. (2016). *Thank you for being late: An optimist's guide to thriving in the age of accelerations*. MacMillan Press.

Fukuyama, F. (1992). *The end of history and the last man*. Free Press.

Fuse, N., Sakurai-Yageta, M., Katsuoka, F., Danjoh, I., Shimizu, R., Tamiya, G., Nagami, F., Kawame, H., Higuchi, S., Kinoshita, K., Kure, S., & Yamamoto, M. (2019). Establishment of integrated biobank for precision medicine and personalized healthcare: The Tohoku medical megabank project. *JMA Journal*, *2*(2), 113–122.

Gaines, A. D., & Juengst, E. T. (2008). Origin myths in bioethics: Constructing sources, motives and reason in Bioethics. *Culture, Medicine and Psychiatry*, *32*(3), 303–327.

Georgetown University Library. (2025). Publications of the national reference center for bioethics literature. Retrieved July 1, 2025, from https://repository.digital.georgetown.edu/collections/10b1ee82-b397-4f2a-99cd-c4b6da134e72

Green, E. D., Guyer, M. S., & National Human Genome Research Institute. (2011). Charting a course for genomic medicine from base pairs to bedside. *Nature, 470*(7333), 204–213.

Green, E. D., Gunter, C., Biesecker, L. G., Francesco, D., Valentina, E., Carla, L., Feingold, E. A., Felsenfeld, A. L., Kaufman, D. J., Ostrander, E. A., Pavan, W. J., Phillippy, A. M., Wise, A. L., Dayal, J. G., Kish, B. J., Mandich, A., Wellington, C. R., Wetterstrand, K. A., Bates, S. A., ... Manolio, T. A. (2020). Strategic vision for improving human health at the forefront of genomics. *Nature, 586*(7831), 683–692.

Guston, D. H. (2014). Understanding 'anticipatory governance.' *Social Studies of Science, 44*(2), 218–242.

Hanna, K. E. (1995). The ethical, legal, and social implications program of the national center for human genome research: A missed opportunity? In R. E. Bulger, E. M. Bobby, & H. V. Fineberg (Eds.), *Society's choices: Social and ethical decision making in biomedicine* (pp. 432–457). National Academies Press (US). https://www.ncbi.nlm.nih.gov/books/NBK231976/

Harari, Y. N. (2024). *Nexus: A brief history of information networks from the stone age to AI*. Fern Press.

Henderson, G. E., Juengst, E. T., King, N. M., Kuczynski, K., & Michie, M. (2012). What research ethics should learn from genomics and society research: Lessons from the ELSI Congress of 2011. *The Journal of Law, Medicine & Ethics: A Journal of the American Society of Law, Medicine & Ethics, 40*(4), 1008–1024.

Herdman, R. C. (1989). The congressional office of technology assessment. *JAMA, 262*(6), 817–821. https://jamanetwork.com/journals/jama/article-abstract/378126

Hilgartner, S., Prainsack, B., & Hurlbut, J. B. (2016). In handbook of science and technology studies. In U. Felt, R. Fouché, C. A. Miller, & L. Smith-Doerr (Eds.), *Ethics as governance in genomics and beyond* (pp. 823–852). MIT Press.

Horgan, J. (1996). *The end of science: Facing the limits of knowledge in the twilight of the scientific age*. Basic Books.

Howley, C., Haas, M. A., Al Muftah, W. A., Annan, R. B., Green, E. D., Lundgren, B., Scott, R. H., Stark, Z., Tan, P., North, K. N., & Boughtwood, T. (2025). The expanding global genomics landscape: Converging priorities from national genomics programs. *The American Journal of Human Genetics, 112*(4), 751–763.

Hudson, K. L., & Collins, F. S. (2015). Bringing the common rule into the 21st century. *The New England Journal of Medicine, 373*(24), 2293–2296.
Indiana University. (2001). "$105 million grant from Lilly endowment fuels genomics initiative at Indiana University". Retrieved April 25, 2025, from https://archives.indianapolis.iu.edu/server/api/core/bitstreams/5d0241a2-7efc-4c7a-bc78-19570e67002e/content
International Electrotechnical Commission (IEC). (2024). *Bio-digital convergence standardization opportunities*. Retrieved April 22, 2025, from https://www.iec.ch/basecamp/bio-digital-convergence-standardization-opportunities
Jasanoff, S., & Hurlbut, J. B. (2018). A global observatory for gene editing. *Nature, 555*(7697), 435–437.
Jerković, A., & Siedschlag, A. (2021). Ethical, Legal, and Social Issues (ELSI). In *Handbook of security science* (pp. 1–19).
Jones, D. P., & Hens, K. (2023). Ethics is everywhere: Human geography, bioethics and the value of interdisciplinary collaboration. *Bioethics, 37*(7), 615–616.
Juengst, E. T. (1991). The human genome project and bioethics. *Kennedy Institute of Ethics Journal, 1*(1), 71–74.
Juengst, E. T. (1994) Human genome research and the public interest: progress notes from an American science policy experiment. *The American Journal of Human Genetics, 54*(1), 121–128.
Juengst, E. T. (2021). Anticipating the ethical, legal, and social implications of human genome research: An ongoing experiment. *American Journal of Medical Genetics. Part A, 185*(11), 3369–3376.
Kakutani, M. (2018). *The death of truth: Notes on falsehood in the age of trump.* Tim Duggan Books.
Karabin, J., Adsit-Morris, C., Lee, S. S. J., Fullerton, S. M., Cho, M. K., & Reardon, J. (2024). A conjunctural analysis of the origins of 'embedded ELSI' in U.S. genomic medicine. *Journal of Responsible Innovation, 11*(1).
Krahn, M. D., Bielecki, J. M., Bremner, K. E., de Oliveira, C., Almeida, N., Clement, F., Lorenzetti, D. L., O´Campo, P., Pechlivanoglou, P., & Tricco, A. C. (2020). Picturing ELSI+: A visual representation of ethical, legal, and social issues, and patient experiences in health technology assessment in Canada. *International Journal of Technology Assessment in Health Care, 36*(1), 40–49.
Kaye, J., Meslin, E. M., Knoppers, B. M., Juengst, E. T., Deschênes, M., Cambon-Thomsen, A., Chalmers, D., De Vries, J., Edwards, K., Hoppe, N., Kent, A., Adebamowo, C., Marshall, P., & Kato, K. (2012). Research priorities. ELSI 2.0 for genomics and society. *Science, 336*(6082), 673–674.
Kosseim, P., & Chapman, S. (2011). Science and society: Some 'made-in-Canada' options for improving integration. *Accountability in Research, 18*(3), 194–216.

Langfelder, E. J., & Juengst, E. T. (1993). Ethical, Legal, and Social Implications (ELSI) program National Center for Human Genome Research, National Institutes of Health. *Politics and the Life Sciences: The Journal of the Association for Politics and the Life Sciences, 12*(2), 273–275.

Lavery, J. V., Green, S. K., Bandewar, S. V. S., Bhan, A., Daar, A., Emerson, C. I., Masum H., Randazzo, F. M., Singh J.A., Upshur, R. E. G., & Singer P. A. (2013). Addressing ethical, social, and cultural issues in global health research. *PLoS Neglected Tropical Diseases, 7*(8), Article e2227.

Lenoir, N. (1999). Universal declaration on the human genome and human rights: The first legal and ethical framework at the global level. *Columbia Human Rights Law Review, 30*(1), 537–587. https://pubmed.ncbi.nlm.nih.gov/12680391/

Lima-Costa, M. F., de Andrade, F. B., de Souza, P. R. B., Jr, Neri, A. L., Duarte, Y. A. O., Castro-Costa, E., & de Oliveira, C. (2018). The Brazilian longitudinal study of aging (ELSI-Brazil): Objectives and design. *American Journal of Epidemiology, 187*(7), 1345–1353.

Lombardo, P. A. (2022). *Three generations, no imbeciles: Eugenics, the Supreme Court, and Buck v. Bell.* Johns Hopkins University Press.

López, J. J., & Lunau, J. (2012). ELSIfication in Canada: Legal modes of reasoning. *Science as Culture, 21*(1), 77–99.

López, J. J., & Robertson, A. (2007). Ethics or Politics?:The Emergence of ELSI Discourse in Canada*. *Canadian Review of Sociology/Revue canadienne de sociologie, 44*, 201–218. https://doi.org/10.1111/j.1755-618X.2007.tb01134.x

Macklin, R. (2010). The death of bioethics (as we once knew it). *Bioethics, 24*(5), 211–217.

Marshall, E. (1996). The genome program's conscience. *Science, 274*(5287), 488–490.

Mathews, D. J. H., Hester, D. M., Kahn, J., McGuire, A., McKinney, R., Jr., Meador, K., Philpott-Jones, S., Youngner, S., & Wilfond, B. S. (2016). A conceptual model for the translation of bioethics research and scholarship. *The Hastings Center Report, 46*(5), 34–39.

McCain, L. (2002). Informing technology policy decisions: The US human genome's project's ethical, legal and social implications programs as a critical case. *Technology and Society, 24*(1–2), 111–132.

McEwen, J. E., Boyer, J. T., Sun, K. Y., Rothenberg, K. H., Lockhart, N. C., & Guyer, M. S. (2014). The Ethical, legal, and social implications program of the national human genome research institute: Reflections on an ongoing experiment. *Annual Review of Genomics and Human Genetics, 15*(1), 481–505.

McLennan, S., Lee, M. M., Fiske, A., & Celi, L. A. (2020). AI ethics is not a panacea. *The American Journal of Bioethics: AJOB, 20*(11), 20–22.

Meslin, E. M. (2022). On the iportance of impact on policy and legacy. In E. S. Dove & N. N. Shuibhne (Eds.), *Law and legacy in medical jurisprudence: Essays in honour of Graeme Laurie* (pp. 127–148). Cambridge University Press.

Meslin, E. M., & Blasimme, A. (2013). Towards a theory of science policy for genetics. *European Journal of Human Genetics., 21*(2), 360.

Meslin, E. M., & Lavery, J. V. (2025). Policymaking as a humanistic endeavour: On the art and science of genomics policy development. In E. Dove, V. Rahimzadeh, & M. Beauvais (Eds.), *Promoting the 'Human' in law, policy, and medicine: Essays in honour of Bartha Maria Knoppers* (pp. 85–108). Brill Publishing.

Meslin, E. M., Thomson, E. J., & Boyer, J. T. (1997). The ethical, legal, and social implications research program at the national human genome research institute. *Kennedy Institute of Ethics Journal, 7*(3), 291–298.

Metzl, J. (2024). *Superconvergence: How the genetics, biotech, and AI revolutions will transform our lives*. Hachette.

Mikami, K., Ema, A., Minari, J., & Yoshizawa, G. (2021). ELSI is our next battlefield. *East Asian Science, Technology and Society: An International Journal, 15*(1), 86–96.

Müller, R., Clare, A., & Ruess, A. K. (2023). STS and bioethics: Forging new synergies for exploring the potentials and pitfalls of public engagement with new biotechnologies. *The American Journal of Bioethics: AJOB, 23*(7), 72–74.

Myskja, B. K., Nydal, R., & Myhr, A. I. (2014). We have never been ELSI researchers – there is no need for a post-ELSI shift. *Life Sciences, Society and Policy, 10*(1).

Nakagawa, Y., Shiroyama, H., Kuroda, K., & Suzuki, T. (2010). Assessment of social implications of nanotechnologies in Japan: Application of problem structuring method based on interview surveys and cognitive maps. *Technological Forecasting and Social Change, 77*(4), 615–638.

National Human Genome Research Institute. (2025). Research areas and sample topics. Retrieved July 1, 2025, from https://www.genome.gov/Funded-Programs-Projects/ELSI-Research-Program/research-areas

National Research Council. (1988). *Mapping and sequencing the human genome Washington*. National Academy Press.

National Institutes of Health. (1989). *NIH guide for grants and contracts, 18*(7), 8–9. https://grants.nih.gov/grants/guide/historical/1989_03_03_Vol_18_No_07.pdf

National Institutes of Health. (1993). *National institutes of health revitalization act of 1993*. Pub. L. No 103–4, 107 Stat. 122. https://www.govinfo.gov/content/pkg/STATUTE-107/pdf/STATUTE-107-Pg122.pdf

Nichols, T. (2017). *The death of expertise: The campaign against established knowledge and why it matters*. Oxford University Press.

NIH-DOE Committee to Evaluate the Ethical, Legal, and Social Implications Program of the Human Genome Project, December. (1996). Retrieved April 22, 2025, from https://www.genome.gov/10001745/elsi-evaluation-report

NIH-DOE Task Force on Genetic Information and Insurance. (1993). Genetic information and insurance. Retrieved April 22, 2025, from https://www.genome.gov/10001750/genetic-information-and-insurance-task-forcereport

Nydal, R., Myhr, A. I., & Myskja, B. K. (2016). From ethics of restriction to ethics of construction: ELSA research in Norway. *Nordic Journal of Science and Technology Studies, 3*(1), 34–45.

OECD. (2024). "Framework for Anticipatory Governance of Emerging Technologies", *OECD Science, Technology and Industry Policy Papers*, No. 165, OECD Publishing, Paris, Retrieved July 7, 2026, from https://www.oecd.org/content/dam/oecd/en/publications/reports/2024/04/framework-for-anticipatory-governance-of-emerging-technologies_14bf0402/0248ead5-en.pdf

Office of Technology Assessment. (1988). *Mapping our genes: Genome projects-how big? How fast?* U.S. Government Printing Office.

Ogbogu, U., & Ahmed, N. (2022). Ethical, legal, and social implications (ELSI) research: Methods and approaches. *Current Protocols*, *2*(1), Article e354.

Ozdemir, V., Faraj, S. A., & Knoppers, B. M. (2011). Steering vaccinomics innovations with anticipatory governance and participatory foresight. *Omics: A Journal of Integrative Biology*, *15*(9), 637–646.

Pang, T. (2002). The impact of genomics on global health. *American Journal of Public Health, 92*(7), 1077–1079.

Patrinos, A., & Drell, D.W. (1997). The human genome project: View from the department of energy. *Journal of the American Medical Women's Association* (1972). 1997 Winter, *52*(1):8–10. PMID: 9033165. https://pubmed.ncbi.nlm.nih.gov/9033165/

Pellegrino, E. D. (1999). The origins and evolution of bioethics: Some personal reflections. *Kennedy Institute of Ethics Journal, 9*(1), 73–88.

Perleth, M., Di Bidino, R., Huang, L.-Y., Jones, L., Mujoomdar, M., Myles, S., Pichon-Riviere, A., Sabirin, J., Schuller, T., & Washington, J. (2022). Disruptive technologies in health care disenchanted: A systematic review of concepts and examples. *International Journal of Technology Assessment in Health Care*, *38*(1), Article e70.

Pew Research Center. (2023, September 23). *Americans' dismal views of the nation's politics.* Retrieved May 1, 2025, from https://www.pewresearch.org/wp-content/uploads/sites/20/2023/09/PP_2023.09.19_views-of-politics_REPORT.pdf

Policy Horizons Canada. (2019). *Exploring biodigital convergence.* Retrieved July 7, 2026, from https://horizons.service.canada.ca/en/2020/02/11/exploring-biodigital-convergence/index.shtml

Potter, V. R. (1971). *Bioethics: Bridge to the future*. Prentice Hall.
Powledge, T. M. (2003). Whither NHGRI? *Genome Biology, 4*, spotlight-20030417-02.
Priest, S. H., Bonfadelli, H., & Rusanen, M. (2003). The 'trust gap' hypothesis: Predicting support for biotechnology across national cultures as a function of trust in actors. *Risk Analysis, 23*(4), 751–766.
Rehm, H. L., Page, A. J. H., Smith, L., Adams, J. B., Alterovitz, G., Babb, L. J., Barkley, M. P., Baudis, M., Beauvais, M. J. S., Beck, T., Beckmann, J. S., Beltran, S., Bernick, D., Bernier, A., Bonfield, J. K., Boughtwood, T. F., Bourque, G., Bowers, S. R., Brookes, A. J., ... Birney, E. (2021). GA4GH: International policies and standards for data sharing across genomic research and healthcare. *Cell Genomics, 1*(2), Article 100029.
Reich, W. T. (1995). The word 'bioethics': The struggle over its earliest meanings. *Kennedy Institute of Ethics Journal, 5*(1), 19–34.
Research Council of Norway. (2014). *Ethical, legal and social aspects of nanotechnology, biotechnology and neurotechnology. Programme description*. Retrieved April 25, 2025, from http://www.forskningsradet.no/servlet/Satellite?c=Page&cid=1224698247089&pagename=elsa%2FHovedsidema
Roberts, L. (1993). Whither the ELSI program?. *Hastings Center Report, 23*, 5. https://doi.org/10.2307/3562916
Sajeer, P. M. (2023). Disruptive technology: Exploring the ethical, legal, political, and societal implications of nanopore sequencing technology. EMBO Rep. 24, Article EMBR202256619.
Sass, H. M. (2008). Fritz Jahr's 1927 concept of bioethics. *Kennedy Institute of Ethics Journal, 17*(4), 279–295.
Schmeck, H. (1988, October 4). DNA pioneers to tackle biggest gene project ever. *New York Times*. https://www.nytimes.com/1988/10/04/science/dna-pioneer-to-tackle-biggest-gene-project-ever.html
Sherkow, J. S., Barker, K. B., Braverman, I., Cook-Deegan, R., Durbin, R., Easter, C. L., Goldstein, M. M., Hudson, M., Kress, W. J., Mathews, D. J. H., McCarthy, C., McCartney, A. M., da Silva, M., Torrance, A. W., & Greely, H. T. (2022). Ethical, legal, and social issues in the Earth BioGenome Project. *Proceedings of the National Academy of Sciences of the United States of America, 119*(4), Article e2115859119.
Shorter Oxford English Dictionary, 6th ed. (2007). Oxford University Press.
Singer, P. A., Taylor, A. D., Daar, A. S., Upshur, R. E. G., Singh, J. A., & Lavery, J. V. (2007). Grand challenges in global health: The ethical, social and cultural program. *PLoS Medicine, 4*(9), Article e265.
Smit, A. K., Gokoolparsadh, A., McWhirter, R., Newett, L., Milch, V., Hermes, A., McInerney-Leo, A., & Newson, A. J. (2024). Ethical, legal, and social

issues related to genetics and genomics in cancer: A scoping review and narrative synthesis. *Genetics in Medicine: Official Journal of the American College of Medical Genetics*, *26*(12), Article 101270.

Sommerfeld, J., & Oduola, A. M. (2007). Health-related biotechnologies for infectious disease control in Africa: Ethical, Legal and Social Implications (ELSI) of transfer and development. *African Journal of Medicine and Medical Sciences*, *36 Suppl*, 1–5. PMID: 17703556. https://pubmed.ncbi.nlm.nih.gov/17703556/

Souppaya, M., Scarfone, K., & Dodson, D. (2022). Secure Software Development Framework (SSDF) version 1.1. *National Institute of Standards and Technology (NIST)*.

Sugarman, J., & Sulmacy, D. P. (2010). *Methods in medical ethics* (2nd ed.). Georgetown University Press.

Ten Have, H. A. M. J. (2012). Potter's notion of bioethics. *Kennedy Institute of Ethics Journal*, *22*, 59–82

Toffler, A. (1975). What is anticipatory democracy? *The Futurist*, *10*, 224–229.

UN Environmental Plan. (2011). Design for sustainability: A step-by-step approach. Retrieved April 22, 2025, from https://wedocs.unep.org/bitstream/handle/20.500.11822/8742/DesignforSustainability.pdf?sequence=3&amp%3BisAllowed=

United Nations. (1993). Convention on biodiversity. Retrieved July 8, 2025, from https://www.un.org/en/observances/biological-diversity-day/convention

United Nations. (1997). Universal declaration on the human genome and human rights. Retrieved April 29, 2025, from https://www.unesco.org/en/ethics-science-technology/human-genome-and-human-rights

US Department of Health and Human Services and US Department of Energy. (1990). Understanding our genetic inheritance: The U.S. human genome project: The first five years. *NIH publication* 90–1590. National Institutes of Health, Bethesda, MD. Retrieved June 25, 2025, from https://www.genome.gov/Pages/About/Understanding_Our_Genetic_Inheritance_1st_5years_HGP.pdf

Vonderschmitt, J., Wöhlke, S., & Schicktanz, S. (2023). Scarce resources, public health and professional care: The COVID-19 pandemic exacerbating bioethical conflicts - findings from global qualitative expert interviews. *BMC Public Health*, *23*(1), 2492.

Vought, R. T. (2025, May 2). Major discretionary funding changes [Letter]. *The white house, executive office of the President*. https://www.whitehouse.gov/wp-content/uploads/2025/05/Fiscal-Year-2026-Discretionary-Budget-Request.pdf

Walker, R. L., & Morrissey, C. (2012). Charting ELSI's future course: Lessons from the recent past. *Genetics in Medicine*, *14*(2), 259–267.

Walker, R. L., & Morrissey, C. (2014). Bioethics methods in the ethical, legal, and social implications of the human genome project literature. *Bioethics, 28*(9), 481–490.

Walters, L. (1987). Ethics and new reproductive technologies: an international review of committee statements. *Hastings Center Report*, *17*(3), S3–S9. https://pubmed.ncbi.nlm.nih.gov/11644029/

Wolfe, A. (2002). Federal policy making for biotechnology, executive branch, ELSI. In *Encyclopedia of Ethical, legal and policy issues in biotechnology*.

Yesley, M. S. (1991). ELSI working group studies genetic bias. *Human Genome News, 3*(3), 12–14.

Yoon, J.-R., Cho, S. K., & Jung, K. W. (2010). The challenges of governing biotechnology in Korea. *East Asian Science, Technology and Society, 4*(2), 335–348.

Yoshizawa, G., et al. (2014). ELSI practices in genomic research in East Asia: Implications for research collaboration and public participation. *Genome Medicine, 6*(39).

Zwart, H., & Nelis, A. (2009). What is ELSA genomics? Science & society series on convergence research. *EMBO Reports, 10*(6), 540–544.

Zwart, H., Landeweerd, L., & van Rooij, A. (2014). Adapt or perish? Assessing the recent shift in the European research funding arena from 'ELSA' to 'RRI'. *Life Sciences, Society and Policy, 10*(11).

Multiple Competition and the Ethics of the Sciences: Toward an Interdisciplinary Dialogue

Anna Kosmützky

1 Introduction

The academic system has always been shaped by forms of competition. Classical sociologists of science such as Merton, Bourdieu, and Whitley highlighted the role of recognition, reputation, and field-specific hierarchies in structuring the generation of knowledge (Bourdieu, 1975; Merton, 1968; Whitley, 1984). Yet the contemporary landscape of science and higher education has undergone a substantial transformation, marked not by the presence of competition per se, but by its multiplication, intensification, and entanglement. The term multiple competition—as used in this contribution—seeks to capture a condition in which diverse actors (individual researchers, universities, states), scarce and increasingly heterogeneous goods (reputation, funding, positions, media visibility), and third parties (funding agencies, scientific elites, governments) interact across a growing number of interrelated and overlapping competitive arenas (Buenstorf et al., 2025; Krücken, 2021).

A. Kosmützky (✉)
Leibniz Centre for Science and Society (LCSS), Leibniz University Hannover, Hannover, Germany
e-mail: anna.kosmuetzky@lcss.uni-hannover.de

J. Hinrichsen et al. (eds.), *Ethics of the Sciences*,
https://doi.org/10.1007/978-981-92-2821-8_4

What makes the concept of multiple competition particularly salient is not just the expansion of what and who is being competed for, but how these competitions interact: actors shift roles, competitions overlap and feed into each other, and outcomes in one arena condition chances in another. Early-career researchers today may find themselves engaged in competition for grants, publications, teaching evaluations, and public engagement—often simultaneously and with little coordination. Universities act not only as enablers but also as strategic actors themselves, competing for cluster funding, top researchers, and institutional prestige. States organize, fund, and evaluate such competitions while simultaneously participating in national and international ranking races. What emerges is a dynamic configuration that places novel demands on academic actors and organizations alike.

The article asks how these competitive dynamics reshape the normative infrastructures of science itself and what consequences they hold for the emerging field of ethics of the sciences. Unlike traditional research ethics—which focuses primarily on conduct within the research process (e.g., informed consent, data integrity, conflicts of interest)—the ethics of the sciences, as proposed in this volume, foregrounds the broader institutional, political, and social conditions under which science is produced, disseminated, and evaluated. This article seeks to contribute to this perspective by offering a sociologically grounded analysis of the contemporary condition of scientific knowledge generation, with particular attention to the systemic dynamics of multiple competition.

The structure of the argument is as follows: The article argues that the contemporary higher education and science system is increasingly shaped by the condition of multiple competition. An analysis of this condition—through a broad and systematic conceptual lens that captures different types of actors and their interrelations—must serve as a basis for reflecting on the practical orientation of an ethics of the sciences. The article is structured accordingly: First (Sect. 2), it introduces a sociological notion of competition that goes beyond market-based exchange relations. Second (Sect. 3), it draws on a conceptual analysis of the constellation of multiple competition, supported by empirical examples from the German higher education and research system. Third (Sect. 4), it demonstrates how multiple competition introduces normative tensions, diffuses responsibilities, and fosters normative pluralism, and outlines aspects of a dialogical perspective that allows both strands—sociologies of competition and the ethics of the sciences—to mutually illuminate one another.

These dynamics, it is argued (in the conclusion, Sect. 5), call for an ethics that is reflexive, procedural, and attentive to the organizational and systemic conditions under which science is practiced today. In this way, the article connects analytical tools from the sociology of science and higher education with a normative agenda for a future-oriented ethics of the sciences.

2 Competition in the Higher Education and Science System

To conceptually grasp competition in the higher education and science system, it is not sufficient to view it solely as a market-based exchange relationship. A significant part of the competition in this field does not take place within market-like structures (Dasgupta & David, 1994; Polanyi, 1962; Stephan, 1996), even though the term "market" is frequently used in higher education research to describe and analyze the contemporary transformation of universities. A large part of competition in the higher education and science system is not coordinated through price mechanisms, and competitive actors are only in rare cases—such as the allocation of study places at prestigious universities with highly selective admissions procedures—linked through direct exchange relationships in the form of supply and demand. While market-like competition is continuous and seemingly seamless, many competitions in the higher education and science system are organized, discrete, and bound to specific places and times (see Arora-Jonsson et al., 2020, 2021; on the distinction between competition as a steering principle and concrete competitions, see Stark, 2020). In this sense, at best one can speak of a quasi-market (Le Grand, 1991).

Instructive reflections to conceptually capture competition in a broader perspective are offered by Simmel's sociology of competition. He describes competition as a dynamic between rival actors who do not directly struggle with each other, but rather strive for the success of their efforts with regard to a third party (Simmel, 2008). Competition can thus be understood as an interaction between three actor positions: two competitors—Ego and Alter—who vie for a limited and desired good, and a third party that decides on the allocation of the scarce good. In this constellation, the relationship between the rival actors is indirect, as Ego's success does not consist in defeating Alter, but in persuading the third party to allocate the scarce good in Ego's favor. Since the good is limited, the competitors mutually restrict each other's access to it, and the success of one party comes at the expense of the other (i.e., a zero-sum game). Competition, therefore, means that Ego and Alter always refer to the third party and maintain an indirect relationship with one another. However, they observe each other to gather information that may help improve their own competitive position (see for further elaboration of Fig. 1: Kosmützky & Meier, 2025). Third parties can, of course, be not only organizations but also individual researchers, such as "star scientists" over whom universities compete as will be elaborated below.[1]

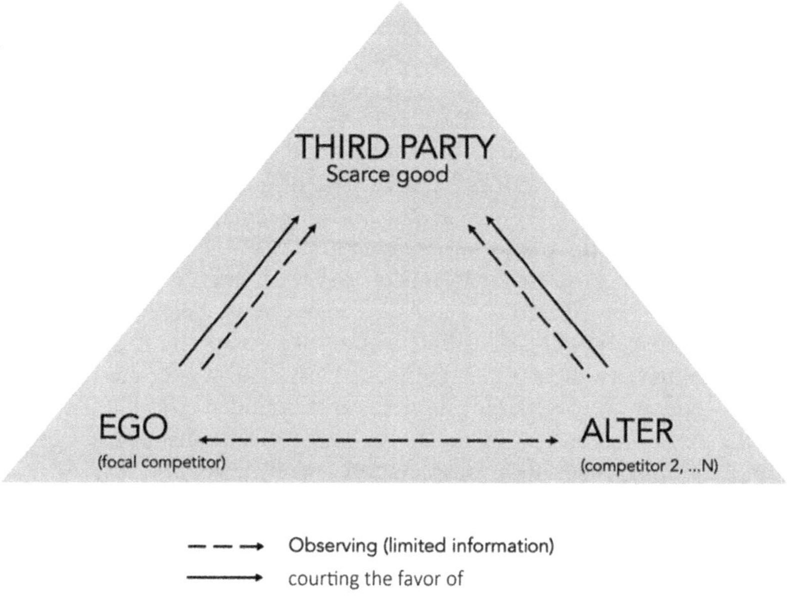

Fig. 1 Triadic competitive constellation

[1] Arora-Jonsson et al. (2020, 2021) furthermore identify actors who define the rules and structure of competition and organize it as so-called fourth parties in the competitive constellation. However, we will not go into the role of fourth parties in more detail here, not least because in many cases they simultaneously act as third parties in competitive constellations. A typical fourth party is the state, which increasingly organizes university funding on a competitive basis. Funding organizations such as the German Research Foundation (DFG) may empirically act as either third or fourth parties.

This foundational concept of a triadic competition constellation underpins the following conceptualization of multiple competition.

3 From Competition to Multiple Competition

There is much discussion about growing competition and increasing pressure on individual researchers and universities. Whether competition has actually intensified over time remains an empirical question. What is clearly observable, however, is a proliferation of competitive arenas and an increasing interweaving of different competitions and competitive actors. Universities—particularly those engaged in both teaching and research—as well as individual academics are now entangled in multiple, overlapping competitions. Moreover, universities, researchers, and even states increasingly depend on one another to succeed in these contexts. This section therefore explores the dynamics of competition in higher education and science through the conceptual lens of multiplicity (Krücken et al., 2021), and illustrates them using examples from the German higher education and science system in order to reflect on their implications for the ethics of the sciences in the following section.

The term *multiple competition* (*multipler Wettbewerb*) is a recent conceptual innovation (Krücken, 2021). In contrast to a genealogical or historical approach to competition—which traces the emergence of competitive regimes through processes of power and institutionalization (Szöllösi-Janze, 2021)—or an ecological perspective, which focuses on the embeddedness of actors and competitive arenas in particular contexts and the diffusion of competitive logics (Krücken, 2024), the notion of multiple competition aims to capture the complex entanglements of actors and competitive arenas that shape contemporary higher education and science. Depending on the metaphor or theoretical framing adopted, different dimensions of the relationships between competition and actors come into view. By adopting the lens of multiple competition, one foregrounds the co-existence and interplay of various forms of competition, the interdependencies among actors at different levels, and the dynamics that emerge from such configurations (Kosmützky & Krücken, 2025).

The concept of *multiple competition* refers to a configuration in which different actors—individual researchers, universities as organizations, and state actors—compete for various types of scarce and desirable goods, such as reputation, funding, talent, or academic positions. These goods are awarded by the abovementioned so-called Third Parties to the competition—individuals, organizations, or collectives—which may include influential scholars, research councils, or collectives of students (see Fig. 2).

Before turning to other layers of multiplicity, it is important to elaborate three interrelated and mutually reinforcing dynamics related to the actors of competition that underpin the multiplication: (1) The multiplication and shift of competitive pressure to earlier career stages in the competition of individual actors; (2) the transformation of universities into organizational competitors (Krücken & Meier, 2006); and (3) shifts at the state level, where higher education and science policy actors increasingly understand themselves as active "competition states" operating in multiple roles (Musselin, 2018) (see Table 1 for a summary).

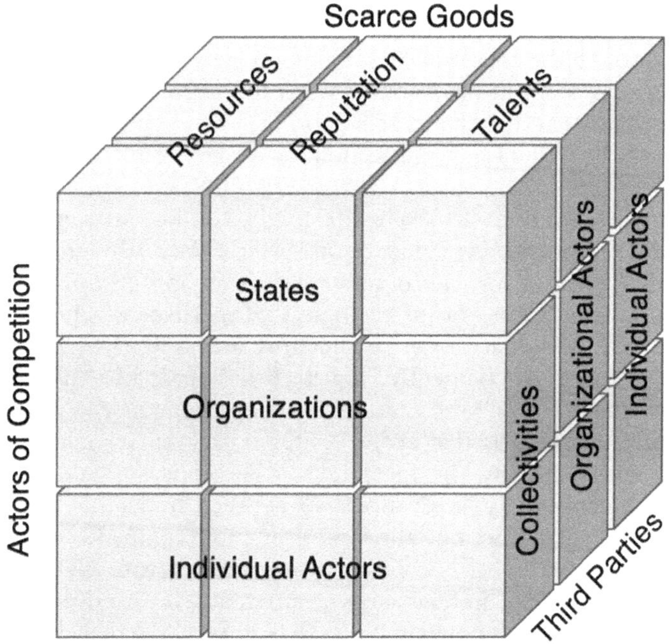

Fig. 2 Visualizing the multiplicity across actors of competition, scarce goods, and third parties

Table 1 Competitive dynamics and examples from the German system

Actors	*Typical scarce goods*	*Dynamics*	*Example(s)*
Individual	Traditional: publications, citations, New: professorships, third-party funding and other scarce goods	– Third-party funding, particular research clusters, becomes a performance criterion and a marker of reputation – Proliferation of social impact, (social) media visibility, internationalization markers as scarce goods: – Competitive pressure shifts to earlier career stages	– Cluster applicants rely on support by their universities (administrative, financial, strategic) – Postdocs apply for ERC Starting or Heisenberg Grants as gateways to professorships, successful candidates then become desirable scarce goods for universities competing for prestige – Third-party funding as performance indicator at the post-doc level
University	(Cluster) third-party funding, excellence status, star scientists	– Transformation into "strategic competitive actors" – Universities actively try to improve the competitive position of "their" university – Dependency on university-affiliated researchers who are clusterable	– Expansion of strategy units and supporting staff (helping professionals) – Internal governance aligned to foster specific researchers and research areas – Depend on state co-funding to support applications and infrastructure
State	Regional reputation, tax revenue	– Competition used as a main governance instrument – Federal states compete for excellence universities – Proliferation of surrogate competitions	– States co-finance excellence initiatives and seek symbolic gains – Mutual dependency between strong states, strong universities, and strong individual researchers

First, for individual actors, one can observe a diversification of competitive goods. Academic competition is no longer limited to the generation of knowledge and the related reputation (the classic "publish or perish," cf. Lofthouse, 1974; Merton 1968) or to securing positions. Today, researchers also compete for citations, third-party funding, societal impact, and public and media attention (Krücken, 2021; Musselin, 2018; Naidoo, 2016). In Germany, acquiring third-party funding has become a key performance indicator in academic competition (Gerhards, 2013; Mayer, 2019). These multiple competitive goods can be tracked and compared using standardized CVs, databases, and platforms—by researchers themselves and by external entities such as appointment committees or funding agencies. The growing relevance of metrics, reflected in the widespread use of digital platforms like Web of Science, Google Scholar, and ResearchGate, significantly contributes to this trend. As a result, diverse academic activities—especially in the area of research—are increasingly reduced to measurable and comparable indicators (Mau, 2019). In addition, the academic job market has become more internationalized, particularly with regard to the competitiveness and availability of professorships at research-intensive universities (Gerhardt et al., 2023).

Second, competition and competitive pressure are shifting to earlier career stages, particularly affecting early-career researchers who have yet to secure a professorship. Specific performance dimensions are gaining importance in the competition for the scarce good of professorships, while others are becoming less relevant (Fochler et al., 2016; Müller, 2014; Waaijer et al., 2018). In Germany, it is now common for scholars from disciplines ranging from history to physics to compete for third-party funding even before attaining a professorship, as this is often interpreted as a sign of exceptional academic standing (Fenkner & Gottschalk-Rayling, 2025). At the same time, universities and funding organizations offer a broad array of support programs and so-called helping professionals—mentoring, coaching, pedagogical training, career advice, writing workshops for doctoral candidates and funding applications—that are meant to strengthen particularly the individual competitiveness of PhD students and postdocs (Hasse et al., 2026). Yet, despite these ongoing initiatives, the systemic conditions surrounding the competition for professorships remain largely unchanged (Krücken, 2023).

Third, the academic competition among individual researchers is increasingly being overlaid by universities acting as competitive organizations. Historically, universities functioned primarily as institutional frameworks and workplaces for individual scholars. They have been described as "sieves, incubators, temples, and hubs" (Stevens et al., 2008: 140). Today, however, they act more and more as autonomous organizations engaged in competition for resources, reputation, and status (Hasse & Krücken, 2013; Krücken, 2021; Musselin, 2018). This competition extends not only to research but also to teaching and so-called third-mission activities such as knowledge transfer and societal impact (Krücken et al., 2021). The modern university is no longer merely the sum of loosely connected subunits—researchers, institutes, faculties—but rather participates in institution-wide strategy and profiling processes that are primarily competition-driven. In this context, the university often serves as a hinge between externally initiated competitions and the individual researchers, whose active participation is essential for organizational success (Kosmützky, 2026).

This transformation is also accompanied by an expansion of science management. This includes departments, personnel, and instruments aimed at monitoring both the external environment and the university's internal structures to optimize competitiveness. However, internal university governance is not simply the result of a linear transformation into a competitive organization. Rather, it is shaped by a web of different organizational models and governance principles (Frost et al., 2016; Kosmützky, 2026; Sahlin, 2012). Research also shows significant variation in universities' strategic capacities (Thoenig & Paradeise, 2016). Nevertheless, institutions in very different countries and regions are increasingly subject to standardized performance measurement and positioning contests. They build appropriate capacities and governance structures, strategically sharpen their organizational identities and profiles, and pursue different competitions and scarce goods—including unique features, rankings, third-party funding, reputation, personnel, and students. In Germany, institutional reputation in research is primarily achieved through prestigious, competitively awarded third-party funding—especially within the framework of the Excellence Initiative, but also through DFG programs such as Collaborative Research Centers and Research Units (Kosmützky & Krücken, 2023).

In addition to these changes at the level of individual and organizational actors, shifts can also be observed at the state level. Traditionally, the state's role in academic competition was limited to defining the financial and legal framework. However, in the past two to three decades—and in line with the global trend toward New Public Management—states have begun to take a more active role. They now use competition as an instrument of steering while simultaneously engaging in political competition with other actors (Krücken, 2024; Mayer, 2019; Szöllösi-Janze, 2021). In short, the state no longer merely sets the rules of the game but actively mandates competition in many forms. The logic of competition has come to shape higher education policy in a variety of national contexts. One can thus view the dissemination and implementation of this paradigm as a development that transcends national boundaries. At the same time, its concrete expression remains tied to specific national traditions and structural features (Krücken, 2024).

In Germany, specific characteristics of the national embedding of the global competition paradigm are evident, which contradict the idea of a "retreat of the state" (Strange, 1996: 85). First, one must consider the federal structure of the German higher education system. This leads to a multiplication of state-run competitions. The use of law as a steering instrument is limited at the federal level due to the educational federalism, whereas financial incentives are increasingly employed to enhance strategic capacity. Furthermore, state regulation is expanding beyond research and teaching to include areas such as third-mission activities, gender equality, and internationalization. This has led to a proliferation of state-induced competitions and high levels of engagement by both federal and state ministries (Seidenschnur et al., 2024). This diversity reflects the balancing mechanisms of federalism and illustrates the commitment of government bodies to strengthening the competitiveness of their universities. These competitive formats target universities as organizations, not only in the realm of research and teaching but also in areas such as gender equality, sustainability, and internationalization. The multiplication of such competitions, in which actors across a wide spectrum of performance levels and profiles can participate and succeed ("everyone can win"), reinforces the legitimacy of an institutional configuration that is structured around competition (Buschkamp & Kosmützky, 2026). Second, it is important to note that Germany's competitive governance of higher education does not entail a shift toward marketization or privatization. Tuition fees have not taken hold, and there is little evidence of significant privatization, even if enrollment in private universities is growing. Rather than diminishing state influence, competitive governance enhances state actor capacities. In contrast, market privatization would ultimately curtail the state's room for maneuver.

To this foundational constellation of multiple competing actors, goods, and third parties allocating scarce resources, we must add another layer of multiplicity: the interrelations among the various competitions in which actors engage, including the associated role shifts—for instance, when actors move from being competitors to acting as third parties. At least three types of such interrelations can be analytically distinguished (for a more detailed discussion of these entanglements, see Buenstorf et al., 2025).

Firstly, actors frequently shift roles within this triadic constellation of competition. For instance, actors who are competitors (Ego) in one context may serve as third parties in another. State actors, for example, depend on high-performing universities to succeed in regional competition for excellence status, while those same universities depend on excellent individual researchers. Conversely, researchers applying for an excellence cluster must often rely on the support of their university and potentially even that of state-level political actors. Another illustrative example concerns academic publishing: while a journal editor initially acts as the third party in a competition when a scholar submits a manuscript, the roles may later reverse—turning the editor into an applicant and the former author into a third party—if the editor later submits a proposal to a special issue or edited volume curated by the original author. In such cases, competitive arenas become entangled, and shifts in positionality are no longer isolated events.

Secondly, most actors—whether as competitors or third parties or both—are typically involved in more than one competition, often simultaneously. Time and resources invested in one competition, such as writing a third-party funding proposal, may be lacking for another, such as publishing in a prestigious journal. At the same time, synergies can emerge: success or failure in one arena can influence outcomes in others. For example, securing an ERC Starting Grant is widely considered a promising step toward a professorship. This also illustrates the aforementioned role shift: researchers initially compete for the grant (as Ego), but once awarded, they become objects of competition themselves (as third parties), as universities compete to recruit researchers with ERC funding.

Thirdly, competition for one scarce good—such as specific third-party funding—may intensify when it also serves as a means to secure another desirable outcome, such as prestige or a permanent position. For instance, universities competing for talented students and top-tier professors are simultaneously competing for budgetary allocations, potential external funding, and the reputational capital tied to both. In this context, previously mentioned instruments such as ERC Starting Grants (or within the German context Heisenberg Grants) exemplify how competitive success for one scarce good (third-party funding) can translate into an advantage in another competition (tenure-track positions), thereby reinforcing cumulative competitive dynamics. Originally designed to support early-career researchers in conducting ambitious projects, they increasingly function as strategic assets in the competition for professorships, providing successful applicants with a significant advantage over their peers.

In sum, the foundational constellation of multiple competition—comprising diverse actors, goods, and third parties—unfolds in a dense web of interdependencies and positional shifts. This multiplicity gives rise to three interrelated dynamics that shape contemporary academic life: role-switching between competitors and gatekeepers, overlapping and strategically entangled arenas of competition, and the cumulative escalation of competition when scarce goods become means for securing further advantages.

These entanglements of competitive dynamics—only briefly sketched here—demonstrate the growing complexity of competition in contemporary higher education and science systems. They are aptly captured by the concept of multiple competition.

4 Multiple Competition and Its Implications for the Ethics of the Sciences

Classical authors in the sociology of science have long emphasized the structured nature of scientific competition and its normative foundations. Merton (1973), for instance, conceptualized science as a normatively regulated enterprise governed by the CUDOS norms (communism, universalism, disinterestedness, organized skepticism). Whitley (1984) underscored the differentiation of scientific fields through reputational structures linked to the social organization of research, while Bourdieu (1975) highlighted symbolic capital and distinction as core dynamics within scientific fields. Yet these foundational approaches did not anticipate the degree of complexity and interdependence introduced by multiple competition in contemporary higher education and science systems. At the same time, philosophy of science has developed a substantial debate on the normative dimensions of scientific practice. As Douglas

(2009) has shown, scientific judgment is intrinsically value-laden, shaped by both epistemic and societal norms (see for a brief overview of the most recent debate: Douglas & Branch, 2024). Ethics is therefore not external to scientific work but embedded in its very processes of reasoning and decision-making. This resonates with arguments emphasizing the entanglement of epistemic and social norms in scientific practice (Carrier et al., 2008).

The emerging program of an ethics of the sciences extends this philosophical debate into a broader, interdisciplinary arena (Hinrichsen et al., and Schicktanz & Reydon, this volume). An ethics of the sciences does not conceive ethics as an external, ancillary, or deficit-oriented oversight mechanism. Rather, ethical reflection is understood as an inherent and constitutive dimension of scientific practice itself. Scientific work is always already shaped by implicit and explicit norms—epistemic norms concerning the quality, transparency, and integrity of research, as well as societal norms concerning responsibility, responsiveness, and trustworthiness. From this perspective, ethics is not merely a reaction to changes in scientific knowledge generation, emerging challenges, or cases of misconduct. Instead, it operates as a constructive, transdisciplinary mode of problem-solving. It provides procedures, forums, and conceptual tools through which researchers, institutions, and societal actors can deliberate on the purposes, consequences, and normative orientations of knowledge generation. Ethical reflection thereby contributes to orienting scientific work within complex institutional and societal contexts.

This ethics-of-the-sciences perspective aligns well with the complexity of the interrelations generated by multiple competition and can enter into a productive dialogue with it, offering avenues for integrating ethical reflection into analyses of competitive dynamics. Dialogue here refers not to a normative ideal of consensus, but to analytical complementarity: each perspective reveals mechanisms and blind spots in the other. Such a dialogue, first, invites closer attention to how different forms of competition intersect with established epistemic and societal norms and create situations in which researchers must navigate partially conflicting evaluative expectations. Second, a dialogical perspective brings into view the redistribution of responsibility across different levels of the scientific system: multiple competition not only shapes individual decision-making but also reconfigures collective, organizational, and system-level responsibilities. Third, it highlights how existing ethical infrastructures—codes of conduct, standards of integrity, community norms—remain essential, yet increasingly require complementary reflexive and procedural forms of engagement to address the systemic effects of competition. These three avenues will be discussed in turn in the remainder of this section.

Tensions Between Epistemic Values and Competitive Logics: A first entry point for such a dialogue concerns the intensified tensions and shifting alignments between epistemic norms and competitive values. Multiple competition does not simply increase the number of evaluation contexts; it also changes the conditions under which epistemic commitments are upheld, negotiated, or strategically adjusted. Examining these tensions therefore provides a starting point for understanding how ethical reflection can engage with the everyday norm conflicts that researchers must navigate.

As scientists are subject to various and sometimes contradictory valuation criteria within different competitions (Kosmützky & Meier, 2025), the alignment between epistemic values such as rigor or originality and institutional rewards becomes increasingly fragile. Merton's (1973) normative ideals, with their emphasis on universalism, communism, disinterestedness, and organized skepticism (CUDOS), may become strained under the weight of strategic behavior induced by competitive pressures.

In systems such as the German higher education and research sector—where third-party funding constitutes the primary, or at least a major, source of research financing (with approximately 50% of university research expenditures derived from such funding; see WR, 2023)—research agendas and epistemic values risk being subordinated to the logic of grant acquisition. This tension between epistemic orientations and funding imperatives has been widely discussed (e.g., Gläser, 2024; Laudel, 2023). Likewise, publication in high-impact journals can incentivize problem choice that prioritizes visibility over relevance or robustness.

The multiplicity of competitions increases the number and heterogeneity of performance expectations to which researchers are subject. Scientific actors must often reconcile divergent normative orders: the epistemic values of their discipline, the strategic imperatives of their institution, and the formalized criteria imposed by third-party evaluations (e.g., excellence schemes, rankings, impact metrics). This leads to what Musselin (2018) describes as a hybridization of academic roles and to strategic action under constraint. As competition multiplies, trade-offs become routine: Should time be invested in publishing, teaching, grant writing, or outreach? Should research agendas be aligned with funding priorities, or pursued for their intrinsic epistemic value?

Furthermore, dimensions of academic work—especially publications and external research grants—that are assessed and compared competitively by both individual academics and external evaluators (such as funding agencies or search committees) tend to be prioritized. Other dimensions, particularly those less legible in competitive frameworks—such as committee work, mentoring, or anonymous peer review—lose visibility and recognition, as does the self-conception of being part of an academic community (Cloete et al., 2023). Such (mis)alignments challenge the integrity of the scientific endeavor and create systemic pressures that contribute to norm erosion and weaken collegiality.

Reconfiguring (ir)responsibility under organizational complexity and shifting academic roles: A second entry point for a productive dialogue between multiple competition and an ethics of the sciences concerns the reconfiguration and redistribution of responsibility across different levels of the academic system. As individual, organizational, and state actors become more interdependent under the condition of multiple competition, the distribution of responsibilities becomes blurred. From an ethics-of-the-sciences perspective, this means that ethical reflection cannot be confined to individual codes of conduct but must address the institutional arrangements that structure opportunity and constrain ethical agency.

One central question is whether contemporary competitive configurations enable or impair collegiality as the modus operandi of academia, and how this affects the collective pursuit of epistemic goals (Cloete et al., 2023). Under conditions of multiple competition, the delineation of responsibilities becomes increasingly opaque. Scientific work is not only more collective and interdependent; actors also alternate between different roles: an individual researcher may act as a competitive applicant in one context and as a third-party evaluator in another. This dynamic triangle (Ego, Alter, Third Party) becomes even more complex when fourth parties—organizers of competition—are included, whose influence often remains insufficiently scrutinized in standard ethical frameworks (Arora-Jonsson et al., 2020, 2021). A dialogical ethics perspective therefore needs to account for these systemic structures of responsibility and consider how accountability can be coordinated across levels rather than assigned exclusively to individuals.

In this sense, Beck's (1992) notion of "organized irresponsibility" can be read as a sociological account of the structural conditions under which responsibility becomes dispersed. A productive ethics of the sciences that takes multiple competition seriously would not seek to contradict this diagnosis, but rather to reframe it: the structural diffusion of responsibility becomes an object of ethical reflection, raising questions about how responsibility can be re-anchored across actors and institutions in practical and context-sensitive ways. Building on Beck's insight, the challenge is to develop procedural and reflexive mechanisms that translate structural forms of irresponsibility into coordinated responsibilities in the day-to-day conduct of research.

Competing Orders of Worth and Reflexive Ethical Engagement: A further area in which a productive dialogue between multiple competition and an ethics of the sciences becomes relevant concerns the increasing pluralization of evaluative standards. Scientific communities have long relied on internal logics of distinction and recognition (Ben-David, 1971; Bourdieu, 1975; Whitley, 1984). Yet across contemporary higher education and research systems, these community-specific norms increasingly coexist with external, often quantitative evaluation devices—such as rankings, audits, and performance-based funding schemes—that introduce additional and sometimes conflicting criteria (Hamann et al., 2022; Sauder & Espeland, 2009). Under conditions of multiple competition, researchers and institutions are thus evaluated simultaneously through heterogeneous and partly contradictory normative expectations.

Such environments resemble what Boltanski and Thévenot (2006) describe as competing "orders of worth," requiring actors to justify their work according to diverse evaluative frameworks. The legitimacy of scientific contributions may hinge simultaneously on bibliometric performance, policy relevance, societal impact, or international mobility. This does not imply the disappearance of shared epistemic standards, but it does make their application more context-dependent and their alignment with institutional reward structures more fragile.

From the perspective of an ethics of the sciences, this raises the need to understand ethical judgment not as the application of a single, unified evaluative grammar, but as a practice that must navigate and reflect on plural and sometimes conflicting normative contexts. Existing ethical infrastructures—codes of conduct, standards of integrity, and community norms—remain indispensable anchors, yet they increasingly require complementary reflexive and procedural forms of engagement to address the systemic effects of multiple competition. Rather than replacing rule-based ethics, such reflexive approaches extend them by offering conceptual and procedural tools to make explicit which evaluative frameworks are mobilized, how they shape scientific decision-making, and where tensions or misalignments arise.

Taken together, these implications illustrate that multiple competition does not merely intensify existing ethical tensions but reshapes the institutional and evaluative environments in which ethical reflection in science takes place. An ethics of the sciences informed by the dynamics of multiple competition must therefore take into account how governance arrangements, evaluation procedures, and inter-organizational dependencies influence the possibilities for ethical agency and collective orientation. Rather than focusing primarily on individual misconduct, it calls for conceptual and procedural tools that enable researchers and institutions to reflect on the normative assumptions embedded in competitive formats and to deliberate on their consequences for epistemic and societal goals. This includes examining how specific evaluation instruments—including widely institutionalized competition formats such as excellence schemes, performance-based funding, or ranking systems—are institutionalized, and considering whether alternative arrangements might better support the epistemic quality, social responsibility, and long-term integrity of scientific work.

5 Conclusion

This chapter has examined multiple competition as a structural condition of contemporary science and higher education and outlined why it poses a central challenge for an ethics of the sciences. Multiple competition reshapes the environments in which epistemic values are pursued, responsibility is allocated, and evaluative standards are applied. It introduces tensions between epistemic and strategic orientations, redistributes responsibility across institutional layers, and generates normative pluralism that requires continual negotiation.

An ethics of the sciences that takes these dynamics seriously therefore needs to extend beyond individual codes of conduct. It must attend to how governance arrangements, evaluation instruments, and organizational dependencies shape the possibilities for ethical agency and collective orientation. Rather than functioning as an external oversight mechanism, ethics becomes a reflexive and procedural resource within academic self-governance—supporting actors in examining the normative assumptions built into competitive formats and in deliberating about how these align with the epistemic and societal missions of science. In this sense, ethics is not an appendix to scientific practice but part of its core and an integral dimension of the process of knowledge generation itself because epistemic practices themselves presuppose judgments about relevance, robustness, and responsibility. The key question is not only how researchers navigate multiple competitions, but also how academic communities can reflect on, and where necessary reshape, the competitive environments that structure contemporary knowledge generation.

Seen together, multiple competition and an ethics of the sciences illuminate one another: their interplay not only clarifies how normative tensions emerge but also offers conceptual resources for imagining alternative competitive architectures. This dialogue reveals not only how science is governed, but also how it might be governed differently.

Competing Interests The author has no conflicts of interest to declare that are relevant to the content of this chapter.

References

Arora-Jonsson, S., Brunsson, N., & Hasse, R. (2020). Where does competition come from? The role of organization. *Organization Theory, 1*(1), https://doi.org/10.1177/2631787719889977.

Arora-Jonsson, S., Brunsson, N., & Hasse, R. (2021). A new understanding of competition. In S. Arora-Jonsson, N. Brunsson, R. Hasse, & K. Lagerström (Eds.), *Competition: What it is and why it happens* (1st ed., pp. 1–25). Oxford University Press.

Beck, U. (1992). *Risk society: Towards a new modernity*. Sage Publications.

Ben-David, J. (1971). *The scientist's role in society in a comparative study*. Prentice-Hall.

Boltanski, L., & Thévenot, L. (2006). *On justification: Economies of worth*. Princeton University Press.

Bourdieu, P. (1975). The specificity of the scientific field and the social conditions of the progress of reason. *Social Science Information, 14*(6), 19–47.

Buenstorf, G., Hottenrott, H., Kosmützky, A., Meier, F., Schaper, T., & Schimank, U. (2025). Analyzing constellations of multiple competition in science and higher education. *Minerva*. https://doi.org/10.1007/s11024-025-09616-9

Buschkamp, L., & Kosmützky, A. (2026). *The construction of (imagined) competitive groups of universities within multiple competition*. Under review in: Soziale Welt.

Carrier, M., Howard, D., & Kourany, J. A. (2008). *The challenge of the social and the pressure of practice: Science and values revisited*. University of Pittsburgh press.

Cloete, N., Côté, N., Crace, L., Delbridge, R., Denis, J.-L., Drori, G. S., Eriksson-Zetterquist, U., Gehman, J., Gerhardt, L.-M., Goldenstein, J., Harroche, A., Jandrić, J., Kosmützky, A., Krücken, G., Lee, S. S., Lounsbury, M., Mizrahi-Shtelman, R., Musselin, C., Gustafsson, H. Ö., ... Walgenbach, P. (2023). "Outroduction": A research agenda on collegiality in university settings. In K. Sahlin & U. Eriksson-Zetterquist (Eds.), *Research in the sociology of organizations* (pp. 181–211). Emerald Publishing Limited.

Dasgupta, P., & David, P. A. (1994). Toward a new economics of science. *Research Policy, 23*(5), 487–521.

Douglas, H. (2009). *Science, policy, and the value-free ideal*. University of Pittsburgh Press.

Douglas, H., & Branch, T. Y. (2024). The social contract for science and the value-free ideal. *Synthese, 203*(2), 40. https://doi.org/10.1007/s11229-023-04477-9

Fenkner, T., & Gottschalk-Rayling, J. (2025). Surviving science—Coping with exit-decisions in physics and history. *Minerva, 63*(3), 449–469. https://doi.org/10.1007/s11024-024-09548-w

Fochler, M., Felt, U., & Müller, R. (2016). Unsustainable growth, hyper-competition, and worth in life science research: Narrowing evaluative repertoires in doctoral and postdoctoral scientists' work and lives. *Minerva, 54*(2), 175–200.

Frost, J., Hattke, F., & Reihlen, M. (2016). Multi-Level governance in universities: Strategy, structure, control. In J. Frost, F. Hattke, & M. Reihlen (Eds.), *Multi-level governance in universities* (Vol. 47, pp. 1–15). Springer International Publishing.

Gerhardt, L.-M., Goldenstein, J., Oertel, S., Poschmann, P., & Walgenbach, P. (2023). The managerialization of higher education in Germany and its consequences: Changes in job advertisements for professorships in German universities, 1990 to 2010. In K. Sahlin & U. Eriksson-Zetterquist (Eds.), *University collegiality and the erosion of faculty authority* (Vol. 86, pp. 59–85). Emerald Publishing Limited.

Gerhards, J. (2013). *Der deutsche Sonderweg in der Messung von Forschungsleistungen*. Interdisziplinäre Arbeitsgruppe Exzellenzinitiative.

Gläser, J. (2024). From effects of governance to causes of epistemic change. *Minerva, 62*(3), 309–337.

Hamann, J., Blome, F., & Kosmützky, A. (2022). Devices of evaluation: Institutionalization and impact—Introduction to the special issue. *Research Evaluation, 31*(4), 423–428. https://doi.org/10.1093/reseval/rvac047

Hasse, R., & Krücken, G. (2013). Competition and actorhood: A further expansion of the Neo-institutional agenda. *Sociologia Internationalis, 51*(2), 181–205.

Hasse, R., Hwang, H., & Mormann, H. (2026) Expanding institutional analysis: Helping professionals and the reconstruction of academic individuals. *Higher Education.* https://doi.org/10.1007/s10734-025-01600-2

Kosmützky, A. (2026). Competition as a catalyst for collaboration: How German universities navigate the paradox of organizational research governance. *Higher Education.* https://doi.org/10.1007/s10734-025-01616-8

Kosmützky, A., & Krücken, G. (2023). Governing research: New forms of competition and cooperation in academia. In K. Sahlin & U. Eriksson-Zetterquist (Hrsg.), *Research in the sociology of organizations* (S. 31–57). Emerald Publishing Limited. https://doi.org/10.1108/S0733-558X20230000086002

Kosmützky, A., & Krücken, G. (2025). *Konkurrenz und Kooperation in Hochschulen und Wissenschaft* (LCSS Working Paper Nr. 20).

Kosmützky, A., & Meier, F. (2025). Competing: An analytical framework and application in higher education. *Studies in Higher Education*, 18–38. https://doi.org/10.1080/03075079.2025.2455425

Krücken, G. (2021). Multiple competitions in higher education: A conceptual approach. *Innovation, 23*(2), 163–181.

Krücken, G. (2023). Die Zukunft der Universitätsorganisation: Forschung, Lehre, Dritte Mission. In Torp, C. (Ed.). *Aufbruchstimmung: Die Universität Bremen und das Projekt Hochschulreform* (pp. 251–75). Wallstein Verlag.

Krücken, G. (2024). The construction of actors in the ecology of competitions permeating science and higher education. In S. Arora-Jonsson, M. Bromgren, J. Pallas, & L. Wedlin (Eds.), *An ecology of ideas permeating science, higher education, and society: Essays for Kerstin Sahlin* (pp. 59–61). Uppsala universitet.

Krücken, G., & Meier, F. (2006). Turning the university into an organizational actor. In G. S. Drori, J. W. Meyer, & H. Hwang (Eds.), *Globalization and organization: World society and organizational change* (pp. 241–257). Oxford University Press.

Krücken, G., Bünstorf, G., Cantner, U., Frost, J., Grebel, T., Hamann, J., Hottenrott, H., Kosmützky, A., Meier, F., Schimank, U., & others. (2021). Multipler Wettbewerb im Hochschulsystem—Interdisziplinäre Perspektiven und wissenschaftspolitische Implikationen. *Das Hochschulwesen, 3*(4), 90–95.

Laudel, G. (2023). Researchers' responses to their funding situation. In B. Lepori, B. Jongbloed, & D. Hicks (Eds.), *Handbook of public funding of research* (pp. 261–278). Edward Elgar Publishing.

Le Grand, J. (1991). Quasi-markets and social policy. *The Economic Journal, 101*(408), 1256–1267.

Lofthouse, S. (1974). Thoughts on "publish or perish." *Higher Education, 3*(1), 59–80.
Mau, S. (2019). *The metric society: On the quantification of the social*. Polity Press.
Mayer, A. (2019). *Universitäten im Wettbewerb: Deutschland von den 1980er Jahren bis zur Exzellenzinitiative*. Franz Steiner Verlag.
Merton, R. K. (1968). The Matthew effect in science: The reward and communication systems of science are considered. *Science, 159*(3810), 56–63.
Merton, R. K. (1973). *The sociology of science*. University of Chicago Press.
Müller, R. (2014). Racing for what? Anticipation and acceleration in the work and career practices of academic life science postdocs. *Forum Qualitative Sozialforschung/Forum: Qualitative Social Research, 15*.
Musselin, C. (2018). New forms of competition in higher education. *Socio-Economic Review, 16*(3), 657–683.
Naidoo, R. (2016). The competition fetish in higher education: Varieties, animators and consequences. *British Journal of Sociology of Education, 37*(1), 1–10.
Polanyi, M. (1962). Tacit knowing: Its bearing on some problems of philosophy. *Reviews of Modern Physics, 34*(4), 601.
Sahlin, K. (2012). The interplay of organizing models in higher education: What room is there for collegiality in universities characterized by bounded autonomy? In Stensaker, B., Välimaa, J., Sarrico, C. S. (Eds.), *Managing reform in Universities: The dynamics of culture, identity and organisational change* (pp. 198–221).
Sauder, M., & Espeland, W. N. (2009). The discipline of rankings: Tight coupling and organizational change. *American Sociological Review, 74*, 63–82.
Seidenschnur, T., Götze, N., & Krücken, G. (2024). Multiple roles of the state–Federal states and their roles in how universities compete in Germany. *Studies in Higher Education, 49*(10), 1753–1762.
Simmel, G. (2008). Sociology of competition. *Canadian Journal of Sociology/cahiers Canadiens De Sociologie, 33*(4), 957–978.
Stark, D. C. (Ed.). (2020). *The performance complex: Competitions and valuations in social life*. Oxford University Press.
Stephan, P. E. (1996). The economics of science. *Journal of Economic Literature, 34*(3), 1199–1235.
Stevens, M. L., Armstrong, E. A., & Arum, R. (2008). Sieve, Incubator, Temple, Hub: Empirical and theoretical advances in the sociology of higher education. *Annual Review of Sociology, 34*(1), 127–151.
Strange, S. (1996). *The retreat of the state: The diffusion of power in the world economy*. Cambridge University Press.

Szöllösi-Janze, M. (2021). Archäologie des Wettbewerbs: Konkurrenz in und zwischen Universitäten in (West-)Deutschland seit den 1980er Jahren. *Vierteljahrshefte Für Zeitgeschichte, 69*(2), 241–276.

Thoenig, J.-C., & Paradeise, C. (2016). Strategic capacity and organisational capabilities: A challenge for universities. *Minerva, 54*(3), 293–324.

Waaijer, C. J., Teelken, C., Wouters, P. F., & van der Weijden, I. (2018). Competition in science: Links between publication pressure, grant pressure and the academic job market. *Higher Education Policy, 31*(2), 225–243.

Whitley, R. (1984). *The intellectual and social organization of the sciences*. Oxford University Press.

Wissenschaftsrat (WR). (2023). *Strukturen der Forschungsfinanzierung an deutschen Hochschulen*. Positionspapier. Drs. 1012-23. Wissenschaftsrat. https://doi.org/10.57674/pms3-pr05

Part 2: The Internal Ethics of Science

The Changing Faces of Freedom: Historicizing Academic Activism as an Ethical Challenge

Marian Füssel

1 Introduction

In a recent article, Swiss historian Caspar Hirschi and Austrian sociologist Alexander Bogner argued that "academic activism is the weakest weapon imaginable in the defense against populist attacks on university autonomy" (Bogner & Hirschi, 2025). On the one hand, they argue, this is due to an "instrumental relationship to academic autonomy"; on the other hand, it is because activism shares fundamental assumptions with its populist opponents. Their claim is clear, activism yes, but only outside the doors of the institution, as a private person, not as a scholar. The "classic topoi of justification in the debate about the politicization of science," they argue, has been reversed compared to the period around the Second World War. Times have obviously changed, and with them the activist discourse.

Regardless of one's own personal position as an academic on activism, historicizing it can be not only a reflective gain in objectivity, but, in the words of Pierre Bourdieu, an act of liberation: "If one needs to know

M. Füssel (✉)
Seminar für Mittlere und Neuere Geschichte, Kulturwissenschaftliches Zentrum, Georg-August-Universität Göttingen, Göttingen, Germany
e-mail: marian.fuessel@phil.uni-goettingen.de

J. Hinrichsen et al. (eds.), *Ethics of the Sciences*,
https://doi.org/10.1007/978-981-92-2821-8_5

history, it is less to feed on it, but to free oneself from it, to avoid obeying it without knowing it or repeating it without wanting to" (Bourdieu 2004, p. 115).

The necessity of historicization can also be clearly illustrated by a simple everyday observation that was made a good decade ago but has hardly lost any of its relevance to highlight the ambiguity of certain slogans. At the time, I came across the image campaign "Free-Space for New Thinking" ("Freiraum für Neues Denken") on my university's website, and at the same time, a banner in front of the history department's students room read "This is a Free-Space" ("Das ist ein Frei-Raum"). The term "free space" referred to two very different claims to validity in knowledge policy: that of the university management for the freedom of the scientific system to think in new ways, and that of student activists within the university to claim freedom for their causes. Could it be that a new semantic cross-frontier has been formed in the spirit of freedom? A more plausible explanation seems to me to be the long-term effects of the complex semantics of academic freedom, which, in my opinion, represents one of the conditions for the tense relationship between university science and activism (cf. Conze et al., 1975; Horn, 1905; Warnach et al., 1972 on freedom).

The question of the "pluralization" of world views, values, and norms and their effects on academic practice is repeatedly confronted with various forms of activism, which makes this a central challenge for an ethics of science. Since universities are "spaces of argument" and "critical reasoning" some scholars conclude that, in the "setting of the university, a legitimate academic activism has to heed certain conditions and hold in view certain responsibilities" (Barnett, 2021, p. 518). It is a multiple challenge because the moral standards of activists and those of the prevailing scientific community do not have to be congruent and the representatives of science see themselves challenged to different degrees (Henning, 2024).

Many of these challenges are ethical in nature. Am I allowed to, or even obliged to, engage in politics as a scientist? Where are the limits of this political activism? Does activism harm academic freedom, or is it a prerequisite for its defense?

The impression that this is a genuinely new situation makes it necessary to historicize the tense relationship between activism, academia, freedom, and science. However, it is precisely the experts in historicization, the historians, who are currently actively involved in the question of activism itself (Nolte et al., 2024). The tension is evident: "Activism and movement knowledge are based on political antagonisms and dramatisations, on emotional attachment, on political plausibility and certainty performance, whereas scientific knowledge is based on critical-rational exchange, communication and doubt, differentiation and complexity of arguments, particular and selective research subjects, time-bound and revision-open areas of validity, and methodically bound evidentialisation. Science can advise and inform politics, but it cannot replace it" (Reichardt, 2024, 14; engl. Trans. MF). But let's look further back.

In the period around 1800, the premodern corporative plural of freedoms, in the sense of privileges, merged with the modern universalist concept of freedom, which continues to have an effect in the concept of academic freedom (Gordon, 2022; Horn, 1905; Lackey, 2018). It was a kind of sublation ("Aufhebung") in the Hegelian sense, because the meanings of the corporative freedoms never completely disappeared. The older concept of freedom, which meant a corporative-exclusive space with its own jurisdiction, also becomes visible in the "free space," marking the "campus" as its own legal territory of the "activist university" (Barnett, 2021). Without this free space, activism would have no stage and no forums for articulation. But academia itself also needs the freedom to think, research, and teach, which many of its representatives see as threatened by activism. In addition, there is a specific historical formatting of the activist persona, which can benefit from the sociology of intellectuals to elucidate it. The aim of this short essay is neither an empirically dense description of individual protest movements nor a sophisticated theoretical framework for their analysis. Rather, it is to show what possible added value questions and approaches of historical studies can provide for a current problem area in the ethics of science. The empirical basis for my argument is the German university of the last two centuries, not least because the historical semantics of academic freedom have been particularly pronounced there since the early modern period (Horn, 1905). Since then, national academic paths have crossed and diverged throughout history. The US for example had no totalitarian experience like universities in Nazi Germany, which caused different path-dependencies. The key event in the recent history of academic activism, the protests of 1968, was a global phenomenon and had an impact in France and the US as well as in Germany. However, the paths from there to the present day have developed quite differently in some respects. This is particularly important to bear in mind when considering the transnational circulation of concepts such as academic freedom or activism, which can encounter different resonances and contexts (Downs & Mannion, 2004). With this in mind, the longue durée is given preference over the transnational comparative perspective in the following.

First, I will illuminate the basic axes of the tense relationship between activism and academia from the perspective of a historical sociology of knowledge (1) Second, I will give an overview of phases in the history of student activism in Germany (2) in order to then, thirdly, identify some key elements for a praxeological analysis of academic activism on this basis (3) My aim is not to provide a magic formula for dealing with activism, but rather to argue that we first need to understand what activism does, how it works, how it argues, and how it changes over time.

2 Historical Sociology of Knowledge and Academic Activism: A Productive Tension

The diagnosis of a return of early modern corporative social formations in late modernity can hardly claim originality anymore. However, in my opinion, it is informative for our question about the relationship between activism and science, at least it helps to explain the apparent contradiction between student free spaces and university image campaigns, which I mentioned at the beginning. Some see activism as a threat to academic freedom; inflammatory terms such as "opinion dictatorship" and "cancel culture" leave no doubt about this (Daub, 2022; Dardan et al., 2023). But the activists also defend a freedom, and a spatially marked freedom at that, even if the campus has long since gone beyond a purely physical space (Nocella & Gabbard, 2013). Thus, structural homologies stand out between an academic freedom, as it was claimed in the premodern corporate society of orders and as it is demanded by activists today: In this respect, the pre-modern university was the quintessential "safe space" (Kaldewey, 2017). Until the nineteenth century, it formed a privileged space for its members and its own legal space, a legal space that, paradoxically, is also claimed by the conservative, attentist side.

But to shed light on the tense dynamics of the relationship between science and activism, we need to consider another, no less ambivalent pole of that "force field" (Bourdieu), the attraction of which I call the promise of the intellectual. I see a structural affinity between activists and intellectuals in the conscious transgression of one's own system boundaries in the sense of a politicization of the science system. This requires certain spokesperson positions in the struggle for the power of interpretation of problems in society as a whole, which are not based on invocation, as with experts, but on personal initiative and epistemic authority. Once again, it quickly becomes clear that academic intellectuals in France, the US, and Germany have developed very different role models, such as Sartre, Foucault, Bourdieu, and Derrida in France, Habermas in Germany, and Noam Chomsky or Judith Butler in the US (Schildt, 2016, 2020). There need not be agreement between their philosophies or their political agendas in order to activate the tension between activism and academia in a similar way. However, the situation is made analytically more complicated by the fact that these intellectual fathers and mothers not only took to the streets themselves, but also provided the theoretical vocabulary with which we analyze activism like discourse or field analysis or texts like Derridas "University without conditions" (Derrida, 2005) or Butlers "Critique, Dissent, Disciplinarity" (Butler, 2009) and this has been the case at least since Marx and Weber.

Analysts of a structural change in the student milieu of the 1990s, for example, rather casually pointed out the old idealistic factor of "respect for scientific authority," with the only difference that this respect "is now shown to the appointed, yet non-conformist critics of a normality in a knowledge- and industrial-society" (Köhler et al., 2001, p. 286). This is a very important point for understanding the dynamics of academic activism. The language of uncompromising advocacy is borrowed from a scientific discourse logic that authorizes criticism of authority itself.

Furthermore, an analytical distinction must be made between the goals of activism that are beyond the reach of the institution, i.e., that affect society as a whole, such as a constitutional issue, or even all of humanity, such as climate change or gender relations, and those that primarily have an intra-institutional focus, such as the reduction of privileges, tuition fees, co-determination, etc. (McCann, 2010; and anti-activist Pluckrose, 2020). In individual cases, the goals may undoubtedly overlap, as in the case of Marxist groups in the 1960s and 1970s, whose criticism was directed as much against "outdated forms of organization ("under the robes, the musty smell of a thousand years"/ "unter den Talaren, Muff von tausend Jahren") as against the 'objectivity of science in the' capitalist state" (Jarausch 1984, p. 229).

In contrast to the intellectual, the activist persona seems to be more strongly influenced by ad hoc movements, or by tactical action in the sense of Michel de Certeau, i.e., acting more fleetingly in time than controlling their own spaces (Certeau, 1988, pp. 34–39). Activism is obviously an attractive proposition for those involved, and the driving force here is the promise of being able to achieve something, to change something.

In the terminology of the sociology of knowledge and the history of science, which are equipped with a historical depth of focus, activists are a relatively new phenomenon. Any search for traces in the times before the term came into political use in the early twentieth century (Cordsen, 2022; Harm, 2024) is at least open to the accusation of being a "productive anachronism." The same applies to the concept of intellectuals. Since Jacques Le Goff's 1957 work *Intellectuals in the Middle Ages*, every epoch has its intellectuals (Le Goff, 1993). But what about the activists? A certain attractiveness of the concept lies in its low threshold, which makes it easier to release historical identification potentials. In contrast to intellectuals, activism does not require academic titles, an academic or aesthetic oeuvre of one's own, or a special institutional status. However, activism also seems more fleeting, although this depends, to a large extent, on the group of people being observed. Many of the actors under discussion could be categorized as both activists and intellectuals, although students often engage in a form of activism that is more a phase in their lives (Bashiri, 2024, pp. 66–67; Germann et al., 2022, p. 436). The following, very abbreviated academic genealogy begins primarily with student activism, but also looks at the professors.

3 Protest on Repeat. Historical Stages of Development of Academic Student Activism

In order to gain a historical perspective on the dynamics of the tension between activism and science, one has to go back well beyond the time when the concept of activism emerged in the early twentieth century (Cordsen, 2022). Historical comparisons are repeatedly used to make the dramatic nature of current developments evident. Herfried Münkler, Professor for Political Science, who was himself the object of activism at the Free University of Berlin, is particularly active in updating historical references when he writes: "The difference between 1968 and today is that the 1968 students wanted to discuss at all costs [...] Today, the radical students strike from the anonymity of the Internet. They try to destroy reputations, silence people—and refuse to engage in any kind of discussion" (Münkler 2019, p. 8, engl. trans. MF). The wording seems to suggest that the scenario of the "new wars" has also been transferred to the lecture halls. But the social logic of this post-Habermas style of articulation is not at issue here; what is more interesting is that, in historical comparison, the much-maligned 68ers appear as the lesser evil, the present as the provisional low point of a spiral of decadence in academic culture. Nevertheless, a historical depth dimension seems to me to be conducive to insight (Dhondt & Boran, 2018).

German student history from 1800 to 1970 has been divided into five stages (Jarausch 1984; Schenke, 2020), spanning a transformation from "Burschenschafter to Sponti" (Schlicht, 1980). For the last fifty years, one or more further phases could be added. The period from 1800 to 1866 was characterized by "national and liberal awakening." The decades saw the founding of the Urburschenschaft (1815), the Kotzebue assassination attempt (1819), the first and second Wartburg festivals (1817 and 1848), and the Hambach festival (1832). It was a phase of a philosophy of action and political professors. In 1837, the Göttingen Seven stood for professorial constitutional activism with student support. With national unification, the focus here was on goals of university-political activism that clearly went beyond the concerns of the university. The student culture of this time still enjoyed the freedom of the academic jurisdiction.

The second phase, from 1867 to 1917, roughly coincides with the abolition of this freedom and marks a "change of tendency among academic youth" (Jarausch 1984: 59), which was characterized, among other things, by the fraternity system and academic antisemitism. A counter-movement emerged from 1900 in the form of the liberal Free Student Body, which saw itself as a rallying point for all non-corporated students. In the academic culture war, the value of "academic freedom" was invoked to suppress non-national students and their associations. A massive shift to the right was unmistakably evident.

The third phase from 1918 to 1932 presented itself as a "permanent crisis." Unlike 1849, the November Revolution of 1918 was a "revolution without students" (Jarausch 1984, p. 118). The political radicalization to the right continued to increase, and the diagnosis of an overcrowding crisis did the rest. Keywords such as massification and proletarianization pointed to unease and fears of a loss of status.

Phase four is characterized by Nazi rule, from now on the National Socialist German Students' League ruled. In May 1933, the "Aktion wider den undeutschen Geist" ("Action to Restore the German Spirit"), initiated by the German Student Body, culminated in the burning of books (Strätz, 1968). "There is no longer any room for apolitical students, just as there can no longer be any room for political groups at our universities that are offshoots of the people-destroying and subversive forces," said Reich-student-leader Scheel in 1937 (Jarausch 1984, p. 174, engl. trans. MF; Grüttner, 1995). The term "activist" had positive connotations in Nazi Germany and after the war it became a category of the group of those incriminated in the Control Council directive (Harm, 2024). There was only isolated resistance, for example in Berlin in the group around Herbert Baum or in Munich with the White Rose.

After World War II, the fifth period of "ambivalent academics" began (1946–1970) (Jarausch 1984: 213). Under the restorative tendencies of the Adenauer era, the corporations were re-established. On July 8, 1953, during the 1000th anniversary celebrations of the city of Göttingen, six hundred members of the corporations marched through the city with a torchlight procession to demonstrate for the wearing of colors. There were no counter-demonstrators. At the same time, the "Asten" (student assemblies) reformed and the parties set up university groups ranging from the SDS to the LSD to the RDCS (Micus et al., 2020). The discussion about the political mandate of the student body became a long-running issue. In Göttingen and Freiburg in 1951/52, films by Nazi movie director Veit Harlan caused political riots, resulting in violent conflicts between students and anti-Semitic youths. I quote at length from the Göttingen City Archives' chronicle of January 25, 1952: "As in other university cities, especially Freiburg im Breisgau and Münster, liberal students gathered in front of the 'Central Theater' on Barfüßerstraße in Göttingen in the afternoon to protest the screening of the film 'Hanna Amon' by Veit Harlan." Target of the demonstrations was the director, who, through his film "Jud Süß" during the Nazi regime, produced anti-Semitic propaganda. "They thus repeated the rallies that had taken place in February 1951 at the same place against Harlan's film 'The Immortal Beloved'. The demonstrating students carried banners with the inscription 'Peace with Israel' and 'We don't want any more Harlans', which was answered by the opposing side with vivid anti-Semitic comments. After brawls broke out in the streets and the third showing was disrupted by stink bombs, the police intervened and banned further screenings of the film to maintain public order and safety. These events were hotly debated for and against by the citizenry for days." Now academic activism came into play: "In a flyer dated January 29, 48 professors and lecturers from the University and the College of Education publicly endorsed the motives of the demonstrations, which they recognized as a sign of the political sense of responsibility of the academic youth with regard to anti-Semitic and undemocratic tendencies. Among the signatories of the declaration were some of the most well-known personalities in Göttingen, such as the president of the Max Planck Society, Professor Otto Hahn, the famous nuclear physicist, Professor Werner Heisenberg, and the expert in constitutional and ecclesiastical law, Professor Rudolf Smend. Heinz Hilpert, the director of the German Theatre ('Deutsches Theater'), was also one of them. The rector of the university, Professor D. Dr. Wolfgang Trillhaas, backed this declaration in a meeting of the student council. Harlan's attempts to negotiate with the rector and student body in Göttingen at the beginning of February failed because neither the rector nor the General Student Committee received him." (Chronik der Stadt Göttingen, https://stadtarchiv.goettingen.de/portal/seiten/chronik-der-stadt-goettingen-900001256-25480.html) [engl. trans. MF]

This period also saw the involvement of the "Göttingen Eighteen," a group of professors who signed the Göttingen Declaration against the arming of the Bundeswehr with nuclear weapons in April 1957 (Lorenz, 2011; Schirrmacher, 2008). Among them were Max Born, Otto Hahn, Werner Heisenberg, Heinz Maier-Leibnitz, and Carl Friedrich von Weizsäcker. Similar to the Göttingen Seven, the ultimately successful auction also received a lot of support from students. On May 20, 1958, fourteen German universities saw demonstrations against "nuclear death," from which the RCDS (=Association of Christian Democratic Students), of course, distanced itself.

A critical student press with organs such as Colloquium or the *Studentenkurier* (1955–1957), which then became the left-wing journal *Konkret* in 1957, ensured a trans-regional discursive network of activist circles. Jarausch views the "political actions" of students in the post-war years as falling between "democratic initiative and conformist adaptation" (Jarausch 1984, p. 221). The visit of the Shah of Iran to Berlin on June 2, 1967 marked the beginning of a radical phase of student activism that is retrospectively referred to as the 1968 movement. It was the "German version of an international youth revolt," which, according to Jarausch, on the one hand stood in the "radical leftist tradition of the German students of 1848, the Finkenschaft or Weimar socialists," but on the other hand was a "local expression of the worldwide protest of intellectuals" against the U.S. Vietnam War (Jarausch 1984, p. 226, engl. Trans MF). However, non-stop activism "did not lead to the expected international youth revolution, but to the disintegration of the student movement from 1969 onwards" (Jarausch 1984, p. 229, engl. trans MF). Protest and activism became a way of life, "one 'demo' chased the other, there was always the need to hang up posters, print newspapers, hold discussions, occupy institutes" (Jarausch 1984, p. 229). The frustration of the 1970s and further radicalization followed, then the No Future mentality. The most recent phase is difficult to analyze. After the loss of a "relative social and socio-moral homogeneity of the student body," political horizons of interpretation have also shifted. Left and right have lost much of their significance in the face of identity policy imperatives, while at the same time an old "educational pathos" continues to feed motivation and language (Schenke, 2020, pp. 485–486). For the 1990s, sociologists then observed a development from "alternative non-conformism to conformist performance and career orientation" (Köhler et al., 2001, p. 285).

But what systematic insights can be gained from these historical highlights? In the heuristic horizon of the social–historical approach, there is usually the expectation of the possibility of a correspondence analysis between the socio-historical development of the education system, both in its frequencies and political controls, and the political radicalization and mobilization willingness of the young, educated classes (Schenke, 2020, pp. 55–61). Simplified to Jarausch's hypothesis that "in phases of growth and opening, the student body liberalized, but in times of crisis of overcrowding or shrinkage, it radicalized, depending on the political circumstances, to the left or right" (Jarausch 1984, p. 247). There is certainly something to be said for this, but in my view, it needs further micro-historical concretion. Jarausch himself rejects the explanatory models of generational conflict and crisis at the university, using the example of the 1968 movement. Although both had the potential to mobilize, they were not sufficient as reasons alone, so that basically a look at the content of the political protest is necessary. Instead, however, I turn here to some aspects of the sociology of knowledge of academic activism.

It is important to specify how general political trends in the Federal Republic, such as democratization and liberalization, manifested themselves in the academic field and changed in the process (Herbert, 2002). None of the key concepts of political struggles are static but are constantly changing as they are appropriated by the respective groups of spokespersons. The call for liberalism, for example, sounds very different when it is directed against a conservative establishment in the 1960s and 1970s than when it is directed against a supposed left-wing "dictatorship of opinion" in the post-2000s. Therefore, the intellectual history of scientific ethics urgently needs to be fine-tuned in terms of the sociology of knowledge and history.

4 Dialectics of Freedom. Elements of a Historical Praxeology of Activism

The opposite pole of activism in terms of action theory is passivism, i.e., a rather passive, wait-and-see behavior. In the polarization between the university and activism, the university usually adopts the role of passivism. In its history, reforms have rarely come from within, but mostly from outside; the institution's decision-making behavior tends to be defensive (Kintzinger et al., 2018; Rohstock, 2010). In this respect, the university as an institution and thus also the social system of science is the ideal point of friction for activist movements.

Activism overlaps to a large extent with the categories of protest and social movements that are well established in historical research. Activism is not limited to one political camp; its long history shows left-wing and right-wing variations. To automatically label protesting students as left-wing would be a historical fallacy (Cordsen, 2022; Schenke, 2020). Activist practices, for example, draw on the same reservoir as those of protest: occupying lecture halls, demonstrations, publications in small formats such as pamphlets, posters, stickers, etc. Especially the campus is a particularly contested space (Nocella & Gabbard, 2013). A transnational comparative perspective is revealing when it comes to the spatial context of activism. For example, American campus universities have a different spatial logic than many German universities, which tend to be spread out across urban areas. Without a clearly defined campus, the path to the lecture hall or the streets of the city may be more obvious. The police autonomy of the campus can also vary in degree.

In the confrontation with the institution, activism automatically takes a position on the tactical, not the strategic side. In Michel de Certeau's sense, it is the institution that has control over space, its critics only over temporary uses—both evident in the dispute over the power of speech in the lecture hall (Certeau, 1988). The activists thus see their activity as legitimized as the "art of the weak." This was symbolically emphasized in 1977, for example, when a "city Indian" in Göttingen referred to himself as a "Mescalero," a group of the Apache tribe, in the newspaper of the Göttingen AStA in a text about the assassination of Federal Prosecutor General Siegfried Buback (Spiller, 2006). Such a self-description would probably no longer be chosen today in the context of an intensive discussion about cultural appropriation from this political milieu.

The Göttingen case attracted nationwide attention at the time, which touches on the spatial dimension of activism in terms of locality and globality. Pre-modern forms of student protest almost always remained locally confined to the space of their own university town. Since the early nineteenth century, however, there has been an increasing tendency toward trans-local networking, and the 1968 protests are considered a global event, from Japan to Europe and the US (Gilcher-Holtey, 2001). In addition to the question of spatial constitution and networking, the academic milieu, on the part of students, is characterized by the limited time spent at university, which brings with it specific challenges for activism in terms of continuity and sustainability, or the question of learning and transferring past experiences (Hoeft, 2021). As a result, generational change tends to occur more rapidly than among professors. On the other hand, this specific academic life cycle also leads to a continuous updating of the activists' programs, which can sometimes lead to inter-activist dissonance.

The aggressive and very successful hip-hop song *Fick die Uni* (*Fuck the University*) by German Band *Antilopengang* can perhaps be read as a symptom of meta-activist anti-academism: "Uni is different from school, because only nerds are left / They're only happy when they're reading and they live their lives to themselves / They only live off cigarettes and coffee / At their parties they dance to electro and ragga / They're politically interested and involved in the AStA / And they all love riding their bikes / They wear second-hand tracksuits and have dreadlocks / During lectures they take notes on their laptops / And they always listen, even though they don't understand a word / In 1933 they would all have been Nazis" (https://lyricstranslate.com/de/antilopengang-fick-die-uni-lyrics.html). From the point of view of the real activists, the academic activists become pseudo-activists, and intellectual promise becomes expression-conformist nonconformism.

The temporary status of academic activists—which, for example, associations with the (the principle of lifelong association ("Lebensbundprinzip") and parties with youth organizations seek to counter—helps to ensure that each generation can fully exploit and experience its own potential for radicalization, even though it is measured against its predecessor generations and almost inevitably differs from them. Jarausch also notes that students were "much more politicized than other young people (56–25%) because the special role of students gave young academics not only a sense of solidarity but also a sense of special responsibility", which I have subsumed under the promise of the intellectual above (Jarausch 1984, p. 235). In the longue durée, this also explains why every student culture, regardless of political orientation, could become a foreign body in relation to its bourgeois environment, be it as a fraternity member of the Vormärz or as a Sponti of the 1970s (Dhondt & Boran, 2018; Schlicht, 1980). And even largely well-adjusted student generations are suspicious of their predecessors. At the beginning of the 1980s, a study of the current student generation concluded: "They are more tolerant, softer, more relaxed than their predecessors, but also more erratic, less driven and more easily frightened" (Jarausch 1984, p. 242, engl. Trans. MF).

Academic activism has always had its polar opposites. Initially, it was the monarchy or the state, but after 1968, it was also private counter-initiatives. Among the best known is the *Bund Freiheit der Wissenschaft* (BFW), which was founded in 1970 but ceased its work in 2015 (Koischwitz, 2017; Wehrs, 2014). Its goal was "to resist any submission to the power claims of individual groups or interests." In 2021, the *Network for Freedom of Science* (*Netzwerk Wissenschaftsfreiheit*) was founded with an initial 70 members, but by September 2022, the network already had 734 members (incidentally, the largest proportion of them were from Berlin; 758 members on Feb. 06.2025). The network's manifesto states: "The Network for Scientific Freedom is an association of scientists with the common concern of defending the freedom of research and teaching against ideologically motivated restrictions and contributing to the strengthening of a liberal scientific environment" (Netzwerk Wissenschaftsfreiheit e.V., 2021) [engl. trans. MF]. This also explicitly includes a reference to the work of activists: "Both university members and external activists scandalize the invitation of unpopular guest speakers in order to exert pressure on the inviting colleagues and the management levels." The network's homepage also contains a collection of cases of Cancel Culture incidents that have produced the first evidence base for a threat to academic freedom (Daub, 2022; Dardan et al., 2023). The question of spaces for discourse and opportunities for articulation ultimately reveals the paradoxical convergence between activists and anti-academics noted at the outset. In the 1968 era, the expansion of speaking positions and dialogue was a key demand. Cancel-culture and the demands of the

Network for Academic Freedom, however, seem to point more toward the stifling of discourse.

While freedom generally enjoys a positive connotation across different eras and political camps, this does not apply to activism, although its negative connotation is now more strongly linked to political camps: "In recent times, there have been signs of a deterioration in meaning in relation to the term activist, particularly in relation to the adjective activist and the abstract noun activism. Primarily on the part of those with more conservative views, activists and activism are sometimes criticised as irrelevant and unscientific [...], and in some contexts, the attribute activist is used to characterise people as non-objective and ideologically driven" (Harm, 2024, engl. trans. MF).

5 Conclusion

The universities were only one part of the academic system that was examined, but historically, universities are particularly influential places when it comes to the question of activism. Universities have always provided forums for articulation, which makes space a contested resource. Debates on contemporary activism always also involve spatial conflicts: Who is allowed to speak where, who speaks on behalf of which institution, and with what scope? The academic free space offers an inherited institutional charisma with corporative roots, in which its critics participate just as much as its advocates.

As communities of teachers and learners, however, universities also train different time regimes of engagement. While professors can pursue certain forms of activism or passivity throughout their lives, students operate in a time frame with more rapid generational succession. This operation in time frames feeds the idea of always being in a completely new situation, while those who are long-term oriented quickly tend to compare the past and the present.

In historical studies, two camps of argumentation are currently emerging: one advocates a systems theory approach to the contrasts between scientific and movement-oriented knowledge, while the other describes "fluid transitions and transformations between historical studies and activism" (Reichardt, 2024, p. 14). While the first position comes across as somewhat dogmatic and exclusionary, the second makes it too easy to always insist on an undifferentiated middle ground.

However, a historicization from the perspective of the sociology and ethics of knowledge means neither "there has never been anything like this before" nor an evocation of a return of the same, as in "that's just the way young people are." Rather, the aim is to make visible the historical formation of patterns that continue to have an effect in the present as reference points, whether consciously or unconsciously. The ethical challenge of historicization is twofold: it is about confronting current political, social, and economic challenges while at the same time drawing on the reflective potential from the long history of these challenges. The encounter with activism makes historians aware that their own ethical standards usually remain implicit, as professional ethics dictate that they refrain from making generalized normative statements. However, this does not mean that they do not take a political stance. On the contrary, this hiatus may be due, on the one hand, to the fact that historians are not trained as ethicists, but on the other hand also to the fact that ethical principles themselves are subject to ongoing negotiation. Our historical tour de force has shown that even very powerful social processes such as democratization are by no means stable, but require continuous work, which also gives rise to ethical responsibility. This responsibility is in turn confronted by numerous shifting boundaries between universalism and particularism, engagement and distancing, etc. History cannot be a teacher with simple formulas for coping with these challenges. Rather, history shows that things were different and can be different from what they are today.

Competing Interests The author has no conflicts of interest to declare that are relevant to the content of this chapter.

References

Barnett, R. (2021). The activist university: Identities, profiles, conditions. *Policy Futures in Education, 19*(5), 513–526.

Bashiri, F. (2024). Conceptualizing scholar-activism through scholar-activist accounts. In P. Mattsson, E. Perez Vico, & L. Salö (Eds.), *Making universities matter: Collaboration, engagement, impact* (pp. 61–97). Springer.

Bogner, A., & Hirschi, C. (2025, September 4). Rollentausch von Aktivisten und Antiakademikern. Die Politisierung der Wissenschaft läuft anders ab als früher. Um sie aufzuhalten, muss man ihre confuse Logik verstehen. Frankfurter Allgemeine Zeitung, No. 508.

Bourdieu, P. (2004). *Schwierige Interdiszplinarität. Zum Verhältnis von Soziologie und Geschichtswissenschaft*. In E. Ohnacker & F. Schultheis (Eds.). Westfälisches Dampfboot.
Butler, J. (2009). Critique, dissent, disciplinarity. *Critical Inquiry, 35*(4), 773–795.
Certeau, M. de (1988). *The practice of everyday life* (S. Rendall, Transl.). University of California Press.
Conze, W., et al. (1975). Art. Freiheit. In *Geschichtliche Grundbegriffe. Historisches Lexikon zur politisch-sozialen Sprache in Deutschland* (Vol. 2, pp. 425–542). Klett-Cotta.
Cordsen, K. (2022). *Die Weltverbesserer: wie viel Aktivismus verträgt unsere Gesellschaft?*. Aufbau.
Dardan, A., et al. (2023). *Canceln: ein notwendiger Streit*. Hanser.
Daub, A. (2022). *Cancel culture transfer. Wie eine moralische Panik die Welt erfasst*. Suhrkamp.
Derrida, J. (2005). The future of the profession or the unconditional university (Thanks to the "Humanities", What Could Take Place Tomorrow). In P. P. Trifonas & M. A. Peters (Eds.), *Deconstructing Derrida* (pp. 11–24). Palgrave Macmillan.
Dhondt, P., & Boran, E. (2018). *Student revolt, city, and society in Europe: From the middle ages to the present*. Routledge.
Downs, J., & Mannion, J. (2004). *Taking back the academy! History of activism*. Routledge.
Germann, P., Held, L., & Wulz, M. (2022). Scientific political Activism—eine Annäherung an das Verhältnis von Wissenschaft und politischem Engagement seit den 1960er Jahren. *NTM Zeitschrift Für Geschichte der Wissenschaften, Technik und Medizin, 30*, 435–444.
Gilcher-Holtey, I. (2001). *Die 68er Bewegung: Deutschland, Westeuropa*. C.H. Beck.
Le Goff, J. (1993). *Intellectuals in the middle ages*. Blackwell.
Gordon, D. (2022). *What is academic freedom? A century of debate, 1915–Present*. Taylor & Francis.
Grüttner, Michael. (1995). *Studenten im Dritten Reich*. Schöningh.
Harm, V. (2024). Art. Aktivist (Aktivismus, aktivistisch). Version 12/2024, zuerst 10/2022. In *Wortgeschichte digital* – ZDL, https://www.zdl.org/wb/wortgeschichten/Aktivist [consulted 10.9. 2025].
Henning, T. (2024). *Wissenschaftsfreiheit und Moral*. Suhrkamp.
Herbert, U. (2002). *Wandlungsprozesse in Westdeutschland: Belastung, Integration, Liberalisierung*. Wallstein.
Hoeft, C. (2021). *Stillstand in Bewegung? Kollektives Lernen in sozialen Bewegungen*. transcript.

Horn, E. (1905). "*Akademische Freiheit*": *historisch-kritische Untersuchung und freimütige Betrachtung nebst einem Anhange über studentische Ausschüsse*. Trowitzsch & Sohn.

Jarausch, K. H. (1984). *Deutsche Studenten: 1800–1970*. Suhrkamp.

Kaldewey, D. (2017). Der Campus als "Safe Space". Zum theoretischen Unterbau einer neuen Bewegung. *Mittelweg 36. Zeitschrift des Hamburger Instituts für Sozialforschung, 26*(4/5), 132–153.

Kintzinger, M., Wagner, W. E. & Crispin, J. (2018) *Universität – Reform. Ein Spannungsverhältnis von langer Dauer (12.-21. Jahrhundert)*. Schwabe.

Köhler, T., Gapski, J., & Lähnemann, M. (2001). Von der alternativen zur konformistischen Revolution? Zum Strukturwandel von "Lebenschancen" und "Lebensführung" im westdeutschen Studierendenmilieu. In E. Stölting & U. Schimank (Eds.). *Die Krise der* Universitäten. Leviathan Sonderheft 20 (pp. 265–294). Westdeutscher Verlag.

Koischwitz, S. (2017). *Der Bund Freiheit der Wissenschaft in den Jahren 1970–1976: ein Interessenverband zwischen Studentenbewegung und Hochschulreform*. Böhlau.

Lackey, J. (2018). *Academic freedom*. Oxford University Press.

Lorenz, R. (2011). *Protest der Physiker: Die »Göttinger Erklärung« von 1957*. transcript.

McCann, B. J. (2010). Borders of engagement: Rethinking scholarship, activism, and the academy. In J. Lee & S. Kahn (Eds.), *Activism and rhetoric: Theories and contexts for political engagement* (pp. 11–25). Routledge.

Micus, M., Pflicke, T., & Scharf, P. (2020). Unterm Brennglas – Facetten des linken Radikalismus in Göttingen. In A. Deycke, J. Gmeiner, J. Schenke, & M. Micus (Eds.), *Von der KPD zu den Post-Autonomen: Orientierungen im Feld der radikalen Linken* (pp. 225–270). Vandenhoeck & Ruprecht.

Münkler, H. (2019). Interview. https://www.welt.de/politik/deutschland/plus189723151/Universitaeten-Wenn-Studenten-zu-Meinungsdiktatoren-werden.html, quoted in Wilhelm Hopf, Meinungsfreiheit verteidigen, Einleitung. In Ibid. (ed.). *Die Freiheit der Wissenschaft und ihre 'Feinde'*. LIT, 7–27, 8.

Netzwerk Wissenschaftsfreiheit e.V. (2021, February). Manifest. https://www.netzwerk-wissenschaftsfreiheit.de/ueber-uns/manifest/

Nocella, A. J., II., & Gabbard, D. (2013). *Policing the campus: Academic repression, surveillance, and the occupy movement*. Peter Lang.

Nolte, P., Reichardt, S., & Rischbieter, L. (2024). Aktivismus und Geschichtswissenschaft, Thematic Issue of *Geschichte und Gesellschaft, 50*(1).

Pluckrose, H. (2020). *Cynical theories: How activist scholarship made everything about race, gender, and identity—And why this harms everybody*. Pitchstone Publishing.

Reichardt S. (2024). Aktivismus und Geschichtswissenschaft. Eine Einleitung/ Activism and Historiography. An Introduction. *Geschichte und Gesellschaft, 50*(1), 8–32.

Rohstock, A. (2010). *Von der "Ordinarienuniversität" zur "Revolutionszentrale"?: Hochschulreform und Hochschulrevolte in Bayern und Hessen 1957–1976.* Oldenbourg.

Schenke, J. (2020). *Student und Demokratie: Das politische Potenzial deutscher Studierender in Geschichte und Gegenwart.* transcript Verlag.

Schildt, A. (2016). *Von draußen: ausländische intellektuelle Einflüsse in der Bundesrepublik bis 1990.* Wallstein.

Schildt, A. (2020). *Medien-Intellektuelle in der Bundesrepublik.* Wallstein.

Schirrmacher, A. (2008). Verantwortung made in Göttingen? Die Erklärung der Göttinger Achtzehn von 1957. In *Göttinger Jahrbuch* (Vol. 56, pp. 57–71).

Schlicht, U. (1980). *Vom Burschenschafter bis zum Sponti. Studentische Opposition gestern und heute.* Colloquium-Verl.

Spiller, S. (2006). Der Sympathisant als Staatsfeind: die Mescalero-Affäre. In W. Kraushaar (Eds.). *Die RAF und der linke Terrorismus* (2nd ed., pp. 1227–1259). Hamburger Edition.

Strätz, M. H. (1968). Die Studentische "Aktion Wider Den Undeutschen Geist" Im Frühjahr 1933". *Vierteljahrshefte Für Zeitgeschichte, 16*, 347–372.

Warnach, W., Pesch, O. H., & Spaemann, R. (1972). Freiheit. In *Historisches Wörterbuch der Philosophie* (pp. 1064–1098).

Wehrs, N. (2014). *Protest der Professoren: der "Bund Freiheit der Wissenschaft" in den 1970er Jahren.* Wallstein.

Digital Sovereignty in Science: Challenges and Opportunities Through Large Language Models in Knowledge Production

Daniel Kurzawe

1 Introduction

The role of digital information and data in society is rapidly expanding, particularly in research. The integration of a diverse array of mobile devices, which provide seamless access to an almost limitless pool of data sources, is transforming how information is engaged with on a societal level and ultimately reshaping our relationship to knowledge itself. Alongside the accessibility of data, its representation is also transforming: Simple factual representations are being complemented by increasingly complex information structures, ranging from discrete data points to semantically interconnected entities and, more recently, to complex model structures, which are used by *Artificial Intelligence* (AI) and *Large Language Models* (LLMs).

The author used Grammarly and AI-powered language correction tools for grammar and stylistic revisions. No generative AI tools were used for content creation, data analysis, or interpretation.

D. Kurzawe (✉)
Research and Development Department, State and University Library, Georg-August-Universität Göttingen, Göttingen, Germany
e-mail: kurzawe@sub.uni-goettingen.de

J. Hinrichsen et al. (eds.), *Ethics of the Sciences*,
https://doi.org/10.1007/978-981-92-2821-8_6

Behind these terms lie concrete technologies and concepts that are already reshaping the way knowledge is produced, accessed, and legitimized. This article examines how LLMs and AI systems challenge scientific knowledge production through infrastructural entanglements and epistemic shifts.

As our societal relationship with information, and consequently, the production, organization, and dissemination of knowledge, becomes increasingly reliant on digital information processing, so does scientific practice itself (Humphreys, 2004, pp. 4–9, 49–55). This concerns not only the frequently cited examples from data-intensive research fields such as astronomy or the life sciences, as Leonelli (2016) describes in detail. The adoption of computational methods begins much earlier and extends across disciplines, encompassing areas that were previously inaccessible with digital research methods. An example is the emergence of *Digital Humanities* and hermeneutical approaches in large datasets (Balzer et al., 2018). With AI-based methods, entirely new domains are now being opened for digital tools: From literature research to concrete data collection and utilization, research is progressively shifting toward incorporating digital working methods. This evolution includes, for example, the simple use of search algorithms, which are increasingly based on static models. It can be found in word processing tools or more sophisticated tools, such as those used for stylistic revision of texts. With the integration of AI-driven functionalities, such as literature management systems, the line is increasingly blurred between writing assistance tools and *intellectual-like contributions*. By generating automatic summaries, these systems engage in a form of *interpretative automation*, performing initial acts of synthesis that reflect a *delegation of epistemic agency*. Moreover, of course, LLM-based chatbots take this to the next level. They all share the commonality that their language models are built upon the underlying intellectual contributions of other publications.

Discipline-specific tools employ a range of computational techniques, from simple algorithms to complex statistical models, particularly in areas such as data analysis and the development of research software as part of the research software engineering process itself. Automation does not merely support interpretation but becomes increasingly embedded within the methodological architecture of research practice. As such, they no longer assist researchers but begin to shape the cognitive process itself. By shifting parts of the human mental process to researcher-assisted tools, a form of *cognitive offloading or intelligence offloading*, as described by Clark (2008, pp. 196–217), is beginning to change the practice of conducting research.

Furthermore, with that, such AI tools are now systematically integrated into all phases of the scientific knowledge production process, finding application across an increasingly broad spectrum of academic disciplines.[1] Moreover, with this, contemporary scholarly communication and research are becoming increasingly dependent on digital research infrastructures, as they depend on distributed systems and information.[2] With this, the information itself is increasingly shifting from singular archives into distributed digital infrastructures. This also brings a wide range of ethical concerns (Hagendorff, 2024). These include questions of authorship and intellectual ownership, the significant energy consumption associated with model training and deployment, and the pervasive biases rooted in the selection and composition of training data. Promises of algorithmic neutrality often prove misleading, as underlying models reflect and reproduce societal asymmetries. The dominance of commercial technology firms further concentrates control over infrastructure and knowledge production, while the overrepresentation of English-language and Western-centric data sources reinforces a limited cultural perspective. Other pressing concerns involve the epistemic reliability of model outputs, e.g., hallucinations and the absence of verifiable provenance. Moreover, the opacity of training processes and the inaccessibility of underlying datasets hinder transparency and accountability. Issues of surveillance, data extraction, and the use of AI in reinforcing existing social inequalities also raise serious normative questions. Finally, the integration of AI into scientific practice prompts concerns about the automation of judgment, the erosion of epistemic agency, and the reconfiguration of research ethics in computationally mediated environments. This article will not cover all these aspects, focusing instead on those related to knowledge production,

[1] First studies show also evidence, that such tools also influence directly the way individuals conduct learning and thinking. With this, LLMs not only change knowledge production methodology, but also the thinking of individuals. An example in the case on the negative effects on learning is discussed in the study Kosmyna et al. (2025). However, it should be noted here that, as with any *new medium or paradigm*, these studies conducted in limited groups over small time periods are of limited significance and future observations are necessary to see all the side effects.

[2] An extensive overview of the history of the development of knowledge infrastructures, particularly about the challenges of data and data sovereignty, is discussed in Borgman (2025).

digital infrastructures, and aspects of Digital Sovereignty. Accordingly, the article focuses on the nexus between knowledge production and digital research infrastructures, with particular attention to how information (as data) and methods (as applications) are relocated into decentralized environments, and how this relocation reshapes provenance, accountability, and institutional control.

2 Epistemic Shifts in Distributed Knowledge Infrastructures

This shift from fixed archival storage to distributed infrastructures and contextual reinterpretation produces hybrid data forms that blur traditional boundaries between source, metadata, and usage. As the data are linked, enriched, and combined with other datasets across diverse platforms and contexts, they are continuously reinterpreted and embedded within new epistemic frameworks. As part of the preservation strategy of archives, data may be transformed to maintain accessibility, and new metadata may be created to compensate for semantic drifts in domain-specific languages. This process not only transforms the original data objects but also generates novel informational artifacts that extend beyond their initial scope and meaning. This form of dispersion and aggregation constitutes a distinct epistemic transformation, fundamentally altering how information is organized, interpreted, and made available for scientific inquiry.

These emergent data constructs reflect complex layers of interpretation, association, and metadata, fundamentally reshaping the boundaries between primary sources, derivative data, and analytical products within contemporary research practices. For example, processes such as data collection, knowledge organization, cooperative writing, peer review, publication, and impact assessment are mediated by interconnected technical systems, distributed services, and data sources. A commonly used tool is, for example, *digital programming notebooks*, which allow the creation of datasets, use analytical processes, and write the publication all within the same framework and, thus, exemplify how analysis, data, and narrative are co-located, often across distributed compute and storage, blurring boundaries between source, method, and publication. Such resources, data, and computing instances can be distributed across several independent infrastructures. Initially used in the context of Data Science, these tools are now also commonly used in other domains, such as digital humanities or Life Sciences.

Infrastructures are not limited to the peripheral organization of research. Still, they are also directly integrated into core analytical processes, influencing data analysis, modeling, interpretation, and the generation of research findings themselves. The concept of decentralized infrastructures is taken even further in so-called *Cloud architectures*. Cloud-based tools transfer control over data and applications entirely to abstract, decentralized infrastructures, fundamentally transforming established models of data management. *Cloud* refers to the abstraction of computational and storage functions from their physical substrates into a virtualized infrastructure, often obscuring the material and institutional conditions under which data is processed. Not only are digital research environments undergoing systematic changes, but discovery systems are as well. E.g., semantic search engines and similar tools are beginning to take over tasks that were once the domain of specialized central catalogs in archives or libraries. And with LLM-based guidance tools, such advanced discovery services will supplement or even substitute information professionals, such as librarians, and local institutional infrastructures, including the mentioned specialized catalogs and archival systems. As a result, scientific information and publications are progressively decoupled from centralized knowledge institutions and embedded within larger, often commercialized, service networks.

The transformation of knowledge production, on the other hand, is driven by this general process of digitalization, which integrates previously isolated resources into interconnected, system-wide constructs. Such infrastructures encompass the technical, institutional, and social systems that organize the production, distribution, and, to some extent, the validation of data and information (Ribes & Lee, 2010). Traditional examples of such infrastructures include scientific libraries and archives. Research libraries, e.g., have undergone a profound transformation from isolated repositories of printed knowledge into integral components of distributed networks. In the context of data-intensive and collaborative research, they serve as both custodians and curators of data, actively contributing to the organization, validation, and dissemination of scientific knowledge (Borgman, 2015, p. 15). Digitalization has become a driving force behind both the epistemic and practical transformation of knowledge infrastructures and the discovery of knowledge resources in recent years (Borgman, 2015; Borgman et al., 2019). This involves not only the technical conversion of analog information into digital formats but also the reconfiguration of the modalities of access, interpretation, and reuse, as in the previously given examples. Digitalization fundamentally reshapes scholarly practice by embedding knowledge production within infrastructures that mediate the discoverability, interoperability, and epistemic operability of data across distributed research environments (Borgman et al., 2019).

3 Digital Infrastructures as the Foundation for AI-Driven Research

Building upon these infrastructural transformations, emerging AI-based systems introduce a new layer of epistemic mediation by not only organizing and curating digital resources but also by actively participating in pattern recognition, inference, and the generation of new knowledge claims, thereby further reshaping the conditions under which knowledge is produced, validated, and communicated. This development reaches a further level of epistemic abstraction with LLMs. These systems do not establish contextual linkages through curated annotations or metadata, which are typically anchored in specific datasets. Instead, they perform large-scale syntheses of heterogeneous corpora, dissolving original contextual frames and recomposing them into unified representational structures based on probabilistic relations. In retrieval-augmented setups, retrieval can reintroduce sources, but synthesis may still obscure attribution and accountability. The result is a form of inferred coherence, a statistical approximation of meaning that emerges from patterns of usage rather than from intentional semantic anchoring.[3] The construction of such models involves aggregating diverse textual resources, ranging from formal academic publications to informal online discourse.[4]

[3] Boleda (2020) highlights in this context the growing dominance of neural networks as tools for modeling semantic similarity based on data-driven approaches.

[4] A commonly used dataset for textual data is, for example, *The Pile* (Gao et al., 2021). It consists of a large, diverse collection of English-language texts compiled specifically for training large language models. Created by EleutherAI, The Pile includes data from academic publications (such as arXiv, a well-known and largely used preprint repository for research publications), web content (like StackExchange), books, code repositories, and more. The aim is to provide a high-quality, heterogeneous corpus that reflects a broad spectrum of language use and knowledge domains. But bigger models, like generic LLMs, rely on much larger data sources for foundational training. Kandpal et al. (2025) released 2025 an alternative data set the *common pile* to addressed legal and ethical issues with providing a standardized set of materials of unlicensed resources.

Within this vast corpus, LLMs model relationships between minimal textual units, known as tokens. Those tokens are then being analyzed by their distributional properties across contexts. Through this process, the models develop *synthetic proximity*, which is a relational structure that reflects the statistical frequency and co-occurrence of linguistic elements. These proximities are then organized into high-dimensional vector spaces, in which *latent semantics* are encoded not as fixed meanings but as emergent statistical regularities. While such structures enable the generation of coherent and context-sensitive outputs, they also displace the epistemic traceability of the sources. What appears as knowledge in LLM-generated text is thus the product of algorithmic pattern recognition and probabilistic inference, not of interpretative synthesis grounded in human judgment (Mugleston et al., 2025).

These structures are designed primarily for computational processing and are not directly interpretable by humans. It is through interpretative interfaces, such as queries executed via large language model-based systems, that the encoded relations and data are translated back into meaningful, human-accessible representations. Many deployed systems are further tuned with human (or human-like, or even completely synthetic) feedback. Both the data and the query are embedded within high-dimensional vector spaces, with the model's internal knowledge represented as distributed patterns of vector relationships and transition probabilities. The final text is not simply retrieved as a pre-existing item but is dynamically synthesized based on inference in response to the specific query context.

An emerging problem in the training of LLMs is not only the need for computing power, energy consumption, and network bandwidth, but also the scarcity of high-quality training materials (Villalobos et al., 2024). By focusing on high-quality, domain-specific data, this can help to enrich generic language models with context. In this context, academic libraries face a profound redefinition of their role. They are now increasingly challenged to navigate the tension between human-curated epistemic infrastructures and the opaque, statistically driven logic of large-scale language models. As LLMs generate text based on patterns rather than provenance, the library's task of preserving context, source transparency, and metadata integrity gains renewed significance, not as a static archival function, but as a dynamic form of epistemic counterbalance.[5] Libraries may thus evolve into critical infrastructures not merely for access provision, but for ensuring traceability, curatorial framing, and normative filtering in a system that otherwise lacks intrinsic mechanisms for validation. Moreover, their potential to contribute high-quality, semantically rich training material through structured metadata, ontologies, and curated corpora positions them as essential stakeholders in shaping the ethical and epistemological contours of AI-driven research environments (Gooding et al., 2025).

Rather than becoming obsolete, libraries are being called upon to facilitate the connection between human-centered models of meaning and machine-oriented architectures of statistical approximation. They offer both infrastructural and intellectual support to ensure that the future of scientific inquiry remains accountable, interpretable, and inclusive. This expands the future role of libraries beyond their traditional curatorial functions, transforming them into proactive architects of scientific information ecosystems (Horstmann, 2022). In addition to safeguarding and improving metadata, they can intentionally provide high-quality, semantically organized collections for training and testing open language models. They may also develop open interfaces and standardized, verifiable access points that ensure both the traceability of AI-generated outputs and their seamless integration into research workflows (Kim, 2025). By building or partnering on non-proprietary LLM infrastructures, libraries can serve as a public counterbalance to commercial platforms. In this way, they become vital players in a digitally sovereign science, protecting the conditions for knowledge creation while actively shaping its future.

Operationally, this would involve curated training corpora with clear licensing, provenance layers for model outputs, and public governance interfaces for auditing and institutional accountability.

[5] Smith (1976) was an early thinker and predicted the potential of the first wave of AI development, without being able to estimate the actual potential and scope or even the change brought about by digitalization itself.

4 Research Infrastructures and the Digital Sovereignty

In the context of scientific knowledge production, digital sovereignty does not denote state sovereignty or complete technological self-sufficiency. Rather, it refers to the capacity of academic communities and public institutions to exercise effective control over the infrastructural conditions of research. This includes (i) governance over data, models, and computational infrastructures, (ii) the ability to define and enforce epistemic standards such as transparency, provenance, and reproducibility, and (iii) institutional accountability for how knowledge is curated, processed, and made visible. Understood in this operational sense, digital sovereignty becomes a prerequisite for maintaining epistemic independence in AI-mediated research environments.

Although digital information infrastructures are often perceived as neutral and technical constructs, they embody normative choices that shape how knowledge is organized and accessed (Bowker & Star, 1999). An example is the presentation of information, which is increasingly controlled by algorithms that influence visibility and findability, thereby impacting also the interpretation of content (Introna & Nissenbaum, 2000). Notably, the public discussion surrounding Google Search and its impact on information and the economy is well-known. The evaluation of the relevance of information is the responsibility of the algorithms or the models used (Mittelstadt et al., 2016). However, the service provider is in control of this algorithm, and users must trust its implementation.

As a result of this debate, the fundamental assumption of *search neutrality* was also discussed and called into question (Crane, 2011). With the inclusion of semantics into search functionalities, the concept of information discovery has undergone significant expansion, and the issue of neutrality has gained another dimension. Modern discovery systems also include presumed semantic context that they infer from the query and its surrounding data. This functionality is particularly beneficial in the context of academic research, for instance, in identifying relevant data or publications across different domains. Subject-specific semantic search engines, which present a range of results based on a search query, also enable the retrieval of relevant information from large and diverse datasets. They do so not only by matching exact wording, but also by interpreting the semantic context, thematic proximity, and conceptual relationships between terms. They offer results that reflect deeper levels of meaning and relevance beyond surface-level keyword matching, sometimes also including metadata of the user themselves. However, this is not always transparent to researchers who use such tools. A core issue, the algorithmic bias and fairness of discovery and recommendation algorithms, accounts not only for specific keywords and rankings, but also for complex semantics.

The implementation of the underlying algorithms reflects choices about what *constitute*s relevant information, and with that, forms the acquired knowledge. These algorithms were developed with a specific intention and therefore reflect the norms inherent in that intention. However, this relationship begins to dissolve when such functionalities are based on statistical models, like those used for advanced semantic search, and especially those used in AI and LLMs. These models rely on statistical transitions that do not arise from predefined intentions but are instead learned from data. While the underlying datasets naturally reflect specific intentions based on their selection, these intentions are now encoded statistically by design, utilizing large, often heterogeneous datasets. This shifts key elements of digital sovereignty, namely control over epistemic standards and infrastructural governance, in two ways: first, epistemic standards are increasingly operationalized through statistical models and their evaluation regimes rather than explicit, contestable algorithms; second, researchers become dependent on the data selection and governance choices of model providers. As a result, normative authority over relevance, meaning, and methodological reliability is partially relocated from transparent, institutionally governed procedures to model- and data-governance regimes that are largely opaque to researchers. Consequently, researchers must rely on trust in upstream choices regarding data selection, model design, and infrastructural governance, which inherently influence any research that depends on such systems.

An even more significant and concise aspect is the ethical design and governance of infrastructures, as they determine the underlying norms and, consequently, which data and information are made accessible, to whom, and under what conditions. The development and deployment of, e.g., large language models are predominantly controlled by commercial organizations and service providers, whose actions are primarily driven by financial interests.

Given that underlying data sources and model governance remain opaque, a portion of methodological responsibility is effectively displaced from researchers and public institutions toward service providers and their proprietary governance regimes. It is occasionally in companies' interests to leave details of methods, particularly the underlying data used to create such models, unclear. Even if such organizations emphasize the appearance of openness, this is often not reflected in their actions (Widder et al., 2024). The previously mentioned shift in processes and methods for handling information, including the adoption of digital research methods and the transition to cloud-based services, underscores its significance in this context. However, this shift in informational authority does not occur in a vacuum. Instead, it intersects with and begins to reconfigure a long-established academic ecosystem within the *research life cycle*.

5 Publishers and Knowledge Production

To understand the ethical and epistemic implications of this transformation, it is essential to revisit the traditional foundations of scholarly knowledge production. An idealized *life cycle* emerged between the generation of knowledge and data, their publication, and their reception in teaching or further research based on them. Historically, research has been embedded in a complex socio-economic system that governs this life cycle with the creation, validation, management, and dissemination of knowledge.[6] This system involves a diverse array of actors and infrastructures, including researchers and authors, founding agencies, peer reviewers, editors, publishers, repositories, and indexing services, each contributing to the credibility, accessibility, and circulation of scholarly information through clearly defined roles and responsibilities (Latour, 1987). Publishing houses have historically functioned as gatekeepers of academic quality and arbiters of visibility within the global knowledge economy.[7] Over time, this structure has solidified into a highly institutionalized framework that not only supports but also shapes how knowledge is recognized, legitimized, and valorized. At the center of this system are publishers, intermediaries, and institutions that mediate between those who produce knowledge and those who consume it. However, this role has increasingly come under scrutiny in the context

[6] The report Fyfe et al. (2017) presents insight into the development over time of various influencing factors and dependencies in the publishing system.

[7] Critics may question whether traditional publishing practices genuinely ensure quality, or whether they primarily reflect reputational economies and entrenched power dynamics within academia.

of the digital transformation of scholarly communication and the rise of the *Open Science* movement, which challenges the monopolistic structures of academic publishing and calls for more equitable, transparent, and accessible forms of knowledge dissemination (Suber, 2012). To open new exploitation channels in the digital space and regain lost markets, publishers offer a variety of digital services in this context. Alongside new business models emerging at the margins of literature and reference management, and a shift toward licensing digital content to public institutions rather than selling it as proprietary products, publishers are increasingly integrating methodological aspects of scientific work itself. This is also applicable to the licensing aspect: With the rising demand for *Open Access* publications and policies, publishers also offer Open Access as a licensing model. While often framed as a democratizing force in scholarly communication, Open Access at commercial services, particularly in its dominant "author-pays" or Article Processing Charge (APC) model, has also become a strategic tool for publishers to monetize scientific output in new ways. Rather than dismantling the economic barriers to knowledge, this model frequently shifts the cost burden from readers and institutions to individual researchers and their funding bodies, reinforcing existing inequalities in access to publishing opportunities (Posada & Chen, 2018).

Major publishing houses have adapted to the Open Science movement not by relinquishing their central position, but by rebranding their role as service providers in a data-driven, metrics-oriented academic economy. They now offer value-added services such as enhanced data analytics, impact tracking, AI-based manuscript tools, and integration with institutional repositories, often bundled into comprehensive publishing platforms. This vertical integration consolidates their influence across the entire research lifecycle, from manuscript submission to post-publication metrics, thereby reinforcing their epistemic authority within digital knowledge infrastructures (Fyfe et al., 2017).

Furthermore, by embedding AI-based tools into editorial and review processes and leveraging algorithmically enhanced discovery systems, publishers are increasingly shaping what counts as visible, credible, and impactful knowledge. The data used to train these AI algorithms, which are part of this business strategy, was derived from scientific output previously published by other researchers. These technologies also introduce new forms of opacity and bias into scholarly workflows, as decisions about relevance, novelty, or peer-review suitability are increasingly mediated by proprietary algorithms rather than transparent academic deliberation (Bender et al., 2021; Hagendorff, 2024). In this evolving environment, the commercialization of openness emerges not as a paradox but as a strategic rearticulation of control, transforming infrastructures of knowledge production into platforms for data extraction, surveillance, and predictive analytics. Thus, the ethical evaluation of AI in academic contexts cannot be separated from a critical inquiry into the political economy of digital publishing and the infrastructural power it entails.

6 From Statistical Abstraction to Intellectual Ownership

Another dimension concerns authorship itself. As previously stated, the training of large language models relies on the intellectual contributions of a vast and diverse body of authors. Beyond encoding formal structures of written language, these models internalize linguistic patterns, argumentative forms, and substantive assertions present in the training data. These constitute original intellectual contributions produced by individual authors.

While the training processes of large language models are often compared to human learning, particularly in their reliance on exposure to large bodies of text, this analogy should not be overstated. The present argument does not hinge on the claim that LLMs learn in a categorically different way from humans. Comparability does not require identity, and similarities in distributional abstraction or pattern recognition do not, in themselves, invalidate such analogies.

The critical distinction instead lies in the institutional and infrastructural conditions under which learning and knowledge synthesis occur. Unlike human scholars, whose engagement with texts remains bounded by individual cognition, limited reproducibility, and explicit norms of attribution, LLMs operate at a scale, speed, and level of extractability that fundamentally alter the relationship between intellectual contribution and its reuse. The capacity of these systems to statistically approximate or reproduce substantial parts of original texts is therefore not primarily an epistemic claim about learning, but a structural feature of infrastructuralized cognition.

Two consequences follow from this asymmetry. First, while LLMs do not store or replicate complete works, generated outputs may, under specific conditions, closely approximate original formulations. Second, the scale and density of relational abstraction involved in model training far exceed the cognitive capacities of individual human readers, enabling forms of synthesis and recombination that are not achievable through human learning alone. Consequently, LLMs constitute a statistically mediated representation of aggregated intellectual labor rather than a direct reproduction of individual works.

This form of representation fundamentally reconfigures established relations between authorship, reuse, and attribution by enabling the large-scale extraction and recombination of intellectual contributions without embedded mechanisms for provenance or consent. These epistemic shifts are further compounded by unresolved ethical and legal questions concerning authors' willingness to make their work available for model training and the legitimacy of rights acquisition by commercial providers. Together, they challenge existing systems of accountability and reinforce concerns about the erosion of digital sovereignty in AI-mediated scientific knowledge production.

7 Conclusion

Despite the challenges discussed in this article and, moreover, in the general discourse, such as the opacity of model-driven infrastructures, the erosion of traditional gatekeepers, the contestation of search neutrality, and the normative weight carried by algorithmic design, these developments should not overshadow the opportunities that AI and LLMs offer for scholarship. Like every foundational technology in the history of knowledge creation, they have the power to spark new questions, accelerate discovery, and connect previously separate areas of data, method, and insight. However, realizing this potential requires more than just routine adoption. It demands a conscious shift in mindset: a willingness to examine the hidden architectures that now shape what we can find and how we understand, an ethical vigilance focused on whose voices are amplified or silenced, and a renewed commitment to transparent and accountable research practices. Mühlhoff (2025, p. 179) writes: "After all, the goal of power-aware AI ethics is not to produce erudite tomes and papers for the library (or training data for the next large language model), but to reach human recipients in academic as well as public, political, and activist debates." Addressing these challenges and fostering open science aligned with ethical principles requires a greater awareness of infrastructural reliance and a cultural shift in how digital infrastructures are integrated into research. Besides the common issue of vendor lock-in to specific closed toolsets, there is an increasing risk of dependence on non-transparent infrastructures and the centralization of power over the methodological foundations of science. While this influence may often remain hidden, the widespread integration of AI tools into research means that even subtle effects can accumulate over time, unintentionally reshaping the scientific system without deliberate intent.

A central dimension of this sovereignty challenge concerns authorship, attribution, and provenance. As argued above, AI-mediated knowledge production reconfigures how intellectual contributions are aggregated, reused, and rendered visible. When large-scale language models synthesize knowledge without transparent links to sources or contributors, established mechanisms of credit, responsibility, and accountability are weakened. This does not merely affect individual authors but reshapes the institutional conditions under which scientific knowledge can be claimed, validated, and governed. Safeguarding digital sovereignty, therefore, requires not only control over technical infrastructures but also renewed attention to epistemic traceability and attribution as foundational elements of scientific integrity.

As infrastructures become abstract and opaque, these questions must take center stage in scientific practice. For this potential to unfold responsibly, it is crucial that guiding frameworks such as the CARE (Collective Benefit, Authority to Control, Responsibility, and Ethics) and FAIR (Findable, Accessible, Interoperable, and Reusable) principles are not merely acknowledged, but meaningfully embedded in the design, governance, and everyday use of research infrastructures and in scientific practice more broadly. These principles serve as ethical and practical anchors, reminding us that the value of data and algorithmic tools extends beyond their purely technical aspects to encompass relational, contextual, and political considerations. If we pair technical ingenuity with this critical reflexivity, AI and LLMs need not merely automate the familiar; they can help us rethink the very contours of inquiry itself, extending the reach of scientific imagination while anchoring it firmly in responsible, human-centered norms.

Competing Interests The author has no conflicts of interest to declare that are relevant to the content of this chapter.

References

Balzer, W., Eleftheriadis, A., & Kurzawe, D. (2018). Digital humanities and hermeneutics. *Philosophical Inquiry, 42*(3), 103–119. https://doi.org/10.5840/philinquiry2018423/415

Bender, E. M., Gebru, T., McMillan-Major, A., & Shmitchell, S. (2021). On the dangers of stochastic parrots: Can language models be too big? In *Proceedings of the 2021 ACM conference on fairness, accountability, and transparency* (pp. 610–623). https://doi.org/10.1145/3442188.3445922

Boleda, G. (2020). Distributional semantics and linguistic theory. *Annual Review of Linguistics, 6*(1), 213–234. https://doi.org/10.1146/annurev-linguistics-011619-030303

Borgman, C. L. (2015). Big data, little data, no data: Scholarship in the networked world. *The MIT Press*. https://doi.org/10.7551/mitpress/9963.001.0001

Borgman, C. L. (2025). *Libraries, digital libraries, and data: Forty years, four challenges*. UCLA. Preprint. Retrieved May 1, 2025, from https://escholarship.org/uc/item/6dn9z9k6

Borgman, C. L., Scharnhorst, A., & Golshan, M. S. (2019). Digital data archives as knowledge infrastructures: Mediating data sharing and reuse: Digital Data Archives as Knowledge Infrastructures: Mediating Data Sharing and Reuse. *Journal of the Association for Information Science and Technology*. https://doi.org/10/gfzhs3

Bowker, G. C., & Star, S. L. (1999). Sorting things out: Classification and its consequences. *The MIT Press*. https://doi.org/10.7551/mitpress/6352.001.0001

Crane, D. A. (2011). Search neutrality as an antitrust principle. *SSRN Electronic Journal*. https://doi.org/10.2139/ssrn.1961742

Clark, A. (2008). *Supersizing the mind: Embodiment, action, and cognitive extension*. Oxford University Press. https://doi.org/10.1093/acprof:oso/9780195333213.001.0001

Fyfe, A., Coate, K., Curry, S., Lawson, S., Moxham, N., & Røstvik, C. M. (2017). Untangling academic publishing: A history of the relationship between commercial interests, academic prestige and the circulation of research. *Zenodo*. https://doi.org/10.5281/zenodo.546100

Gao, L., Biderman, S., Black, S., Golding, L., Hoppe, T., Foster, C., Phang, J., He, H., Thite, A., Nabeshima, N., Presser, S., & Leahy, C. (2021). The Pile: An 800GB dataset of diverse zext for language modeling (Version 1). arXiv. https://doi.org/10.48550/ARXIV.2101.00027

Gooding, P., Terras, M., & Ames, S. (2025). *Library catalogues as data: Research, practice and usage*. Facet Publishing.

Hagendorff, T. (2024). Mapping the ethics of generative AI: A comprehensive scoping review. *Minds and Machines, 34*(4), 39. https://doi.org/10.1007/s11023-024-09694-w

Horstmann, W. (2022). Library technology—Between local tools and global solutions. *Bibliothek Forschung und Praxis, 46*(2), 261–270. https://doi.org/10.1515/bfp-2022-0011

Humphreys, P. (2004). *Extending ourselves: Computational science, empiricism, and scientific method*. Oxford University Press.

Introna, L. D., & Nissenbaum, H. (2000). Shaping the web: Why the politics of search engines matters. *The Information Society, 16*(3), 169–185. https://doi.org/10.1080/01972240050133634

Kandpal, N., Lester, B., Raffel, C., Majstorovic, S., Biderman, S., Abbasi, B., Soldaini, L., Shippole, E., Cooper, A. F., Skowron, A., Kirchenbauer, J., Longpre, S., Sutawika, L., Albalak, A., Xu, Z., Penedo, G., Allal, L. B., Bakouch, E., Pressman, J. D., ... Murray, T. (2025). The Common Pile v0.1: An 8TB Dataset of Public Domain and Openly Licensed Text (arXiv:2506.05209). arXiv.

Kim, J. (2025). Academic library with generative AI: From passive information providers to proactive knowledge facilitators. *Publications, 13*(3), 37. https://doi.org/10.3390/publications13030037

Kosmyna, N., Hauptmann, E., Yuan, Y. T., Situ, J., Liao, X. H., Beresnitzky, A. V., Braunstein, I., & Maes, P. (2025). Your brain on ChatGPT: Accumulation of cognitive debt when using an AI assistant for essay writing task. arXiv preprint arXiv:2506.08872.

Latour, B. (1987). *Science in action: How to follow scientists and engineers through society*. Harvard University Press.

Leonelli, S. (2016). *Data-centric biology: A philosophical study*. The University of Chicago Press.

Mittelstadt, B. D., Allo, P., Taddeo, M., Wachter, S., & Floridi, L. (2016). The ethics of algorithms: Mapping the debate. *Big Data & Society, 3*(2), 2053951716679679. https://doi.org/10.1177/2053951716679679

Mugleston, J., Truong, V. H., Kuang, C., Sibiya, L., & Myung, J. (2025). Epistemology in the age of large language models. *Knowledge, 5*(1), 3. https://doi.org/10.3390/knowledge5010003

Mühlhoff, R. (2025). *The ethics of AI: Power, critique, responsibility*. Bristol University Press. https://doi.org/10.51952/9781529249262

Posada, A., & Chen, G. (2018). Inequality in knowledge production: The integration of academic infrastructure by big publishers. In *22nd International conference on electronic publishing*.

Ribes, D., & Lee, C. P. (2010). Sociotechnical studies of cyberinfrastructure and e-Research: Current themes and future trajectories. *Computer Supported Cooperative Work (CSCW), 19*(3–4), 231–244.

Suber, P. (2012). *Open access*. MIT Press.

Smith, L. C. (1976). Artificial intelligence in information retrieval systems. *Information Processing & Management, 12*(3), 189–222.

Villalobos, P., Ho, A., Sevilla, J., Besiroglu, T., Heim, L., & Hobbhahn, M. (2024). Will we run out of data? Limits of LLM scaling based on human-generated data (arXiv:2211.04325). arXiv.

Widder, D. G., Whittaker, M., & West, S. M. (2024). Why 'open' AI systems are actually closed, and why this matters. *Nature, 635*(8040), 827–833.

Determination of the Unknown? A Spotlight on Current Challenges in Research Data Management from the Data Protection Perspective

Friederike Knoke and Margrit Seckelmann

1 Introduction: Annoyance and Restrictions on Academic Freedom or Spaces for Free Play?

Presumably, every researcher will agree on the importance of ethically sound science. Since ethics lay the foundation for the protection of personal data and privacy (see Horn, 2017, p. 3), one can assume as well that this agreement, in principle, expands to the following of data protection rules applicable to the processing of personal data for scientific research purposes. In addition to this inner dimension, which is connected to epistemic norms, a focus on data protection-compliant data handling in research contexts entails establishing a sound basis for the processing of data and provides for a transparent communication toward externals such as research participants. Thus, it is a significant sign of

F. Knoke (✉) · M. Seckelmann
Institut für Rechtsinformatik, Leibniz Universität Hannover, Hannover, Germany
e-mail: friederike.knoke@iri.uni-hannover.de

M. Seckelmann
e-mail: margrit.seckelmann@iri.uni-hannover.de

J. Hinrichsen et al. (eds.), *Ethics of the Sciences*,
https://doi.org/10.1007/978-981-92-2821-8_7

quality of research endeavors, which ensures reliability of scientific practices and can therefore be expected to enhance individual, and, in the long term, societal trust in science (regarding the role of trust in science in modern democracies, see in detail Hönnige and Nyhuis in this volume). It might even give those researchers who know how to manage them effectively a competitive edge in the various facets of competition in scientific contexts (see regarding these in detail Kosmützky in this volume). However, interviews with researchers, conducted in the course of the "FODAHEMM" research project[1] reveal a rather skeptical perception of the legal rules related to the protection of personal data in the course of scientific research: Interviewees have stated that data protection is a topic with a 90% grade of annoyance at least (001_Forschende, #12:48),[2] and is outbraking projects by requiring researchers to concern themselves with it in a very time-consuming manner (020_Forschende, #42:04; 023_Forschende, #36:54). Furthermore, doubts about data protection obligations affecting the research process have been expressed: One researcher reflected that one of the toughest lessons learned in the course of time had been to realize that because of data protection rules, one has to already think the research process through to the end at a very early stage in the process (023_Forschende, #42:27). Another researcher expressed their impression that data protection (law) feels disproportionately complicated and even seems to be a shot off target at times (005_Forschende, #20:00).

Quotes from interviews with personnel specialized in data protection at tertiary institutions like universities, such as data protection officers or offices dealing with research data management thereof (conducted in the same FODAHEMM project mentioned above), are inclined to draw a different picture: From the standpoint of one of the interviewed specialists, to perceive data protection as a "big, black cloud hovering

[1] Further details on the project as well as the empirical data are provided in Sect. 2 of this chapter. The sociological part of the project is led by Professor Dr. Stefanie Büchner, Institute for Sociology, Leibniz Universität Hannover; the interviews have been conducted by Tim Albrecht, MA, research associate in FODAHEMM's sociological aspects. The authors wish to thank both of them for consenting to the use of some of the interview material for this text.

[2] Explanatory note on the citations from the interviews (interview key): The three-digit number indicates the number of the interview, the term following the underscore refers to the group of interviewees ("Forschende" meaning researchers, "Spezialstelle" meaning data protection specialists in tertiary institutions), and the information following the comma ("#") indicates the position of the statement in the interview in minutes.

above research projects," is, in fact, not necessary (024_Spezialstelle, #37:00). On the contrary, one of the specialists claims that permissions of someone processing personal data for research purposes extend very far compared to those of someone processing data for other purposes (010_Spezialstelle, #41:02). Even in cases where there are legal obligations imposed upon those responsible for the processing of the data, data protection specialists have expressed many things to be legally justifiable once one has a knowledge of existing spaces for free play, allowing for deviances facilitating the needs of a specific project (019_Spezialstelle, #20:12; 024_Spezialstelle, #18:00). Furthermore, in the eyes of one of these interviewees, researchers sometimes tend to think something not to be possible from a legal perspective, whereas in fact it is (014_Spezialstelle, #50:55), and another stated that the perception of data protection to be something of a hindrance is a misperception of the legal situation, which actually requires efforts with regard to the reflection and establishment of processes (019_Spezialstelle, #26:36). The fact that the assessment of those interviewees with a data protection specialization differs so widely from that of the cited researchers may advise against taking the critical view on data protection at face value.

Nonetheless, the concerns researchers have uttered in the interviews about restrictions on their academic freedom through legal obligations (for example, regulations such as the General Data Protection Regulation (GDPR)), and the fact that the observations differ significantly from those of data protection specialists point toward a conflict between research interests and data protection obligations where the handling of personal data is involved. This calls for a further investigation of the legal situation to shed light on the matter, to which we will dedicate ourselves in this chapter. Doing so will include seeking answers to the following questions: Which legal status on the researcher's part do the data protection rules potentially interfere with, and can researchers hold that position against these rules? Can the feeling of complexity be connected to the legal regime in place and, if so, does this legal complexity only work against the researcher's interests? How are the characteristics of the research process reflected in the legal rules and to what extent do they have the potential to redress the expressed contradictoriness?

To address these questions and thus investigate the conflict between research interests and data protection rules, the chapter is structured as follows: First, we expand upon the empirical data initiating our dive into the subject of this text (Sect. 2) before outlining the two competing rights and interests mainly involved in this field (Sect. 3). We subsequently provide an overview of the legal regime in place, its challenges and the timing of obligations related to the management of research data (Sect. 4), followed by an analysis of the existing privileges for scientific data handling, focusing on the GDPR and its capacity to address the specific situation of this data-handling scenario (Sect. 5). In the final section, we summarize the findings and present an outlook on the problem in a conclusion (Sect. 6).

2 Methodological Note on the Empirical Database

The interview statements cited above, which have given us the motivation to investigate the friction between the research interests and data protection rules, stem from a research project entitled "FODAHEMM: Forschungsfreiheit vs. Datenschutz—Wie können Hemmnisse beseitigt werden?" In this project, we examine potential barriers to research through data protection and what could be done to reduce them. The project is carried out at Leibniz Universität Hannover in cooperation between the Institute for Legal Informatics and the Institute for Sociology of the Leibniz Universität Hannover, and is funded by resources from zukunft.niedersachsen, the Ministry of Science and Culture of Lower Saxony, and VolkswagenStiftung's joint research funding program in Lower Saxony, Germany. The authors of this chapter are responsible for the legal aspects of the project, while the sociological part of the project is led by Professor Dr. Stefanie Büchner at the Institute for Sociology, and Tim Albrecht, MA works on the project's sociological aspects as a research associate. To acquire knowledge of the current situation and engage in a dialogue with actors of the academic and research landscape on how to advance the status quo in a most suitable and interest-preserving way, FODAHEMM included, among other interactive formats like discussion forums, the collection of empirical data in the form of qualitative interviews.

The aforementioned empirical dataset consists of 28 problem-centered, guideline-based interviews focused on the perception and handling of problems related to data protection in the context of research. Driven by a descriptive and normative epistemological interest, the collection of data aimed at the reconstruction of the variety and typical challenges of dealing with increasing complexities in research data protection, and thus was of a qualitative nature, without being geared toward a representativity of the sample. The acquisition of interviewees concentrated on areas in which these challenges are recurring regularly, so that the expression of perspectives relevant for the research questions could be expected. Between May and November 2023, the interviews have been conducted with researchers in the field of social sciences (11 interviews in total), specialists in data protection at tertiary institutions like universities, such as data protection officers or offices dealing with research data management (eight interviews in total), personnel in hybrid positions like research data centers (six interviews in total), and with members of ethical committees at universities (three interviews in total). The material has been analyzed with the software MaxQDA, following the methodology of focused interview analysis (Kuckartz & Rädiker, 2021). The interviews have been conducted in German, and to be able to use them in this collected volume, we paraphrased the statements in English, sticking to the original wording as closely as possible.

3 Research Data Management from the Scientist's Perspective and the Role of Academic Freedom

3.1 *Academic Freedom as a Fundamental Right*

Article 13 of the Charter of Fundamental Rights of the European Union (CFREU) guarantees that scientific research shall be "free of constraint" and that academic freedom shall "be respected." These guarantees form the freedom of sciences, a fundamental right that entails, among other things, to protect all activities related to research, including actions of all kinds, even when their nature is preparatory or supporting (Jarass 2019, Article 13 marg. no. 8 with further references). Similar guarantees are to be found in national regulations. In Germany, the right to academic freedom (guaranteed in Article 5(3) of the German Constitution—Grundgesetz) goes even further, as it is not subject to any explicit restrictions (whereas the CFREU contains a general limitation clause, which can be found in Article 52(1)). Nonetheless, also in Germany, the right to academic freedom has to be balanced with conflicting fundamental rights of other persons.

3.2 *The Right to the Protection of Personal Data*

One (and the most famous) of those conflicting fundamental rights is the right to informational self-determination, which the German Federal Constitutional Court (BVerfG) derived from the general right of personality according to Article 2(1) in conjunction with Article 1(1) of the Grundgesetz (in its judgment of 15 December 1983—1 BvR 209, 269, 362, 420, 440, 484/83, BVerfGE 65, pp. 1 et seq.). Under the regime of the CFREU, the right to the protection of personal data (Article 8) and the right to respect for private life (Article 7) are in conflict with Article 13 CFREU. To harmonize the legal rules for the handling of personal data and resolve conflicts of interest of the public and the parties involved, the European law-making bodies have enacted the GDPR. Given that Article 1(1) of the German Constitution guarantees human dignity, the derivation of the right to informational self-determination from, among other sources, Article 1(1) of the German Constitution shows the strong connection between the right to the protection of personal data and human dignity. From a philosophical perspective, with respect to the GDPR, it was suggested that "human dignity is the fundamental concept that provides the framework within which one needs to interpret what the GDPR ... understand[s] by informational privacy" (Floridi, 2016, p. 1). In its Article 89, the GDPR contains certain facilitations for research (see Sect. 5.1 below in detail).

3.3 *The Inherent Conflict of Interests*

Both the aforementioned regulations serve to guarantee specific interests: The scientific freedom safeguards an unhindered pursuit of scientific knowledge and therefore secures the underlying research interests, and the data protection law aims at the protection of the right to privacy and data protection to serve the interests of those to whom data can be connected by preventing the handling of the data from getting out of hand. While both of these interests are important and legally recognized, they are—as outlined above—not necessarily in accordance with each other. Their conflict seems to be inherent, as the aims of science and data protection interfere with each other—for scientific purposes, on the one hand, it is beneficial to gain as much knowledge as possible, whereas data protection rules, on the other hand, can be understood as mechanisms of control put in place to keep the knowledge about individuals to a minimum. Hence, researchers, from their epistemological point of view, tend to be interested in generating and working with datasets which are as complete as possible, whereas it is in the data subject's (e.g., subject groups or interviewees) interests to restrict the datasets and the processing, such as storing of data, so that no compromising information about them is included any longer. Consequently, the conflict between these two cannot be solved or prevented completely. Instead, one can see data protection law as—metaphorically speaking—the definition of the outer edge or margin of the researcher's quest for knowledge and the associated collection and handling of data, which means that data protection, on a functional level, *does* actually *limit* research interests. Nevertheless, this functional limitation does not necessarily imply an actual *violation* of the research interests. Rather, to draw the line between these two, the question to be asked is if data protection rules compromise the research interests, which—to keep up the metaphorical language—is only the case if the outer edge is drawn in a way that does not leave enough space for the research interests to be pursued, thus setting the margins or limits too narrowly.

While reflecting on this, a philosophical debate comes to mind. For some years, the question has been put forward as to whether it is advisable to put constraints on the freedom of scientific inquiry in the light of moral grounds or social values like the equality of ethnic backgrounds or genders, should those presumably be affected by the research (see in particular Kitcher, 2003, pp. 93 et seq.). As an effect to this thinking, "scientists ... must recognize the ways in which social values may hinder their ability to discern significant truths and how such facts show up in the first place" (Victor, 2014, p. 1048). The line drawn to the researcher's position or academic freedom by the research participant's interests and fundamental rights can be interpreted in this regard: As much as it might be considered objectionable to investigate, for example, differences between ethnic backgrounds or genders nowadays, it might be considered unwanted to conduct research that is not compatible with data protection requirements. As a consequence, the researcher's obligations arising from data protection law might limit their freedom of inquiry with respect to those truths and facts that only materialized themselves should the collection and processing of the data be unlimited, and this would have to be tolerated for the sake of social values and on moral grounds.

Nonetheless, there is a significant difference between these two forms of limitations to scientific inquiry: The constraints debated by Kitcher and Victor concern *research topics* (not) to be investigated by scientific means, and the right to the protection of personal data results in *process-related* limits without disallowing a concrete kind of content addressed by research: Other than in the case of the EU's Artificial Intelligence Act (Regulation (EU) 2024/1689 of 3 June 2024), which legally prohibits a listed set of artificial intelligence (AI) *practices* (see Article 5(1) of the AI Act), the GDPR does not enlist particular areas of research which will be banned from investigation. This does not mean that aspects which call for a sensitive and delicate, or, so to speak, value-oriented handling are disregarded in data protection law. On the contrary, Article 9(1) GDPR stipulates as a rule that the processing of personal data revealing, among other things, ethnic origin or political opinions, and the processing of genetic data, data concerning health, or data concerning a natural person's sex life or sexual orientation is prohibited. The processing of such "special categories of personal data," as the GDPR puts it, is only allowed if one of the exceptions enumerated in Article 9(2) GDPR applies. This shows the legislative approach taken in the GDPR: The processing of certain, more sensitive personal data is not completely prevented, but is subjected to specific and stricter requirements laid down in the regulation. As data are defined as the central concept through which a natural person's fundamental rights and freedoms are protected (see Article 1(1) and (2) GDPR), these particular conditions are attached to categories of this data. If special categories of data, as enumerated in Article 9(1) GDPR, are being processed, the processing will have to follow the conditions for sensitive data applicable to them. This differs from the idea to refrain from certain inquiries in general in order to prevent sensitive social values from being affected. Instead, processing necessary for scientific or historical research purposes or statistical purposes is indeed part of one of the exceptions in Article 9(2) GDPR (concretely, Article 9(2) lit. j), so processing in research contexts is acknowledged as one of the cases in which an exception from the rule can be legally accepted. Regarding the specific conditions to which such processing is subject, Article 9(2) lit. j refers to Article 89(1) GDPR, where special rules for processing in scientific research contexts are stipulated (see Sect. 5.1 below in detail), but highlights with respect to the special categories of data that their processing "shall be proportionate to the aim pursued, respect the essence of the right to data protection and provide for suitable and specific measures to safeguard the fundamental rights and the interests of the data subject." Regardless of the data category, Article 89(1) GDPR stipulates safeguards for all cases of processing of personal data for scientific research purposes. It seems that, as far as the protection of personal data is concerned, the European legislator took the needs of those affected by the processing into account, and sought to address the conflict between the researcher's and the data subject's interests in the GDPR. This draws attention to the concrete rules rather than thinking about installing or establishing uncodified limits to scientific inquiry.

Thus, the existing legal rules related to research involving the handling of personal data will now be examined in more detail.

4 Legal Obligations for Research Involving Personal Data and the Question of Timing

4.1 *The Legal Regime for the Protection of Personal Data Processed for Research Purposes and its Challenges*

Since many research projects involve the processing of data ("data-driven research"), which can often be regarded as personal data upon their generation, research data, like any other data relating to a natural person, may fall under the jurisdiction of data protection law, such as the GDPR and national data protection laws. This includes, for example, data stemming from experiments involving human participants or produced in laboratory or clinical surroundings, such as names or birth dates of interviewees, patients or other study participants, as often found in social sciences or medical research (see Knoke, 2023, p. 19 et seq.). Besides, this may include any scientific field up to those involving big data, Large Language Models or AI applications, as long as, as the definition in Article 4(1) GDPR stipulates, the data processed there include information relating to an identified or identifiable natural person. Moreover, according to recital 26 of the GDPR, it includes data that has undergone a pseudonymization process, and excludes only anonymous or anonymized data (see in more detail Knoke, 2023, pp. 131 et seq.). In general, this means that the specific terms and definitions of the GDPR apply to such actions carried out in research projects, whether these are related to actors involved, which in the GDPR are referred to, among other things, as "controllers" and "processors" at the processing actor's end and "data subject" at the individual's end, or whether they are related to the subject matter of "personal data" and "processing of data" itself (see the legal definitions in Article 4 GDPR). Moreover, the general principles relating to the processing of personal data, laid down in Article 5 GDPR, like the lawfulness, fairness, and transparency of the processing, including the requirement of a legal basis for the processing of data as further stipulated in Article 6 GDPR, are to be fulfilled by those responsible for the processing.

The wide-ranging nature and scope of the GDPR already recognized in its title ("General Data Protection Regulation") and the fact that the rules of the GDPR, being adopted in the form of a regulation, are directly applicable throughout the EU and within its territorial (and material) scope notwithstanding, there are a number of cases in which national and, depending on the national legal system, possibly regional regulations are to be considered as well. The legal mechanism issuing this effect is the use of opening clauses. The term refers to clauses in EU regulations, which allow for the EU Member States to enact their own national rules concerning certain aspects of data protection. Such clauses can be found in several provisions of the GDPR (Kühling & Martini, 2016, p. 449, have counted a total of 45 opening clauses throughout the norm), such as in the provisions regarding the lawfulness of processing (Article 6(2), (3), and (4) GDPR), to the processing of special categories of personal data (Article 9(4) GDPR), and in the article setting safeguards and derogations specifically for processing for research, statistical, or archiving purposes (Article 89(2) GDPR). At the same time, such national rules may not replace the provisions laid down in the European regulation, but may only clarify them without providing additional requirements or changing their scope (see Court of Justice of the European Union (CJEU), judgment of 24 November 2011, C-468/10, C-469/10, *ASNEF*, marg. no. 35). In addition, the general data protection rules of the GDPR and at the national level may be complemented by special legal rules applicable in specific areas of processing of data, such as in the health sector. Consequently, general and special legal rules potentially overlap with one another in some cases, such as research projects dealing with medical data. All of this leads to a necessity to consult more than one regulation when in search of relevant provisions to be followed in a concrete research project, as, to a certain extent, they may apply cumulatively, and one needs to find out, if this is the case, to what effect they do so.

Once the relevant provisions for a case have been identified, the next step is to ascertain the relevant norms' regulatory content. As legal norms are of a general nature and are supposed to cover different real-life situations, they all require some interpretation when applying them to a concrete case. Nonetheless, the room for interpretation and concretization of legal norms varies between the different norms, depending on the used terms, which are likely to increase the room and need for concretization the more normative a legal prerequisite is, and are likely to decrease it the more descriptive it is. In the norms relevant for research projects, a number of rather normative and vague legal concepts are to be found: For example, the question of data being regarded personal or not, and if, for example, this (still) is the case after research data have been pseudonymized or anonymized; or the determination of responsibility for the processing and thus for fulfilling the legal obligations coming along with the role of the "data controller," through legal subsumption under that term. Both concepts pose challenges when they need to be applied to research data and to the circumstances in which they are generated. Notably, this step of interpreting the norm(s) may already have to be taken within the aforementioned step of identification of the relevant norms—for example, if one has to determine if data are to be regarded as personal data to know if the data protection regime is applicable at all, or one needs to determine if a certain research project processes data for scientific research processes or not in order to see if Article 89 GDPR with its specific rules for scientific data handling is applicable to it.

All in all, in order to obtain knowledge of the legal rules applicable to a concrete research project and of their specific content, one needs to navigate through the legal landscape thoroughly and skillfully, and by means of analysis, following an approach that can be seen sensu stricto juridical. This explains why particularly laypeople have the impression that they face a complex legal situation with regard to the data protection framework.

4.2 Legal Rules Requiring an a Priori Perspective on Processes

Assuming that a research project includes the processing of data to be considered as personal data and that the GDPR applies to the handling of that data, the GDPR imposes a number of legal obligations on those responsible for the processing. For the planning of a research project that fulfills these obligations and therefore is in line with data protection rules, it is crucial to ascertain not only what has to be done, but also at what time. In many projects, the data protection obligations to be met come to the researcher's attention at the latest at the moment in which they prepare the collection of research data. They want to obtain consent from, for example, the participants in their study or from persons they wish to interview, and become aware that this step includes legal aspects. In order to be able to prepare the documents with which consent shall be obtained from participants, they need to find out what has to be described and included in those documents. In this phase of the project, the scientific process has begun, some literature and/or theoretical review has probably taken place, and also some methodological choices—for example, on which data shall be collected and in what form—will have been made, but all in all, the project is, very likely, still at a fairly early stage, running for approximately a few months. At this point in time, the legal rules come into play.

Firstly, the principle of lawfulness, laid down in Article 5(1)(a) GDPR, requires that the processing of personal data be based on a legal ground. These legal grounds are enumerated and specified in Article 6 GDPR. The consent of the persons whose data are being processed in the course of the project, which is listed as a possible legal basis in Article 6 GDPR (specifically, Article 6(1), subparagraph(1)(a)), so the common practice to obtain consent from research participants is in conformity with the law. Article 4(11) GDPR defines "consent" as "any freely given, specific, informed and unambiguous indication of the data subject's wishes by which he or she ... signifies agreement to the processing of personal data relating to him or her." In this context, the elements "informed" and "specific" are particularly relevant, since they enable the participants of a study to assess the consequences of their declaration and the kind of processing to which they are agreeing. In recital 42 of the GDPR, we learn something about what "informed" means concretely. In its fourth sentence, it is stipulated that "[f]or consent to be informed, the data subject should be aware at least of the identity of the controller and the purposes of the processing for which the personal data are intended." In addition, the type of data to be collected and used, and the existence of the right to withdraw consent are necessary to make sure the consent is informed (see Article 29 Data Protection Working Party, 2017a, p. 13). The element of "specific" consent, for its part, is "closely linked to the requirement of 'informed' consent" and includes the determination of "a specific, explicit and legitimate purpose for the intended processing activity" (2017a, pp. 11 et seq.). As a consequence, the requirement of "specificity" adds "specific, explicit and legitimate" as further criteria to the necessity to inform about the purposes of the processing. What does that imply with regard to the timing of legal obligations? It is hard to imagine how a data subject should be able to agree in a specific and informed way to a form of processing that has not been described to them in a detailed way beforehand. Hence, in order for the data subject's declaration to be informed, all the necessary information needs to be provided before the beginning of the data collection (Kühling and Buchner in Kühling & Buchner, 2024, Article 7 DS-GVO marg. no. 59b). Likewise, for it to be specific, the determination of the purposes must precede the collection of consent (Article 29 Data Protection Working Party, 2017a, p. 12). This means that the researchers need to describe specific, explicit and legitimate purposes of the intended processing, name the type(s) of data, provide the identity of the controller, and indicate that the participants have the right to withdraw their consent, in the documents they present to their study participants.

Secondly, the principle of purpose limitation is of great importance: Only by means of a concrete case- and person-related determination of the purpose of the data processing, the factual possibilities of those controlling the data become assessable and checkable for the data subject as well as for supervisory authorities (Herbst in Kühling & Buchner, 2024, Article 5 DS-GVO marg. no. 22). As such, it is one of the fundamental data protection principles, and has been referred to as "a cornerstone of data protection" (Article 29 Data Protection Working Party, 2013, p. 4), and an "essential element of data protection law" ("datenschutzrechtliche[s] Wesenselement"; see Benamor, 2024, p. 268). Purpose limitation requires, among other things, that personal data shall be "collected for specified, explicit and legitimate purposes" (Article 5(1) lit. b GDPR). The Article 29 Data Protection Working Party, the predecessor of the European Data Protection Board, has stressed that a purpose "that is vague or general ... will—without more detail—usually not meet the criteria of being 'specific,'" and listed the wording "for 'future research'" as one of the examples for a purpose which is only vaguely or generally specified and thus not sufficiently identified (Article 29 Working Party, 2013, p. 16).

The wording in Article 5 GDPR indicates that the specification and explication have taken place before the beginning of the data collection—otherwise the purposes might not be seen as specified and explicit at the time of the collection. The Article 29 Working Party argued that it follows from the formulation "collected" in Article 5(1) lit. b that the purposes must be "specified prior to, and in any event, not later than, the time when the collection of personal data occurs" (Article 29 Working Party, 2013, p. 15). The CJEU decided in 2022 that it follows from recital 39 of the GDPR that purposes have to be identified and clearly stated at the time the data are collected at the latest (CJEU, judgment of 24 February 2022, C-175/20, *Valsts ieņēmumu dienests*, marg. no. 64 et seq.). In addition, a different interpretation would come into conflict with the fact that the determination of the purpose is so closely linked to the requirements for a valid consent. An explicit mentioning of the provision of information on the purposes of the processing before the respective processing takes place can be found in Article 13(3) GDPR: In cases in which a controller "intends to further process the personal data for a purpose other than that for which the personal data were collected, the controller shall provide the data subject prior to that further processing with information on that other purpose with any relevant further information" (see also the almost identical recital 61, sentence 3 of the GDPR). This makes clear the most important consequence of the principle: After being defined and communicated to the data subject, the purposes of the processing cannot be changed without informing the data subject (again) or finding another (new) legal basis as a legal ground for further processing. In other words, researchers have to explain in a detailed way what they plan to use the data for and how they will process it, and are obliged to do so at the time they ask participants to agree to become part of their research, which is rather early in the research process. Following the principle of purpose limitation as explained above, they will be bound to that specification during the rest of the data handling, otherwise they would have to undergo the information and consent process again.

Another important obligation of those responsible for the processing of personal data is to make sure that they provide the necessary information to the data subject. The provision of information to the data subject serves to guarantee the fairness and transparency of the processing (see CJEU, judgment of 1 October 2019 C-673/17, *Planet49*, marg. no. 79). Apart from the provision of information about the purposes of the processing (Article 13(1) lit. c GDPR), this includes information about the recipients of the personal data (Article 13(1) lit. e GDPR) and the period for which the personal data will be stored (Article 13(2) lit. a GDPR). In Article 13(1) and (2) GDPR, it is explicitly stated that in cases where data are collected directly from the data subject, the controller is obliged to provide information on the circumstances of the processing "at the time when personal data are obtained." In recital 42, sentence 1, it is stated that "[t]he information in relation to the processing of personal data relating to the data subject should be given to him or her at the time of collection from the data subject." Both sentences are phrased in a way that seems to support the interpretation that the information could take place together with the data collection, instead of prior to it. Yet, not only is it hard to imagine that researchers inform their participants in the middle of carrying out an actual interview or of another method to collect research data, but also, and even more so, the purpose of the obligation and its link to the requirements for informed consent need to be considered: As has been stressed by the German Federal Court of Justice (BGH), the provided information is supposed to serve as a basis for the data subject's decision as to whether or not to consent to a processing, and this aim would be missed if the controller provided the information after the beginning of the data flow (BGH, decision of 10 November 2022—I ZR 186/17, GRUR 2023, 193 (196, marg. no. 31)). Consequently, and in the same case, the CJEU ruled that "the validity of the consent given by the data subject depends, inter alia, on whether that person has previously obtained the information in the light of all the circumstances surrounding the processing of the data in question ... under Articles 12 and 13 of the GDPR, and which allow him or her to give consent in full knowledge of the facts" (CJEU, judgment of 11 July 2024, C-757/22, *Meta Platforms Ireland*, marg. no. 59 f.). The Court uses the term "previously" here, and expressly confirms the existing link between the information and the decision on agreeing to the processing, even spelling it out as a condition for a valid declaration of consent. This states very clearly that the information enumerated in Article 13 GDPR shall be provided before the actual processing of the data begins, which means probably at the time when consent is collected from participants or other data subjects. Earlier, we indicated that the fundamental right safeguarded by data protection law has been derived from the general right of personality, which itself is based, among other things, on human dignity (see Sect. 3.2). Further discussion of how to defend privacy through philosophical anthropology leads to the conclusion that "our dignity rests in being able to be the masters of our own journeys and keep our identities and our choices open" (Floridi, 2016, p. 4). Such a human exceptionalism viewpoint entails an identity-related mastership and the openness of choices. This supports the importance of the provision of information to participants prior to the processing and as a prerequisite for their granting of consent, and of the principle of purpose limitation, binding the researchers to what they have communicated to the research participants.

Following these elaborations, researchers are obliged to determine and communicate at the latest upon the collection of consent and thus before having started the process of analyzing the research data, not only the specific, explicit and legitimate purposes of the processing, the type(s) of data, the identity of the controller, and that the participants have the right to withdraw their consent, but also the persons or institutions that will receive the data, as well as the period of time it will take them to analyze and handle the data in the course of their research process.

5 Privileges for the Processing of Personal Data in Research Contexts and Their Potential to Solve a Knot

The principles and obligations described so far have been general. As mentioned earlier, the GDPR knows different scenarios in which personal data are handled, and in some cases addresses them by specific rules, whether by including further restrictions for the processing, such as for special categories of personal data, or by moderating the obligations to be followed. One such "specific processing scenario," as the GDPR calls it, is the processing of data for the purpose of scientific research. Whereas the term "scientific research" is not defined in the GDPR, the Article 29 Working Party interprets it as "a research project set up in accordance with relevant sector-related methodological and ethical standards, in conformity with good practice" (Article 29 Working Party, 2017a, p. 28). This interpretation can serve as guidance for determining the meaning of the term in the GDPR.

5.1 Article 89 GDPR and Derogating Regulations

With Article 89 GDPR, a complete article is dedicated to the processing for scientific research and other related purposes, namely archiving in the public interest, historical research, and statistical purposes. According to its paragraph 1, sentence 1, processing for these purposes shall be "subject to appropriate safeguards … for the rights and freedoms of the data subject." Sentence 2 of this paragraph stipulates that these safeguards "ensure that technical and organizational measures are in place," and highlights the aim of ensuring "respect for the principle of data minimization" in particular. Pseudonymization is stated explicitly as an example for safeguarding measures (Article 89(1), sentence 3 GDPR). Moreover, the norm obliges researchers to further process data in a manner "which does not permit or no longer permits the identification of data subjects" whenever it is possible to do so and at the same time to fulfill the purposes pursued (sentence 4).

If such safeguards are in place, the possible negative impact on the rights of data subjects can be prevented or limited, and the fact that the included purposes typically go hand in hand with processing data in pseudonymized or anonymized form, and that access to such data is limited, suggests that high standards are set in these scenarios (Herbst in Kühling & Buchner, 2024, Article 5 marg. no. 52). This has been a factor for the German Federal Constitutional Court (BVerfG) in its earliest verdict on the fundamental right to informational self-determination as well (BVerfG, judgment of 15 December 1983—1 BvR 209, 269, 362, 420, 440, 484/83, BVerfGE 65, p. 1 (p. 62)): In this judgment, the Court ruled that it is key for statistical data to "be treated confidentially and rendered anonymous as soon as possible." In contrast, "[i]f data is not collected exclusively for statistical purposes, it is imperative that the sharing and use of such data be subject to strict and specific purpose limitations" (p. 62). In other words, safeguards like anonymization minimize the risks for the data subject and therefore justify a granting of privileges for the processing for research purposes—for example, allowing a less strict application of the principle of purpose limitation, as in Article 5(1) lit. b, 1st half sentence of the GDPR (see Sect. 5.2 below in detail). With Article 89 GDPR obliging researchers handling personal data for scientific purposes to put such safeguards in place, some space is provided for them in return—for example, for the reuse of data collected for other purposes. Their implementation also offers opportunities for further privileges, as stipulated in Article 89(2) GDPR: According to this paragraph, EU or national Member State law may provide for derogations from the data subject rights guaranteed in Articles 15 et seq. GDPR, should those rights be "likely to render impossible or seriously impair the achievement of the specific purposes," and the derogations be necessary to achieve the purposes. Significantly, this possibility is bound explicitly to the observance of the conditions and safeguards stipulated in Article 89(1)—which means that the high data protection standard acts as a counterbalance to the privileges granted to processing of data in, inter alia, scientific research contexts.

5.2 *Identification and Compatibility of the Purposes in Research Contexts*

Recital 33 of the GDPR acknowledges in its first sentence that "[i]t is often not possible to fully identify the purpose of personal data processing for scientific research purposes at the time of data collection." In sentence 2, it is laid down that "[t]herefore, data subjects should be allowed to give their consent to certain areas of scientific research when in keeping with recognized ethical standards for scientific research" (recital 33, sentence 2 of the GDPR). From this, the German Data Protection Conference (DSK) concludes that a so-called "broad consent" can be permissible under certain conditions, making a slightly less determined description of the purposes of the processing acceptable when consent is given prior to the beginning of the processing (Datenschutzkonferenz, 2019, p. 1). The DSK points out that it is not compliant with recital 33 to broaden the use of the data wholesale to "certain areas of research," but that mechanisms need to be developed to limit the use of the collected data in a way that is comprehensible to the research participant; also, the research project in question has to be described as precisely as possible, and specific measures for the protection of the data have to be spelled out in the consent form (Datenschutzkonferenz, 2019, p. 1). However, it should be noted that the existence of recital 33 does not exempt researchers from obligations related to the requirement of specific consent (see Sect. 4.2 above). Instead, the principle remains applicable that scientific research projects "can only include personal data on the basis of consent if they have a well-described purpose"; only for cases in which the purposes of the processing cannot be specified before the collection of data does the recital make it possible to describe the purposes "at a more general level" (Article 29 Working Party, 2017a, p. 28).

In addition, in Article 5(1)(b), 2nd half sentence GDPR, there is a special regulation related to the principle of purpose limitation, or, more specifically, to the requirement of compatibility of the purposes: "[F]urther processing for archiving purposes in the public interest, scientific or historical research purposes or statistical purposes shall, in accordance with Article 89(1), not be considered to be incompatible with the initial purposes ('purpose limitation')." By addressing "further processing," the exemption is limited to cases in which research data are not processed for the purposes defined before the initial data collection, but for other, new purposes compared to the purposes of the initial processing. It is an element of the principle of purpose limitation that further processing is not permitted if it is incompatible with the initial purposes (see Article 5(1) lit. b, 1st half sentence GDPR). Therefore, for further processing of data aiming at other goals than the preceding processing, it needs to be verified in each individual case that the new purposes are compatible with the initial purposes (Heberlein in Ehmann & Selmayr, 2024, Article 5 marg. no. 25). By the 2nd half sentence, it is implied in law that further processing for scientific research purposes is compatible with the initial purposes, with the effect that a compatibility check is not required in these cases, in that way privileging the enumerated purposes over other purposes (Herbst in Kühling & Buchner, 2024, Article 5 DS-GVO marg. no. 50).

5.3 Derogations to the Obligation to Provide Information

As described in Sect. 4.2 above, the provision of information to research participants is of high relevance for the protection of their rights and interests. In principle, this is also true in cases in which data are not collected directly from the data subject, which is why Article 14(1)–(4) GDPR addresses this situation and imposes obligations on those responsible for the processing to inform the data subject in these situations as well. From the data protection perspective, constellations in which the controller obtains personal data from other sources—for example, other controllers, publicly available sources, or data providers are most sensitive, since they take place out of the range of influence and cognizance of data subjects and are prone to errors (Knyrim in Ehmann & Selmayr, 2024, Article 14 marg. no. 2 with further references). Nonetheless, the distance between the controller and the data subject may make it difficult to fulfill the information obligation. In the GDPR, this is taken into account by including exceptions to the duty in paragraph 5 of Article 14. Its subparagraph (b) concerns circumstances in which the provision of the required information is impossible or involves a disproportionate effort. It is recommended that, as a basic principle, these exceptions be interpreted and applied narrowly (Article 29 Working Party, 2017b, p. 28). Whereas the impossibility is given mainly when the data subject is unknown to the controller, who thus cannot identify the person (Bäcker in Kühling &

Buchner, 2024, Article 14 DS-GVO marg. no. 54 with further references), Article 14(5) lit. b, sentence 1 GDPR enlists processing scenarios in which the involvement of a disproportionate effort can be given in particular, and the processing for scientific research purposes is one of them. As there are no exemptions from the obligation to inform the data subject when data are collected directly from a data subject (Article 13 GDPR), the Article 29 Working Party concludes that "disproportionate effort typically arises by virtue of circumstances which do not apply if the personal data is collected from the data subject"; they therefore "must be directly connected to the fact that the personal data was obtained other than from the data subject" (Article 29 Working Party, 2017b, p. 30). On the other hand, there might be constellations which demand for the data subject's information anyway, because they affect data protection interests intensively, such as medical research dealing with health data, which is why this exemption is understood as constituting only a rebuttable presumption (Bäcker in Kühling & Buchner, 2024, Article 14 DS-GVO marg. no. 56).

According to recital 62 GDPR, the "number of data subjects, the age of the data and any appropriate safeguards adopted" should be taken into consideration in relation to the assessment of the question whether the information of the data subjects would involve effort that is disproportionate or not. An example of a case that justifies lifting the information obligation is the reuse of a 50-year-old dataset on surnames with 20,000 data subjects, which does not contain any contact details, by historical researchers (see Article 29 Working Party, 2017b, pp. 60 et seq.). The circumstances and numbers in this example illustrate that the bar for an exemption from the provision of information to participants in research projects is, although limited to the reuse of datasets, set rather high and will not be reached on a regular basis. That said, data controllers will have to be aware that they need to inform data subjects if the data are further processed, even if it is "old stock" data (Knyrim in Ehmann & Selmayr, 2024, Article 14 marg. no. 45, giving the example of an old set of customer data, prospectively to be used for statistical or research purposes). Moreover, even if the requirements for the exemption are met, those responsible for the processing need to take appropriate measures for the protection of the data subjects, which includes "making the information publicly available" (Article 14(5) lit. b, sentence 2 GDPR). In other words, in cases in which the provision of information directly to those to whom the data relate is lifted, the researchers need to take protective measures and particularly consider publishing the information on the processing of the data by other means.

6 Conclusion

One of the data protection specialists interviewed in the FODAHEMM project claimed that there is actually no area of tension between researchers or the freedom of research and data protection, and that it is rather a problem of understanding and communication (019_ Spezialstelle, #43:00). Whereas at the end of our assessment we would not assent to the first part of the statement, since we did identify an area of conflict between the academic freedom and the right to the protection of personal data, we tend to be able to relate to its second part. Ideally, we have contributed to alleviating the problem in understanding by having shed some light on the subject through the legal analysis provided in this chapter. The findings of this analysis can be summarized as follows.

Researchers do face legal complexity if their research project involves the processing of personal data. This complexity is traceable to two factors in particular: The interaction of regulations of different levels, and the vagueness of legal terms and concepts applicable in such case. However, it is noteworthy that instead of having a monodirectional negative effect on the researcher's interests by imposing additional obligations upon them, the confirmed legal complexity can also be found in the special regulations for the processing of personal data for research purposes, and hence has a potential to work in their favor as well.

The situation is aggravated by the fact that a number of data protection provisions require an a priori perspective on the processing, as the provision of information to data subjects and the determination of the purposes of the processing are crucial to enable a proper decision by data subjects as to whether or not to consent to processing. Since the processing of data extends over a large period of the research process, in the context of scientific data processing, this entails that researchers are required to take an a priori perspective on their whole research project before they start to collect data. This is all the more challenging as scientific processes, aimed at generating new knowledge as they are, to a certain extent do involve bringing things forward in an inventive way, which implicates an incremental modus operandi (see Roßnagel, 2019, 158) that is being developed or adapted over the course of time and thus not being defined beforehand and strictly adhered to afterwards. Confirming the inference put forward in one of the interviews conducted in the FODA-HEMM project and cited in the introduction, this indeed puts researchers in a position of being obliged to determine something they do not have certainty about at the time they are supposed to do so—i.e., to think the research process through before beginning to collect data.

The right to academic freedom and research is guaranteed, among other things, by the CFREU, and has to be regarded in the implementation and wording of legal rules aiming at the protection of individuals' data and privacy. It has found its way into the GDPR through the inclusion of special rules for data processing in scientific contexts.

The examples of the exceptions from the principle of purpose limitation (Article 5(1) lit. b GDPR and recital 33) and from the obligation to provide information show that the legislator was aware of the specialties of the research process and has sought to address this in the legal regime.

Alas, the privileges to be found in the GDPR only address certain constellations in the processing for research purposes and need to be interpreted in the light of the general principles for data protection. In fact, so couched in nebulous terms are the special regulations for processing data for scientific research purposes that the actual privileges for this processing scenario are difficult to pin down, even if looked at closely. On the other hand, researchers are bound to the additional requirements set out in Article 89 GDPR, regardless of whether their processing enjoys legal privileges or not. It thus seems that the lifting of the obligations for this processing environment falls short of the additional obligations placed upon them instead of lifting the requirements for this processing environment in equal measure.

Although the importance of data protection compliance in scientific research, as outlined earlier, is not contradicted, the analysis in this chapter suggests the other side of the coin of this topic: On the one hand, researchers should integrate data protection aspects and requirements into their research process; on the other hand, the codification and interpretation of data protection rules should ensure the integration of the researchers' interests, too. For the sake of the exercise of the right to academic freedom, and, as the German Council for Scientific Information Infrastructures (RfII) put it, to achieve a "research-conducive data protection" (RfII, 2017, p. 2, translation by the authors), it would be wise for legislators and jurisprudence to keep being aware of the circumstances of data handling in scientific contexts. It needs to be taken into account that "the conditions under which researchers work with research data, act as producers of data themselves or actively utilize commercial data for research are subject to unique considerations" (RfII, 2024, p. 3). Hence, it is to be hoped that attention will be paid to this particular processing scenario, facilitating a further concretization of the interpretation of the regulations, whether by the CJEU or by expert and professional associations. Also, the actors involved in the further shaping of research data infrastructures, whether on an international, national, or local level, would be well advised to appreciate and consider both the researchers' and the data subjects' interests. However, all of that cannot supersede the necessity to assess the legal situation on a case-by-case basis. Given the results of the analysis above, advisory support for researchers provided by judicial experts is key in terms of enabling them to meet the challenges of legal complexity as well as to meet their responsibility toward the research participants. Ideally, such expert service is not only acquainted with the legal terms and possibilities, but also with the researchers' perspective and the performance demands that are irrefutably brought about in the wake of scientific endeavors.

Competing Interests This study was carried out in the course of the research project "Forschungsfreiheit vs. Datenschutz—Wie können Hemmnisse beseitigt werden? (FODAHEMM)," funded by means from zukunft.niedersachsen, the Ministry of Science and Culture of Lower Saxony, and the joint research funding program of the VolkswagenStiftung, Germany (grant no. 76202-3172/2021 (76ZN2005)).

References

97th Datenschutzkonferenz. (2019). Resolution relating to the interpretation of the term "certain areas of scientific research" in recital 33 GDPR (Beschluss der 97. Konferenz der unabhängigen Datenschutzaufsichtsbehörden des Bundes und der Länder zu Auslegung des Begriffs „bestimmte Bereiche wissenschaftlicher Forschung" im Erwägungsgrund 33 der DS-GVO). https://www.datenschutzkonferenz-online.de/media/dskb/20190405_auslegung_bestimmte_bereiche_wiss_forschung.pdf (accessed 01/13/2025).

Article 29 Data Protection Working Party. (2013). *Opinion 03/2013 on purpose limitation (WP 203).* https://ec.europa.eu/justice/article-29/documentation/opinion-recommendation/files/2013/wp203_en.pdf (accessed 01/13/2025).

Article 29 Data Protection Working Party. (2017a). *Guidelines on consent under regulation 2016/679 (WP 259 rev. 01).* https://ec.europa.eu/newsroom/article29/items/623051 (accessed 01/13/2025)

Article 29 Data Protection Working Party. (2017b). *Guidelines on transparency under Regulation 2016/679 (WP 260).* https://ec.europa.eu/newsroom/article29/items/622227 (accessed 01/13/2025).

Benamor, Sofiane (2024). Speicherdauer biometrischer und genetischer Daten (Anm. zu EuGH, U. v. 30.1.2024—C-118/22), ZD, pp. 267–269.

Ehmann, E., & Selmayr, M. (Eds.). (2024). *DS-GVO. Datenschutz-Grundverordnung. Kommentar* (3rd ed.). C.H. Beck.

Floridi, L. (2016). *On human dignity and a foundation for the right to privacy.* https://papers.ssrn.com/sol3/papers.cfm?abstract_id=3839298 (accessed 11/12/2025).

German Council for Scientific Information Infrastructures (RfII—Rat für Informationsinfrastrukturen). (2017, März). *Datenschutz und Forschungsdaten.* Aktuelle Empfehlungen. https://rfii.de/?p=2253 (accessed 11/20/2025).

German Council for Scientific Information Infrastructures (RfII—Rat für Informationsinfrastrukturen). (2024). *Federated infrastructures for scientific use.* NFDI, EOSC, Gaia-X, and the European Data Spaces: Comparison and Recommendations for a Committed Engagement to Shape the European Research Data Ecosystem. https://rfii.de/?p=11424 (accessed 11/20/2025).

Horn, N. (2017). *Denkimpuls digitale Ethik: Grundlagen der digitalen Ethik—Eine normative Orientierung in der vernetzten Welt.* https://initiatived21.de/uploads/03_Studien-Publikationen/Denkimpulse-Ethik/01-Grundlagen/d21-denkimpuls-ethik01-grundlagen.pdf (accessed 11/10/2025).

Jarass, H. D. (2019). *Charta der Grundrechte der Europäischen Union unter Einbeziehung der sonstigen Grundrechtsregelungen des Primärrechts und der EMRK. Kommentar* (4th ed.). C.H. Beck.

Kitcher, P. (2003). *Constraints on free inquiry, in: Science, truth, and democracy* (pp. 93–108). *Oxford Academic* (online ed.). https://doi.org/10.1093/0195145836.003.0008 (accessed 11/17/2025).

Knoke, F. (2023). *Subjektive Rechte an Forschungsdaten. De lege lata und de lege ferenda.* Mohr Siebeck.

Kühling, J., & Buchner, B. (Eds.). (2024). *Datenschutz-Grundverordnung/ BDSG. Kommentar* (4th ed.). C.H. Beck.

Kühling, J., & Martini, M. (2016). *Die Datenschutz-Grundverordnung: Revolution oder Evolution im europäischen und deutschen Datenschutzrecht?* (pp. 448–454). EuZW.

Kuckartz, U., & Rädiker, S. (2021). *Fokussierte Interviewanalyse mit MAXQDA.* Springer VS.

Roßnagel, A. (2019). *Datenschutz in der Forschung. Die neuen Datenschutzregeln in der Forschungspraxis von Hochschulen, ZD* (pp. 157–164).

Victor, E. (2014). Scientific research and human rights: A response to Kitcher on the limitations of inquiry. *Science and Engineering Ethics, 20*, 1045–1063.

Model Transfer as a Case of Interdisciplinary Research in Science

Catherine Herfeld and Edoardo Peruzzi

1 Introduction

Scientific research is characterized by pronounced disciplinary specialization, which frequently results in highly abstract models tailored to the representational and explanatory demands of particular target systems. At the same time, a key characteristic of contemporary science is that there is increasing interaction across different scientific domains (e.g., Herfeld & Lisciandra, 2019). One way the resulting tension between domain specialization and cross-domain interaction emerges is in the transfer of models across scientific disciplines—a phenomenon that has become a key characteristic of modern science.

There are numerous examples of such model transfers, often taking place between different scientific disciplines. They range from the natural to the social and the life sciences, with some of the most representative cases being the Ising model, the Lotka-Volterra equations, network models, and rational-choice models. All of these models have been prominently applied in various disciplines, with some achieving such broad

C. Herfeld · E. Peruzzi (✉)
Institute of Philosophy, Leibniz University Hannover, Hannover, Germany
e-mail: edoardo.peruzzi@philos.uni-hannover.de

C. Herfeld
e-mail: catherine.herfeld@philos.uni-hannover.de

J. Hinrichsen et al. (eds.), *Ethics of the Sciences*,
https://doi.org/10.1007/978-981-92-2821-8_8

applicability that they have become common tools across otherwise distinct areas of scientific research. To understand how science is organized and how we can ensure its progress, it is therefore informative to study cross-disciplinary model transfers and the ways in which such transfers shed light on interactions between different disciplines, despite increasing specialization, in contemporary science.[1]

In this chapter, we discuss how model transfer across different disciplines help overcome tensions emerging from parallel specialization and increasing cross-disciplinary interaction. The motivation for this focus is that cross-disciplinary model transfer raises questions that typically do not arise in intra-disciplinary model transfer. Cross-disciplinary model transfer involves moving models between distinct epistemic cultures, rather than within a shared one. While intra-disciplinary transfer usually presupposes substantial agreement on the core methodological commitments, key background assumptions, epistemic values, standards of research, and evidential norms, cross-disciplinary transfer frequently cannot rely on shared presuppositions. Instead, these presuppositions themselves become objects of scrutiny during and after the transfer. More concretely, within a single discipline, key concepts and variables tend to have relatively stable meanings, or at least meanings governed by shared tacit understandings. In contrast, when a model is transferred across disciplinary boundaries its core concepts are likely to be reinterpreted or partially redefined to fit a new target domain. Studying these transfers raises questions about whether a model must be modified—or even fundamentally transformed—to be applicable in a new discipline, and about how such modifications help resolve tensions between disciplinary specialization and cross-disciplinary interaction.

Against this background, we approach the question of how model transfers help overcome the tensions emerging from parallel specialization

[1] Note that in the literature on model transfer, the focus is typically on transfers between what has generically been labeled "domains" (see Herfeld & Lisciandra, 2019). While domains can refer to different disciplines, they need not be entirely distinct fields; a domain can also denote a subfield within a particular discipline. This broader understanding allows for the analysis of model transfer within a single discipline—for example, the transfer of the Ising model from its original use in statistical mechanics to describe ferromagnetism in lattice systems, to its application in condensed matter physics to study phase transitions in more general systems. In what follows, we restrict our attention to cases where "domain" refers to a discipline in order to meaningfully discuss cases in which such transfers manifest in cross-disciplinary interaction.

and increasing cross-disciplinary interaction by sketching a taxonomy of four different ways in which models are modified to enable model transfer between different disciplines. This taxonomy shows the different ways in which cross-disciplinary research manifests itself through various modification procedures. Our taxonomy provides a basis for more concrete analyses of whether specific disciplines—for example, economics, which has often been criticized for its apparent isolationist or imperialist attitudes—have become more interdisciplinary through such model transfers. It also contributes to a more general analysis of how contemporary science is organized and how cross-disciplinary research leads to epistemic benefits in science overall.

Our goal is to explore how model transfer can illuminate the multiple ways in which cross-disciplinary interaction occurs. Focusing on model transfer, we take cross-disciplinary interaction to occur when models, modeling practices, or representational tools move across disciplinary boundaries. However, the extent to which they are reinterpreted, repurposed, or materially reconfigured in the process—and thereby adapted to the receiving discipline—varies. In this way, cross-disciplinary model transfer gives rise to different forms of interdisciplinarity.

The term interdisciplinarity is notoriously overdefined in the existing literature. This is in large part because what we take to be cross-disciplinary interaction varies in its details from case to case. As Andersen (2022, p. 376) observes with respect to interdisciplinary interaction:

> [Interdisciplinarity] varies from case to case, and ... it can vary in many different ways. ... It may vary from the focused import into one discipline of a particular method from another discipline in order to solve a specific problem, to the import of a broad methodological repertoire that also opens for the definition of a whole range of new problems. It may vary from a narrow interlocking of a few selected concepts across disciplinary boundaries, to a major integration of conceptual resources across disciplines. It may vary from feeding a particular problem from one discipline to another, to the recognition of problems that can only be defined and addressed by drawing on multiple disciplines. It may vary from one discipline calling upon the mere services of another discipline, to a full-fledged and equal collaboration between the two.

We take our taxonomy to reflect this variety of ways in which interdisciplinarity can occur in science. To do so, we will further elaborate on our definition of interdisciplinarity in Sect. 4 below.[2]

2 The Template-Based Account of Model Transfer

Let us begin by introducing the kinds of models mostly studied to illustrate model transfers. The Ising model offers a classic case of such a model. It was originally developed in statistical physics to explain ferromagnetism. In its simplest form, it consists of a core mathematical structure that describes a system of binary variables arranged on a graph or lattice, where each variable interacts with its neighbors. In the original physical context, these variables represent particle spins, whereas in other domains they may represent different kinds of binary states. Moreover, the model mathematically specifies the strength of neighbor interaction and the overall system's behavior is described probabilistically. It is a paradigmatic example of model transfer in that its mathematical structure has been transferred broadly into various disciplines.

While in physics the aim has been to understand how local interactions lead to collective phenomena such as phase transitions, in neuroscience it has been used to describe the collective behavior of neurons, where spin states represent firing or resting states. In biology, the Ising model has been applied to population dynamics, genetic networks, and protein folding, where the binary nature of spin states can model key switches in biological systems. The questions emerging around such transfers are which parts of the model are transferred, what explains their transfer, and what justifies such transfers.

Philosophers of scientific modeling currently address these questions through either an analogy-based or a template-based account of model

[2] As we understand it, *interdisciplinarity* is mostly defined as often involving some degree of integration of conceptual, methodological, or analytical tools from two or more disciplines. By contrast, *transdisciplinary* interaction is typically defined as involving forms of engagement between science and society—such as the inclusion of non-academic actors or the co-production of knowledge—which are not present in our case. *Multidisciplinarity*, in turn, is mainly characterized by the cooperation of multiple disciplines without substantial integration. While this last feature is common in many instances of model transfer, our discussion in this chapter is deliberately limited to interactions between two disciplines only. For these reasons, we refrain from using these labels in a more prominent way here.

transfer (see Herfeld, 2024, pp. 271 f.). The analogy-based approach, following Mary Hesse (1966), highlights the role of analogical reasoning and the identification of shared features of source and target system in enabling model transfer. In contrast, the template-based approach, developed by Paul Humphreys (2002, 2004, 2019) and further elaborated by Knuuttila and Loettgers (2014, 2016, 2023), sees contemporary science as organized around formal, theoretical and computational templates that, to varying degrees, operate independently of specific target systems. Their generality and tractability make them valuable across disciplines. For instance, mathematical game schemes from economics, originally used to analyze strategic decision-making in conflict situations, have been adapted in evolutionary biology to explain sex ratios and the evolution of aggression in animals (e.g., Grüne-Yanoff, 2011).

Philosophers of science currently tend to lean toward a template-based account, although there are prominent attempts to add an element to it by re-introducing the role of analogies when explaining the transfer. Knuuttila and Loettgers (2023), for example, propose the concept of a "model template" to explain such model transfers, which consists not only of abstract formulas or structures. Rather, a model template also includes a minimal ontology, a set of prototypical properties and behaviors, focal conceptualizations, and the paradigmatic questions it addresses. One important motivation for Paul Humphreys to propose templates as the unit of analysis to study contemporary science is their central role in the organization and application of present scientific practices. According to Humphreys, templates are theoretical or computational structures that serve as generalized frameworks for understanding and representing various systems. They can include mathematical equations, algorithms, and simulation structures, which scientists adapt to represent and analyze particular phenomena. Their generality makes templates flexible, which is fundamental to the ability of models to be transferred between scientific fields.

Some philosophers, especially Humphreys, view templates as sometimes being subject-independent. As such, they are also not inherently tied to any specific scientific domain, which in turn allows them to be applied to diverse sets of fundamentally distinct phenomena. By providing a reusable and adaptable mathematical structure, templates streamline the process of model-building and problem-solving across disciplines. For instance, features of computational models like differential equations or network models act as templates that scientists can tailor to fit the specific requirements of their target systems. Because templates can sometimes be formulated independently of any specific target system—what Humphreys (2019) calls "formal templates"—they can be understood as part of the toolbox for constructing computational models and, as such, transferred across domains.

For mathematical disciplines such as economics, the template-based account of models provides a useful foundation for understanding model transfer, as it aligns with the kinds of mathematical models commonly used in those fields. However, it could benefit from a clearer analysis of the ways in which such templates serve to construct applicable models in a new domain. This would help us understand the different ways in which such models are modified during the transfer process, thereby ensuring their transferability in cross-disciplinary interaction while allowing them to serve as applicable models in highly specialized new target domains. In particular, the template-based account of model transfer treats this transfer process mainly as a process of model construction (see Herfeld & Peruzzi, 2026). Moreover, a number of recent case studies have questioned the view that models transferred remain more or less unchanged. Those studies not only suggest that model modifications are crucial for their successful transfer, but also that such modification processes come in different degrees (e.g., Bakeev, forthcoming; Grüne-Yanoff, 2011; Herfeld & Lisciandra, 2019; Herfeld & Peruzzi, 2026; Knuuttila & Loettgers, 2013, 2014, 2023).

Humphreys specifies the modification path from a template to an applicable model in the form of a stepwise model construction process. According to Humphreys, we start with either a theoretical or formal template; the former is part of a larger theoretical framework and comes with a general theoretical interpretation, while the latter is a purely mathematical object that is accompanied only by a mathematical interpretation. As those templates are mathematical objects that function more as schemas than as models, the goal of model construction on the basis of those templates is to ensure their computational tractability—i.e., to turn them into what Humphreys calls a "computational template"—and ultimately into a computational model that can serve empirical purposes of explanation and prediction through its function as an accurate representation. A computational template is turned into a computational model by adding a set of construction assumptions, i.e., idealizations, abstractions, ontological assumptions, constraints (which include initial and boundary conditions as well as laws), and a model interpretation. To allow changes to the model in case the model cannot fulfill its functions as an accurate representation in the target domain, a correction set is added. The result of this construction process is a computational model.[3]

On this account, templates function as tools for model construction. When a model is transferred from a source domain to a target domain, it is not the model itself that is directly transferred, but rather the underlying template used in the source domain. This template is then employed in the construction of a new model tailored to the target domain. What explains the transfer on this account is the tractability of the constructed model; what justifies the transfer is that the construction assumptions are selected in such a way that the model applies to the new target system.

3 A Taxonomy of Model Modification Strategies

On Humphreys' account of models, model transfer seems to boil down to a form of model construction process in which the template transferred serves as a basis for constructing a computational model that is tractable and applicable in a new target domain. Selecting both the construction assumptions and the interpretation is, of course, constrained by the envisaged target system. Yet the question of how—on Humphreys' account—this model construction process also amounts to a model modification process is not straightforward. First, the degree to which a model gets modified in the transfer process varies; second, the step in the construction process at which this modification takes place and the extent to which such modification is informed by domain-specific features, properties, and knowledge can also vary.

Although our starting point is Humphreys' framework of model transfer as model construction, we maintain that model transfer is not merely the process of constructing a model from a template. Rather, we propose that model transfer often requires what we characterized elsewhere as a two-fold modification process of *deconstruction* and *reconstruction* (Herfeld & Peruzzi, 2026). In our view, a model—to be

[3] Of course, a model can be accepted due to other functions than its representational function, particularly because it is debatable whether models (always) have a representational function and what that representational function exactly amounts to. However, given that we adopt Humphreys' account of scientific models here, we do assume that models are transferred also because they will have some representational function in the target domain.

applicable in some domain—needs to be constructed based on domain-specific features—including target system specific properties, theoretical frameworks, and considerations about model effectiveness. As a consequence, since such domain knowledge informs the construction of a model, applying it successfully in a new domain frequently requires an initial step of deconstruction to strip away features specific to the source domain and enrich the model with domain knowledge of the target domain. Only in this way can the model be reconstructed in a form that is tractable and applicable in the new target domain.

This two-step modification process that a model undergoes in a transfer—albeit to different degrees—forms the basis of our proposal for the following taxonomy of model modification strategies, focusing on four distinct ways in which model modification occurs in the practice of model transfer. This taxonomy is not intended to be exhaustive but aims to capture the most significant ways in which models are modified through deconstruction and reconstruction during transfer, as well as the varying degrees to which such modifications are undertaken.

3.1 Model Imperialism

The first kind of model transfer process has already been discussed in the literature under the label "model imperialism" (Bradley & Thébault, 2019) as one way in which scientific imperialism, or more generally "scientific expansion" manifests itself (Dupré, 1994; Mäki, 2013; Mäki et al., 2017). The idea is that scientific imperialism occurs when a particular scientific discipline expands its concepts, methods, and explanatory frameworks beyond its original source domain, whereby this can occur at the expense of alternative approaches and a clear demarcation of disciplinary boundaries. A reason why the phenomenon of expansion is sometimes framed in terms of an imperialistic tendency is because expansion can occur without any attempt to adapt to the new domain; in case of a model transfer, the model becomes imposed on the target domain to address target-domain-specific problems without any degree of deconstruction and reconstruction.

To put this in slightly more technical terms, in the case of model imperialism, a model that has been used in the source domain A is transferred for use in the target domain B without any modification whatsoever. As such, it primarily leads to an "extension of the original model via an expansion of the domain of phenomena it is taken to adequately describe"

(Bradley & Thébault, 2019, p. 81). Because nothing specifically about the model, particularly neither its underlying conceptual apparatus nor its assumptions, has been altered in the process of transfer, model imperialism requires only an expansion of the justificatory sanctioning that licenses the same idealizations in a broader context. In effect, the scope of the model's sanctioning has been extended in that the model's "certificate of permissible applications" has been broadened to encompass a wider range of circumstances (Bradley & Thébault, 2019, p. 81).

Economics offers probably the most prominent set of examples of such a model imperialism, namely the transfer of models grounded in Chicago price theory to fields such as sociology, social psychology, anthropology, among other domains. The economist Gary Becker—who is most famous for his controversial attempt to explain a wide range of human behavior with a specific type of rational-choice model—has, for example, developed a well-known model of reproductive choice based on economic models of optimization under constraints (Becker, 1960). The model was calibrated to account for the value of a child and the opportunity costs characteristic of different countries. Becker employed the model to explain differentials in family size and parental age in different countries—*explananda* (or "problems" as we call them here) that were widely considered to belong to the domain of sociology. In this case, the transfer involves virtually no modification of the model itself: Becker's behavioral assumptions and underlying conceptual framework remain unchanged as the model is applied to a new domain.

Interestingly, in the case of model imperialism, it is often the scientists from the source domain A who begin to address a problem in domain B. For example, Gary Becker employed an economic model (developed in domain A, economics) to address a sociological problem (application to a problem in domain B, sociology) (Mäki, 2009). Subsequently, the model and its refinements were eventually adopted and further modified by sociologists (e.g., Coleman, 1990, 1993; Coleman & Fararo, 1992). But the initiative clearly came from Becker and his colleagues from economics departments, without much initial extensive involvement from the side of sociology. The lack of significant conceptual and theoretical modification, combined with the fact that model transfer from economics into other fields has often been initiated solely by economists, has led to the phenomenon known as "economics imperialism" (Mäki, 2009).

To sum up with respect to our framework: In cases of model imperialism, the transfer is primarily initiated by scientists from the source domain, who bring the original computational template into a new context. Although the template remains largely unchanged, it acquires a slightly different interpretation when adapted into a computational model in the target domain. Interestingly, the motivation for this transfer is often to tackle a problem that was originally addressed using models or frameworks native to the target domain—yet the imported model is seen as potentially offering a better solution. As a result, the model's scope of application is extended beyond its traditional boundaries. Notably, this extension occurs without substantial modification to the underlying computational structure.

3.2 *Model Migration*

Model transfer as "model migration" has been discussed by Bradley and Thébault (2019). It describes a modification strategy in which the model's assumptions added in the source domain A do not directly apply to a fundamentally different target system in the target domain B. Because the application of the model as constructed in the source domain A to the new target system in target domain B would have called into question the model's original idealizing assumptions, the model must be modified in the transfer such that new idealizing assumptions of the model allow for its application to the system in the target domain. This is done by what Bradley and Thébault (2019) call re-sanctioning, i.e., the re-establishment of the validity or justification of a model's assumptions for a domain different from the source domain. This modification process legitimizes new idealizations in the model and thereby ensures the model's applicability to a different class of target systems.

One example of this modification strategy is the transfer of models from physics into economics, which has given rise to the new field of econophysics. This field adapts statistical physics models to explain and predict economic behavior, particularly in financial systems. A striking case is the migration of Boltzmann's kinetic theory of gases into the economic context. In the original model, molecules are treated as ideal hard spheres that collide in pairs, exchanging kinetic energy through a scattering process. Over time, this leads to a well-known equilibrium distribution—the Maxwell–Boltzmann distribution—which matches experimental data for real gases. The Drăgulescu–Yakovenko model reinterprets this structure for economic systems (Dragulescu–Yakovenko, 2000). The concept of molecules is reinterpreted as human agents, and kinetic energy as money. Instead of physical collisions, agents randomly trade: their wealth is pooled and redistributed according to a random fraction. Although the structural form of the model is preserved, the representational content shifts significantly when its application shifts from physical particles to economic actors.

Bradley and Thébault use this example as a paradigm case of model migration, where a model is not simply applied to a new domain as in the case of model imperialism, but is actively modified by reinterpreting it and thereby ensuring the re-sanctioning of its assumptions. Not all assumptions carry over seamlessly. For instance, the assumption of binary interactions is well-grounded in the physical context because the density regime of gases justifies the dominance of two-particle collisions. This supports the model's legitimacy in that domain—it can be sanctioned with a certificate of permissible applications. But this justification does not apply in the economic domain. There is no direct analogue of a "density regime" for agents in a financial system, so the assumption that agents interact only in pairs lacks an equivalent empirical grounding. As a result, the binary interaction idealization requires re-sanctioning in the new context. Without this re-sanctioning, skepticism about the model's legitimacy is warranted.

By considering the importance of re-sanctioning, we can understand why scientists in the target domain sometimes resist the transfer of models. Economists, for example, are justified in questioning whether the idealizations supporting kinetic exchange models of wealth can be meaningfully applied to economic systems, especially when the assumptions lack adequate justification in the new context. That said, when applying the strategy of model migration, the degree of modification in terms of deconstruction and reconstruction of the original and subsequently the new model can be considered moderate. This is because successful application in the target domain often requires only new justificatory arguments—different from those used in the source domain—for the model's idealizing assumptions, rather than altering or replacing the assumptions themselves.

3.3 Model Translation

In Sect. 2, we briefly outlined our proposed generalization of Humphreys' account in which model transfer involves a two-step modification process: deconstruction and reconstruction. This process is often necessary when transferring a model because constructed models are shaped, to different degrees, by domain-specific features. To apply a model successfully in a new domain, it must therefore first undergo deconstruction to remove some features that enabled the application of the model in the source domain. It can then be reconstructed using features specific to the target domain, yielding a tractable model that is applicable to the new target system. In cases of model imperialism, this process is largely absent, as models are transferred with almost no modification whatsoever. In model migration, only limited deconstruction and reconstruction occur, typically involving the re-sanctioning of certain domain-specific idealizations. However, in many instances of model transfer, a more substantial act of what we call "translation" is required which entails both the deconstruction of an existing model (or its components) and the reconstruction of a new, modified version. This is especially important when even the most abstract elements of the model, most notably its formal template, are shaped by features specific to the source domain.

Elsewhere, we identified three types of domain-specific properties that shape model construction: (1) properties of the target system, (2) the theoretical and conceptual framework of the domain, and (3) considerations about model effectiveness within that domain (Herfeld & Peruzzi, 2026). The transfer of game-theoretic models from economics to political science by William Riker (1962) illustrates how these dependencies call for a process of deconstruction and reconstruction during model transfer. In the 1950s, economists began applying n-person games from von Neumann and Morgenstern (1944) to the study of markets. An important example is the class of "market games," where formal templates were employed as the basis to construct models of exchange economies (e.g., Shapley 1953, 1955, 1959; Shubik 1959). The construction assumptions underlying these models were justified by features of the target system (such as complete information and the use of money for side payments) and by their alignment with the prevailing economic framework centered on rationality, utility, and equilibrium. This framework supported certain tractability assumptions, like the continuity and concavity of utility functions. Considerations of model effectiveness further guided choices such as using the characteristic function form to represent n-person games and selecting a well-known solution concept for such games, namely the so-called "core."

When these models were transferred into political science, Riker had to engage in a process of deconstruction and reconstruction to modify the model (Herfeld & Peruzzi, 2026, sect. 3–5). In particular, he had to strip away from market models specific properties that originated from the model construction process in economics. Since the goal was to model coalition formation processes and their outcomes in political settings, Riker had to deconstruct the model by, for example, "stripping away" assumptions that were unrealistic or irrelevant to the political domain, leading also to the rejection of the core as a solution concept. He then had to reconstruct the model by even altering the formal structure, i.e., the formal template itself, by imposing additional assumptions, such as a zero-sum condition, modifying key constraints, and introducing new domain-specific considerations to obtain model results (the so-called "size principle"). By drawing on existing concepts from political theory and domain-specific knowledge about actual coalition behavior, this activity of reconstruction enabled him to introduce specific properties of the target domain into the model construction process and thereby ensure the model's applicability in political science.

To sum up, successful model transfer often requires such a two-step translation process. First, deconstructing the model to remove domain-specific elements and, second, reconstructing it as a tractable model applicable to the new target domain. This translation process ensures a model's applicability in fundamentally different domains. However, translation comes in degrees, as the extent to which deconstruction and reconstruction are needed varies across cases. It is in this sense that translation can be understood as an activity that underlies all kinds of modification processes sketched here.

3.4 Model Integration

A fourth kind of model transfer concerns cases in which models initially developed and used by a minority of scientists in domain A are later adopted by scientists in domain B for problems in the latter domain, but only after undergoing such extensive modification that they bear little resemblance to the model category to which they originally belonged in the source domain. In those cases, the degree of deconstruction of the original model A is extremely high in that the model is usually stripped of every domain-specific property—assumptions, domain-specific concepts, theoretical elements, epistemic and methodological values—that might have been part of the original model. In this sense, the original model from one domain becomes fully integrated into the theoretical, conceptual, and methodological context of the new target domain.

An exemplary case of model integration is evolutionary game theory. Originally developed within economics, classical game theory was later applied in evolutionary biology to study strategic interactions among organisms. In this new context, however, the models underwent significant transformation. As Grüne-Yanoff (2011) details, biologists not only reinterpreted the conceptual underpinnings to suit biological phenomena but even went so far as to profoundly adapt the formal structures constituting the game-theoretic template (see also Rice & Smart, 2011). They introduced new solution concepts, developed novel methodologies, and reinterpreted model variables according to a distinct biological ontology.

Model integration can also take a more transformative form. In some cases of model transfer, researchers from two distinct domains A and B collaborate to develop an entirely new class of models that draw on existing models from both domains. These new models (which we might call A&B-models) are sometimes applied within both original domains, or—more importantly—they give rise to an entirely new domain, C, emerging from repeated cross-disciplinary model transfer.

A prominent example is the domain of cultural evolution, which largely emerged out of an attempt to integrate theoretical frameworks, templates and concepts from a number of different fields to ground a new domain of modeling cultural change (e.g., Fogarty et al., 2024). Game-theoretic models have played a central role in this process. Over time, these models have been reconfigured and enriched through insights from evolutionary biology, sociobiology, and cultural studies, incorporating concepts such as fitness landscapes, cooperation, kin selection, group selection, population-level dynamics (e.g., replicator dynamics, drift, mutation), transmission biases, social learning, and norm dynamics (e.g., Boyd & Richerson, 1985; Henrich, 2015; McElreath & Boyd, 2007).

Such cases represent ideal forms of integration: we observe a genuine fusion of elements from domains A and B to create models that are not simply exported from one field to another, but instead are jointly constructed from the outset. However, these cases also raise a conceptual question: does the notion of model transfer still apply? Strictly speaking, there is no unidirectional transfer into domain B; rather, there is a mutual contribution from both A and B into a newly constituted domain C—or even the creation of C itself. What emerges is not just a new model, but a new epistemic space within which a new set of models is used. Given the depth of cross-disciplinary collaboration involved, one might question whether this process should still be described as model transfer at all, or whether we should instead understand this step as a form of model construction, in which two different disciplines contribute equally to the set of newly constructed models.

4 Model Transfer as a Case of Interdisciplinarity

Philosophers of science have only recently begun to take a sustained interest in interdisciplinarity (MacLeod et al., 2019; Mäki, 2016), especially when compared to the extensive literature on the topic in fields such as sociology, education, management, and scientometrics.

Broadly speaking, at least three main research questions guide the study of interdisciplinarity, each corresponding to a different approach. The first is empirical and concerns how interdisciplinarity can be measured. This line of inquiry has developed alongside a rich set of empirical techniques, particularly within the sociology of science. A substantial body of work is dedicated to evaluating interdisciplinarity (e.g., Pierce, 1999; Porter et al., 2007; Rafols & Meyer, 2010; Repko & Szostak, 2020; Wagner et al., 2011). One key insight from this literature is that interdisciplinarity is inherently multifaceted, even at the operational level. It can be assessed in various ways: *interdisciplinary crossing* (how often scientists publish in disciplines other than their own), *cooperation* (the proportion of papers co-authored by researchers from different disciplines), *borrowing* (the extent to which scientists cite works from other fields), and *integration* (the degree to which there is conceptual and/or theoretical integration on the level of epistemic instruments).

The second approach to interdisciplinarity is explicitly normative, seeking to articulate the conditions under which interdisciplinarity is desirable—rather than assuming its value by default. As Mäki (2016, p. 339) cautions, philosophers of science should resist generalized normative claims about the virtues of interdisciplinarity. The widespread view that disciplinarity is inherently conservative and restrictive, while interdisciplinarity is necessarily progressive and emancipatory, is often considered misleading because it is a simplistic view. The third approach is conceptual, aiming to clarify what interdisciplinarity actually is and how it differs from related notions such as multidisciplinarity and pluridisciplinarity (see, e.g., Klein, 2010).

To address the question of how model transfers help overcome the tension between specialization and cross-disciplinary interaction, we are concerned here only with the third, more narrowly defined, conceptual issue of the relation between model transfer and the degree of interdisciplinarity. One might expect that the transfer of scientific models across disciplines would foster more interaction between researchers from different scientific communities and thus a higher degree of interdisciplinarity. The intuition is that both communities, by transferring and then using the same models, would have more in common, would need to increase their interactive and integrative efforts before and during the transfer, as well as exchange ideas and possibly even engage in cooperation that is aimed at improving the explanatory and predictive power of these models in old and new domains. This intuition has been questioned, however. Some philosophers of science have suggested that model transfer can occur without necessarily increasing the degree of interaction between the communities involved, let alone the integration between frameworks or modeling tools (Grüne-Yanoff, 2011, 2015; Grüne-Yanoff & Mäki, 2014). Thus, the relation between model transfer and the degree of interdisciplinarity, as defined here, is not straightforwardly one of positive correlation.

We propose to use the resources provided by our taxonomy to shed light on the relationship between model transfer and interdisciplinarity in a more systematic way. Instead of opting for a single definition of interdisciplinarity, we treat it as a polysemic concept with multiple meanings (cf. Andersen, 2022). We suggest that each of the four types of transfer processes can be considered interdisciplinary in different ways. In doing so, we show that model transfer often exemplifies different forms of interdisciplinarity and suggest that our taxonomy reflects this diversity.

One meaning of interdisciplinarity relates to the movement (either import or export) of scientific products (theories, models, methods, concepts, etc.) between scientific domains considered distinct. Grüne-Yanoff and Mäki (2014, p. 55), for example, claim that "an interdisciplinary exchange occurs if objects employed or developed in one discipline are used to solve problems of another discipline." Similarly, Lisciandra (2025, p. 6) argues that "an interdisciplinary exchange occurs when objects—such as models, theories, concepts, instruments, or methods—'move' from one domain to another, by importation or exportation, to address issues in one of the domains involved." We call this form of interdisciplinarity an instance of "interdisciplinarity-as-exchange." From this perspective, several types of transfer processes in our taxonomy exemplify this kind of interdisciplinarity-as-exchange. For example, consider model migration and model translation. In the former, the modification of a model in its transfer consists of a re-sanctioning of idealizing assumptions; in the latter, a deeper modification process of the model is required. In both cases, models are transferred from one domain to address problems in another. These are clear instances of "interdisciplinarity-as-exchange."

Importantly, viewing model transfer as a model construction process à la Humphreys also highlights where to look for interdisciplinarity. In model migration, for instance, it is crucial to re-sanction idealizing assumptions using knowledge of the target domain. Thus, a minimal source of interdisciplinarity comes from understanding how to re-sanction idealizations such that the model can be adapted to the target domain. In the case of model translation, even more so: to deconstruct the model to remove domain-specific elements, and then reconstruct it as a tractable model applicable to the new target domain, one must be familiar with both the source and target domains. In this sense, our approach to understanding model transfer as model construction—and the taxonomy we propose—helps make explicit the connection between interdisciplinarity and the level of knowledge required in both the source and target domains by the scientists carrying out the transfer.

A different, and broader, sense of interdisciplinarity also exists—one that relates to the degree of integration between two distinct fields (let us call this "interdisciplinarity-as-integration"). Lisciandra (2025, p. 7), for example, asserts that "interdisciplinary activities lead to the integration of different domains, or specific aspects of them, such as models or concepts, to tackle novel problems that require contributions from both sides." This definition represents a stronger form of interdisciplinarity than interdisciplinarity-as-exchange, under which most types of transfer in our taxonomy would fall short. In processes such as model migration and translation, there is typically no high degree of integration between domains, which remain largely distinct. Even when scientists from both domains interact and collaborate closely, there is often no genuine attempt to contribute equally to the modeling process (Grüne-Yanoff, 2011, 2015; Grüne-Yanoff & Mäki, 2014). Furthermore, the problems addressed in these types of transfer tend to be highly domain-specific (for instance, wealth distribution in the case of model migration discussed earlier, or political coalitions in the case of model translation). Only a small subset of transfer cases, perhaps only those we classify as strong cases of model integration, would satisfy the criteria for interdisciplinarity-as-integration.

Finally, we have so far set aside the case of model imperialism, which appears to involve no interdisciplinarity at all. Within our framework, imperialistic transfer does not entail any processes of model deconstruction or reconstruction; the model is not modified in any substantive way apart from a shift in its interpretation. Consequently, scientists engaging in this form of transfer need little to no specific knowledge, or even familiarity, with the target domain beyond identifying a problem to which the model from the source domain can be applied. There is neither genuine exchange involved nor any meaningful interaction between researchers or disciplinary communities. Therefore, it seems to exemplify neither what we called interdisciplinarity-as-exchange nor interdisciplinarity-as-integration.

5 Normative Issues and the Expertise Required to Address Them

Each of the four kinds of model transfer distinguished in our taxonomy gives rise to questions concerning the kind and degree of interdisciplinarity involved. In addition, each gives rise to normative issues that are specific to the particular form of interdisciplinarity manifested in the modification process and that pose distinct demands for ensuring a smooth transfer. Generally, all four kinds of model transfer share an overarching risk of a model being wrongly applied in the new domain. However, because these risks arise in different ways, their underlying reasons differ systematically across the distinct kinds of transfer. As a consequence, and given researchers' responsibility to carry out model transfers appropriately, questions arise concerning the kinds of expertise required to address and overcome these issues.

In cases of model imperialism, the primary risk arises from the possibility that a model is applied to a target system to which it does not in fact apply. Here the danger lies in overextending a model beyond the range of phenomena for which its assumptions, mechanisms, or explanatory structure are appropriate. The normative concern is thus one of *misapplication*: researchers bear the risk of imposing an ill-suited framework onto a target system in a way that obscures relevant features of the target phenomena or marginalizes alternative approaches that might be better suited. Because appropriate modification or even transformation of the model would in principle be required to align it with the target domain, model imperialism might often not be a useful strategy to begin with unless the possibility of misapplication is anticipated and transfer is avoided accordingly. This, in turn, requires appropriate expertise in both the source and the target disciplines.

Model migration involves a distinct kind of risk. Rather than straightforward misapplication, as in the case of model imperialism, the danger here lies in how idealizations are re-sanctioned to sustain a model's justification over time. While a model may be formally adaptable to a new domain, the idealizations that supported its original use may no longer be defensible. In response, newly introduced idealizations meant to secure the model's continued justification can themselves prove inappropriate or misleading, especially as the model is repeatedly applied to the new target system. The normative issue is therefore not whether the model applies at all, but whether its re-sanctioning through newly selected simplifying assumptions is undertaken in such a way that it does not distort the target system in ways that are epistemically misleading.

The risks associated with model translation are more complex and, in some respects, more severe. Because translation involves substantial deconstruction and reconstruction, including reinterpretation, of a transferred model's components, there are multiple stages in the construction process at which misunderstandings or errors can occur. As the Riker example illustrates, extensive deconstruction of the original model and its reconstruction may furthermore raise the question of whether the resulting model in the target domain still bears a meaningful relation to the original model in the source domain. The normative concern here centers on *transparency and continuity*: researchers have a responsibility to make clear how much of the original model has been preserved, altered, or abandoned, and to avoid overstating the degree of continuity between the models used in the source and target domains respectively.

Similar concerns apply to model integration, but often in an even more pronounced and problematic form. Integration requires the coordination of multiple models, assumptions, and standards drawn from different disciplines, increasing both the complexity of the epistemic task and the potential for error. The risk of "getting it wrong" here stems not only from misapplication, inappropriate idealization, or errors occurring at the different stages of transformation, but from failures of coherence, incompatibility between modeling frameworks, or unresolved conflicts between disciplinary standards of validation agreed upon in different disciplines.

Responsibility in such cases includes careful management of these tensions on the conceptual, theoretical, and/or methodological level and explicit acknowledgment of unresolved uncertainties.

Across all four kinds of model transfer, then, such kinds of normative issues can arise from the need for researchers to be explicit about the risks involved in model transfers and about the limits of the transferred models. These concerns can become especially pressing when models are applied in disciplines with direct real-world consequences, such as economics, political science, or public policy. In such contexts, the responsibility of the scientist to communicate across disciplines when transferring a model, including the potentials and limitations of a model transferred in light of domain-specific limitations, and the awareness of different possibilities of error is not merely epistemic but also normative.

Ideally, cross-disciplinary model transfers produce novel research in the target domain while simultaneously establishing some degree of compatibility with existing bodies of work in that domain. This demands not only a high level of cross-disciplinary communication, but often discipline-specific knowledge about model use, epistemic values and methodological standards, theoretical frameworks and the concepts involved in both source and target domains. This requires a specific kind of expertise. This is why the scientists undertaking this model transfer can be understood in terms of what Collins and Evans (2002, 2007) call "interactional experts". Interactional experts acquire fluency in the conceptual distinctions, inferential norms, problem-framing practices, and evaluative standards that structure expert discourse in both disciplines. This fluency enables them to participate meaningfully in technical conversations, to assess claims, and to mediate between communities with differing epistemic cultures. Crucially, their competence does not rest on the ability to perform the relevant experiments, calculations, or technical operations, but on the ability to understand and deploy the field's discourse in a way that is recognized as competent by "contributory experts."

In the case of model transfer, because such expertise consists of effectively combining knowledge from the source and target domains, it requires some form of cross-disciplinary engagement. In this sense, interactional experts occupy a distinctive epistemic position: they are neither laypersons nor full participants in the specialist practice, but intermediaries capable of translating, aligning, and integrating knowledge across disciplinary boundaries. Their role is especially salient in contexts in which cognitive progress depends on coordinating heterogeneous bodies of expertise, preserving continuity with established research traditions while enabling conceptual or methodological transfer. It is precisely this combination of novelty and compatibility that characterizes the epistemic labor performed by scientists that engage in successful model transfer.

6 Conclusion

Scientific research is characterized by strong disciplinary specialization that often results in highly abstract models tailored to particular target systems. At the same time, a key characteristic of contemporary science is that there is increasing interaction across different fields. One way in which the resulting tension between specialization and domain interaction manifests itself is in the transfer of models across different scientific domains, which—as we have seen—has become a pertinent feature of modern science (e.g., Humphreys, 2004; Knuuttila & Loettgers, 2016, 2023, 2014). While models are typically designed for specific systems and sometimes even taken to represent them, they are also used as flexible tools that can be applied across disciplines, especially when detached from general but field-specific theories. At first glance, such model transfers reflect both the structure of contemporary science and the potential shape of successful interdisciplinary research. To understand how science is organized and how we can ensure its progress, it is therefore informative to study model transfers and how they relate to what we understand as successful interdisciplinary exchange more generally. This paper examined how we can learn from processes of model transfer, particularly the movement of mathematical models between fields, about the variety of ways in which interdisciplinary research is possible alongside increasing specialization

Acknowledgements We thank the ERC MODEL TRANSFER team for their valuable comments on an earlier draft of this article. We also gratefully acknowledge the financial support provided by the European Research Council (ERC). This work is part of a project that has received funding from the ERC under the European Union's Horizon 2021 Research and Innovation Programme (Grant Agreement No. 101043071), entitled Model Transfer and Its Challenges in Science: The Case of Economics.

Competing Interests The authors have no conflicts of interest to declare that are relevant to the content of this chapter.

References

Andersen, H. (2022). Scientific progress and interdisciplinarity. In *New philosophical perspectives on scientific progress* (pp. 374–391). Routledge.

Bakeev, M. (Forthcoming). Model transfer, the challenge of tractability standards and methodological embedding. *Philosophy of Science.*

Becker, G. (1960), An economic analysis of fertility. In G. B. Roberts (Ed.), *Demographic and Economic Change in Developed Countries* (pp. 209–240). Princeton University Press.

Bradley, S., & Thébault, K. P. Y. (2019). Models on the move: Migration and imperialism. *Studies in History and Philosophy of Science Part A, 77*, 81–92.

Boyd, R., & Richerson, P. J. (1985). *Culture and the evolutionary process.* University of Chicago Press.

Coleman, J. S. (1990). *Foundations of social theory.* Harvard University Press.

Coleman, J. S. (1993). The impact of Gary Becker's work on sociology. *Acta Sociologica, 36*, 169–178.

Coleman, J. S., & Fararo, T. J. (Eds.). (1992). *Rational choice theory: Advocacy and critique.* Sage.

Collins, H. M., & Evans, R. J. (2002). The third wave of science studies: Studies of expertise and experience. *Social Studies of Sciences, 32*(2), 235–296.

Collins, H. M., & Evans, R. J. (2007). *Rethinking expertise.* The University of Chicago Press.

Dragulescu, A. & Yakovenko, V. M. (2000). Statistical mechanics of money. *The European Physical Journal B, 17*(4), 723–729.

Dupré, J. (1994). Against scientific imperialism. *PSA: Proceedings of the Biennial Meeting of the Philosophy of Science Association, 1994*(2), 374–381.

Fogarty, L., Kandler, A., Creanza, N., & Feldman, M. W. (2024). Half a century of quantitative cultural evolution. *Proceedings of the National Academy of Sciences of the United States of America, 121*(48), Article e2418106121.

Grüne-Yanoff, T. (2011). Models as products of interdisciplinary exchange: Evidence from evolutionary game theory. *Studies in History and Philosophy of Science, 42*, 386–397.

Grüne-Yanoff, T. (2015). Models of temporal discounting 1937–2000: An interdisciplinary exchange between economics and psychology. *Science in Context, 28*(4), 675–713.

Grüne-Yanoff, T., & Mäki, U. (2014). Introduction: Interdisciplinary model exchanges. *Studies in History and Philosophy of Science, 48*, 52–59.

Henrich, J. (2015). *The secret of our success: How culture is driving human evolution, domesticating our species, and making us smarter.* Princeton University Press.

Herfeld, C. (2024). Model transfer in science. In T. Knuuttila, N. Carrillo, & R. Koskinen (Eds.), *The Routledge handbook of philosophy of scientific modeling.* Routledge.

Herfeld, C., & Doehne, M. (2019). The diffusion of scientific innovations: A role typology. *Studies in History and Philosophy of Science Part A, 77*, 64–80.
Herfeld, C., & Lisciandra, C. (2019). Special issue on 'Knowledge Transfer and Its Contexts'. *Studies in History and Philosophy of Science Part A, 77.*
Herfeld, C., & Peruzzi, E. (2026). Translation as a modification strategy in model transfer. *Synthese.*
Hesse, M. B. (1966). *Models and analogies in science.* University of Notre Dame Press.
Humphreys, P. (2002). Computational models. *Philosophy of Science, 69*(S3), S1–S11.
Humphreys, P. (2004). *Extending ourselves: Computational science, empiricism, and scientific method.* Oxford University Press.
Humphreys, P. (2019). Knowledge transfer across scientific disciplines. *Studies in History and Philosophy of Science Part A, 77*, 112–119.
Klein, J. T. (2010). A taxonomy of interdisciplinarity. In R. Frodeman, J. T. Klein, & C. Mitcham (Eds.), *The Oxford handbook of interdisciplinarity* (pp. 15–30). Oxford University Press.
Knuuttila, T., & Loettgers, A. (2013). Synthetic biology as an engineering science? analogical reasoning, synthetic modeling, and integration. In H. Andersen, D. Dieks, W. Gonzalez, T. Uebel, & G. Wheeler (Eds.), *New Challenges to Philosophy of Science. The philosophy of Science in a European Perspective* (vol. 4). Springer.
Knuuttila, T., & Loettgers, A. (2014). Magnets, spins, and neurons: The dissemination of model templates across disciplines. *The Monist, 97*, 280–300.
Knuuttila, T., & Loettgers, A. (2016). Model templates within and between disciplines: From magnets to gases—And socio-economic systems. *European Journal for Philosophy of Science, 6*, 377–400.
Knuuttila, T., & Loettgers, A. (2023). Model templates: Transdisciplinary application and entanglement. *Synthese, 201*, 200.
Lisciandra, C. (2025). Explanatory norms and interdisciplinary research. *Synthese, 205*(2), 1–19.
MacLeod, M., Merz, M., Mäki, U., & Nagatsu, M. (2019). Investigating interdisciplinary practice: Methodological challenges (Introduction). *Perspectives on Science, 27*(4), 545–552.
Mäki, U. (2009). Economics imperialism: Concept and constraint. *Philosophy of the Social Sciences, 39*(3), 351–380.
Mäki, U. (2013). Scientific imperialism: Difficulties in definition, identification, and assessment. *International Studies in the Philosophy of Science, 27*(3), 325–339.
Mäki, U. (2016). Philosophy of interdisciplinarity: What? Why? How? *European Journal for Philosophy of Science, 6*, 327–342.

Mäki, U., Walsh, A., Pinto, M. F. (Eds.). (2017). *Scientific imperialism: Exploring the boundaries of interdisciplinarity*. Routledge.
McElreath, R., & Boyd, R. (2007). *Mathematical models of social evolution: A guide for the perplexed*. University of Chicago Press.
Pierce, S. J. (1999). Boundary crossing in research literatures as a means of interdisciplinary information transfer. *Journal of the American Society for Information Science, 50*(3), 271–279.
Porter, A. L., Cohen, A., Roessner, J. D., & Perreault, M. (2007). Measuring researcher interdisciplinarity. *Scientometrics, 72*(1), 117–147.
Rafols, I., & Meyer, M. (2010). Diversity and network coherence as indicators of interdisciplinarity: Case studies in bionanoscience. *Scientometrics, 82*(2), 263–287.
Repko, A., & Szostak, R. (2020). *Interdisciplinary research: Process and theory*. 4th ed. Sage.
Rice, C., & Smart, J. (2011). Interdisciplinary modeling: A case study of evolutionary economics. *Biology & Philosophy, 26*, 655–675.
Shapley, L. S. (1953). A value for n-person games. In H. W. Kuhn & A. W. Tucker (Eds.), *Contributions to the theory of games, Volume II* (Annals of Mathematics Studies, Vol. 28, pp. 307–317). Princeton University Press.
Shapley, L. S. (1955). *Markets as cooperative games*. RAND Corporation Technical Report P-629.
Shapley, L. S. (1959). The solutions of a symmetric market game. *Annals of Mathematics Studies, 40*, 145–162.
Shubik, M. (1959). Edgeworth market games. *Annals of Mathematics Studies, 40*, 267–278
Von Neumann, J., & Morgenstern, O. (1944). *Theory of games and economic behavior*. Princeton University Press.
Wagner, C. S., Roessner, J. D., Bobb, K., Klein, J. T., Klein, K. W., Boyack, J., Keyton, I., Rafols, I., & Börner, K. (2011). Approaches to understanding and measuring interdisciplinary scientific research (IDR): A review of the literature. *Journal of Informetrics, 5*(1), 14–26.

Part 3: Science in the Societal and Political Sphere

Three Problems for Predictive Policy Advice

Simon Lohse and Philippe van Baßhuysen

1 Introduction

Scientific policy advice offers a systematic method to enhance the rationality, trustworthiness, legitimacy and (societal) responsiveness of policymaking, especially when faced with multifaceted problems, such as climate change, or global epidemics. Today, it can be seen as an integral element in progressive legislation, government work and policymaking in general and is used as a matter of course in almost all policy areas. For instance, the German Council of Economic Experts, already set up in 1963, is "[…] mandated with the task of providing an impartial expert view in the form of periodic assessments of macroeconomic developments in Germany, thus helping economic policymakers and the general public

S. Lohse (✉)
Institute for Science in Society, Radboud University, Nijmegen, The Netherlands
e-mail: simon.lohse@ru.nl

Faculty of Humanities, University of Johannesburg, Johannesburg, South Africa

P. van Baßhuysen
Centre for Ethics and Law in the Life Sciences, Leibniz University Hannover, Hannover, Germany
e-mail: philippe.van.basshuysen@cells.uni-hannover.de

J. Hinrichsen et al. (eds.), *Ethics of the Sciences*,
https://doi.org/10.1007/978-981-92-2821-8_9

to make informed decisions".[1] It is considered to be one of the federal government's most important advisory boards. In the Netherlands, the Scientific Council for Government Policy (WRR) advises the government "on strategic issues that are likely to have important political and societal consequences"[2] in the form of policy briefings, reports and other publications since 1976. These "strategic issues" include environmental pollution and nuisance, artificial intelligence and migration policy, among many other things.

Scientific policy advice can be realized in different institutional arrangements. These include departmental research within ministries and other government bodies (e.g. "*Ressortforschung*" at the Robert-Koch-Institute in Germany), expert commissions that provide policy advice on a specific issue (often used in educational policy), councils of experts that are established on a more long-term basis (see above), more informal ad hoc expert hearings for emerging issues (COVID-19 expert consultations around the world), and policy advice by NGOs, such as think tanks (for a comprehensive overview, see Weingart & Lentsch, 2008). In most of these settings, scientific policy advice can take different forms (Sanderson, 2002). First, it can provide *input for objective setting*, e.g. when legislators are still uncertain regarding the regulatory target in environmental protection policy. Second, policy advice can have the goal to *monitor* a specific policy intervention, say after introducing a new type of tax incentive for businesses. Third, it can serve as a *legitimization strategy* for political decision-makers and civil servants, e.g. through the "epistemization" of political issues or via cherry-picking of scientific evidence (Bogner, 2021; Strassheim & Kettunen, 2014). Finally, policy advice can be *predictive* in nature to guide policy decisions in a specific field. This can be done, for example, through quantitative forecasts of economic growth rates that are regularly conducted around the world to inform economic policy, qualitative risk assessments of new biotechnologies (e.g. gene drive technologies) to guide environmental policy, or in the form of conditional scenario projections. Epidemiological model projections during the COVID-19 pandemic provide numerous examples for the

[1] https://www.sachverstaendigenrat-wirtschaft.de/en/about-us/objectives.html [31 October 2024].

[2] https://english.wrr.nl/about-us [31 October 2024].

latter. Modelers often aimed to project different scenarios for transmission dynamics, depending on certain baseline assumptions and possible policy interventions. Another mode of this type of policy advice aims to predict the sometimes long-term impact of different policy measures in order to provide decision-makers with guidance on which measures to implement, for instance, in climate policy. This mode of policy advice is most clearly aligned with the ideal of evidence-based policy, aiming to make policy more rational through science (Cartwright & Hardie, 2012).

In this article, we seek to make progress on *predictive* policy advice by analyzing three key problems associated with it and discussing their interplay as well as possible ways of coping with them. We will do this using examples from recent *external scientific* policy advice. We will thus not discuss policy advice that is based within state institutions, or non-scientific forms of policy advice (although we assume that the problems and issues identified here are also relevant in other institutional constellations). The first problem revolves around epistemic pluralism and the question which disciplines should—and indeed can—be drawn upon in providing input for predictive policy advice. The second problem concerns navigating inherent but often implicit value influences in predictive policy advice, which can hardly (if ever) be avoided. Third, we discuss the issue of model performativity, namely the fact that some models do not only predict but influence the world, which presents us with deep epistemic and ethical problems. While each of the three problems has been discussed in the literature, our contribution is novel as we explore connections between these issues, which haven't to date been considered. As we will argue, each of the three issues is interwoven with the others in intricate ways, which may reinforce the challenges they pose for practitioners seeking to provide reliable and trustworthy policy advice. Scrutinizing these intricate connections will not only serve to deepen the philosophical analysis, but it will also open up ways to better deal with the problems and thus to better position scientific policy advice for the future.

2 Lack of Epistemic Pluralism Generates Imbalanced Input for Policy Advice

There is broad agreement that intra- and interdisciplinary pluralism is useful for policy advice.[3] (Bschir & Lohse, 2022; Mormina et al., 2024). Here we give three arguments that focus on the specific advantages for *predictive* policy advice. First, a multitude of perspectives avoids myopia and facilitates a rich description as well as a better prediction of real-life problems that are complex and allow for several framings. The climate crisis, for instance, has many physical, biological, economic, and social aspects that will need to be considered to guide policy-makers. In context like these, it is also important to gain a better understanding of the possible unintended side-effects of policy measures in the spirit of a more comprehensive risk assessment, which can be better-achieved through a variety of perspectives.[4] Further, every scientific approach is fallible, i.e. although successful in predicting certain aspects of reality, it will mislead or fail in predicting others (Popper 1992 [1992]). In light of this, the triangulation of different scientific approaches can increase the robustness of predictions for policy. During the COVID-19 pandemic, modelers frequently applied this methodological insight by triangulating different models (e.g. combined SIR models with Bayesian parameter inference, see Dehning et al., 2020) in an attempt to increase the robustness of their predictions. Finally, alternative approaches can reveal shortcomings of established perspectives and methods by bringing to light problems that would remain invisible without these alternatives (Feyerabend, 1999 [1999]). To use the pandemic as an example again, social network researchers argued that the prevalent epidemiological predictive models systematically underestimated interaction network effects in the spread of the virus (Bellingeri et al., 2024; Herrmann & Schwartz, 2020; Manzo, 2020)—a limitation that only becomes apparent against the background of alternative (in this case: network-based) models.

[3] Extra-scientific perspectives can also be important. However, we do not address this point in this paper. Furthermore, a complementary line of work focuses on the importance of pluralism in science education and communication, as discussed by Nerhing et al. in this volume.

[4] This also applies to unintended and undesirable effects of policy measures based on scientific predictions that have turned out to be false—a scenario that must be considered fairly typical given the error-prone nature of predicting complex processes, see our next point (many thanks to Karim Bschir for making us aware of this aspect).

A certain degree of epistemic pluralism is thus a clear advantage for predictive policy advice and should be encouraged. However, this recommendation faces several challenges (see Bschir & Lohse, 2023). With regard to predictive policy advice, there is one question in particular that has hardly been discussed, but, in our view, deserves critical reflection, namely the question which disciplines should—*and indeed can*—be drawn upon in providing input for predictive policy advice. The arguments outlined above seem to suggest that, depending on the problem policy-makers face, we should strive for a relatively broad epistemic pluralism, i.e. one that includes a variety of intra- and interdisciplinary perspectives and approaches. However, in addition to pragmatic difficulties in implementing such a form of policy advice, field-specific differences in predictive modes and capabilities, pose a problem for the realization of this ideal. If we look at natural and social science fields, for example, it becomes clear that their capabilities for prediction are quite different. At least to date, complex social processes (e.g. social movements) can hardly be reliably predicted, and certainly cannot be predicted with quantitative accuracy, whereas this is possible, at least to a limited extent, for certain physical systems (e.g. large scale climate patterns). Differences in predictive capabilities in different fields are not limited to natural vs. social sciences, but also occur within these scientific domains, and they are related to several factors that we can only hint at here (cf. Lohse & Greve, 2017): We have a fairly stable knowledge of regularities and mechanisms with regard to certain physical and biological phenomena in the world, whereas we lack this knowledge with regard to other phenomena (including many social and ecological processes). In addition, there may be differences in the degree of causal complexity and relevant contextual factors responsible for a phenomenon that are either manageable or present us with insurmountable epistemic hurdles; and there may be specific interaction effects between model and target that limit our predictive success, especially in social domains (see below).

But there are not only object-related aspects and interaction effects that are relevant here. There are also differences in research traditions in different fields and in *assumptions* about certain fields. Taking the social sciences as an example, economics is much more focused on quantitative predictions than large parts of political science and sociology, which make their raison d'être much less dependent on the ability to predict. This tendency is also reflected in a different focus on (possibly fundamental) limitations of and systematic biases in the prediction of complex socio-economic processes, which are typically taken more seriously in political science and sociology than in economics—although this may also vary with different sub-fields and orientations in the extremely multi-paradigmatic social sciences. For example, rational choice approaches in higher education research are more open to making predictions than hermeneutic and practice-theoretical paradigms in the sociology of culture, which may also result in certain self-selection effects with regard to the willingness to provide predictive policy advice.[5] Finally, a lack of causal knowledge due to methodological restrictions may also be responsible for differences in predictive capacity: While in certain fields it is possible to isolate and test causal factors in experimental setups, in other areas this approach is hardly possible for ethical and/or practical reasons, making the identification of causal mechanisms and regularities that much harder.

It should be noted that existing differences in predictive abilities may sometimes be overemphasized to such an extent that it is assumed, e.g. by policy-makers looking for advice, that social scientific approaches (including qualitative paradigms) are *completely* incapable of predictions. Although this is incorrect, it is nevertheless effective in terms of social attribution processes, which can lead to the marginalization of certain sub-fields in predictive policy advice.

What do these remarks imply? At least three implications seem salient. *First*, given the differences in predictive capabilities in different fields, predictive policy advice for multifaceted problems may lead to imbalanced input for policy-makers, as it will only predict certain aspects of such problems and highlight only predictable consequences, while excluding others from the spectrum. This will often (but not only) affect societal aspects and consequences. An example for this issue are predictions of the effects of the COVID-19 pandemic in many countries that tended to focus on numbers of infected people and ICU requirements, while there were no equally detailed projections for effects on educational inequality and social disintegration.[6] It is safe to assume that such an imbalance has (had) an influence on policymaking and can lead to a specific kind of "prediction bias", i.e. greater attention to those aspects that can be predicted, even if

[5] Sub-field specific differences regarding predictions can also be found in the life sciences (Hoyningen-Huene, 2013, p. 78ff).

[6] See Lohse and Canali (2021) for other factors that led to a certain marginalization of the social sciences during the pandemic.

other aspects of a problem might be of equal or even greater importance for public welfare. *Second*, such an imbalance can also be expected when there are predictions from different fields that concern different aspects of a phenomenon, but these differ fundamentally in level of detail. Consider cases in which, on the one hand, certain economic effects of a technological innovation (say effects of novel AI tools on productivity: Noy & Zhang, 2023) are predicted quantitatively (whether these predictions are reliable or not) and, on the other hand, there are only scenario projections on possible social disintegration effects that are to be expected but cannot be predicted with specific likelihoods. *Third*, different predictive capabilities will have an impact on policy advice especially when it comes to building "what-if" machines. As Nancy Cartwright and Jacob Stegenga convincingly argue (2011), we need more than general principles, mechanisms and extrapolations from other contexts if we are interested in what really works in evidence-based policy, i.e. in determining the likely effects of a policy instrument to be implemented in a specific context. Building what-if machines requires not only knowledge of a specific causal mechanism that can lead to a desired effect, but also knowledge of a variety of contextual factors that ultimately co-determine this mechanism and its effects. Using a stylized example, scientists may be able to model the emission reduction and economic effects of incentives to buy and use e-cars with reasonable precision—but, of course, *ceteris paribus*: cultural factors also play a role here, making a switch to the e-car in a given society more or less likely. However, it may be considerably more difficult to capture these contextual factors and include them in an overall predictive assessment of this policy measure, either because the corresponding modes of action are not sufficiently understood or we do not have sufficient empirical data on these aspects.

This last challenge relates to an additional issue we want to highlight, namely the challenge of knowledge synthesis in the forward-looking assessment of policy interventions. In view of the multifaceted real-world problems focused on here, it would be desirable if scientific policy advice could achieve a certain level of knowledge integration. This would prevent a scenario where various advisory panels from the natural, social sciences etc. predict the expected effects of policy measures in isolation from each other, highlighting different aspects and standing side by side without any kind of epistemic synthesis. Although this will at times probably be unavoidable for various reasons, including political ones, we believe that this "juxtaposition" of predictive policy advice is not ideal. For one thing, policy-makers would have to perform an extremely complex epistemic synthesis themselves, and for another, relevant interaction effects at different levels may be missed in this way. However, if we want to come closer to an ideal of pluralistic knowledge integration, many challenges must be addressed. To name just two (see Bschir & Lohse, 2023 for more): How should we deal with different epistemological affinities and standards of different fields when integrating knowledge? How should different modes of prediction from different fields (quantitative prediction, qualitative scenario projects, etc.) be assessed and weighted? Answers to these questions may be developed by learning from best-practice approaches, such as the sophisticated integrated assessment modeling of climate scientists that feed into the IPCC reports (Rogelj et al., 2018). However, this kind of synthetic assessment modeling is extremely resource and time intensive and will not be a viable model for many other policy contexts.

As an interim conclusion, it can be said that our discussion brings to light a number of epistemic and practical challenges for (desirable) pluralistic predictive policy advice that have not yet been adequately addressed. These challenges have normative relevance as they can lead to imbalanced policies due to systematic predictive blind spots.

3 Problematic Values Influences May Jeopardize the Legitimacy of Predictive Policy Advice

It is widely accepted in the philosophy of science that scientific policy advice cannot be completely value-free (Douglas, 2009; Steele, 2012). Rather, it always has to deal with and navigate the influence of non-epistemic (i.e. political, cultural, and ethical) values.[7] This includes predictive policy advice, as can easily be shown from three brief examples. (1) As indicated above, it will frequently not be clear how a complex policy problem needs to be framed. Is the nitrogen crisis in the Netherlands (the extremely high environmental pollution with nitrogen in large parts of the country) primarily an environmental crisis—or should we also understand it as an economic, political and/or social problem? With regard to the question of value influences, this observation now takes on

[7] In the following, "values" will refer to non-epistemic values, unless otherwise indicated.

a different meaning. Depending on how the problem is articulated and framed, other sub-problems may or may not come into focus, such as the extent to which effects of the nitrogen crisis on farming or the housing market will be part of scenario projections for different policy options. However, which of these issues should be given (particular) consideration in predictive policy advice requires at least some kind of prioritization—and therefore: value judgments. These value judgments will only rarely be specified by politicians since they will often lack a sufficiently deep understanding of the problem space, at least in cases of multifaceted problems. (2) Values will often have an implicit influence on predictive policy advice, for example when they influence modeling assumptions; say, when, due to political assumptions of (ir)relevance, the effects of unpaid care work are neglected in economic forecasts or when only aggregated but not distributed effects of the climate crisis are predicted. Another good example can be found in the projection of ICU capacity in Switzerland during the COVID-19 pandemic. The models assumed that elderly people in nursing homes would transfer to hospitals if their breathing was too severely affected by a COVID-19 infection. However, it turned out, through social surveys, that many elderly people wanted to remain in their nursing homes instead of being transferred to intensive care units. This decision was not anticipated by the modelers precisely because the value of "protecting life" was integrated into the modeling as an unquestioned assumption (Bschir & Lohse, 2022). (3) In practice, scientific facts often have to be enriched with ethical, political and pragmatic considerations when they are passed on to political decision-makers (Bschir & Lohse, 2023). Consider the policy briefings of the IPCC (Intergovernmental Panel on Climate Change). For one thing, it must be decided when prediction models are considered sufficiently robust to be included in a report to decision-makers; in other words, when an adequate epistemic threshold has been reached. The determination of this threshold depends on the seriousness of possible predictive errors (cf. Douglas, 2000 on "inductive risks"): Which would be riskier for political action, a prediction that may overestimate or underestimate certain risks? When should scientists consider a prediction model as corroborated *enough* for policy advice given the drastic costs that false predictions could have (e.g. inaction where massive action would have been required)? It is hard to see how one could answer these questions without relying on non-epistemic value judgment—whether explicitly or implicitly. In addition, in the face of complex problems such as the climate crisis, certain aspects must always be selected as relevant in the projection of future scenarios (this issue is related to the framing problem mentioned above). Which ones will be selected, however, depends, among other things, on considerations regarding the political and pragmatic addressability of identified challenges and, in this example, tacitly on questions of intergenerational justice. If such a value-laden emphasis on certain aspects were to be completely avoided, this would not only be extremely difficult to realize—arguably, it would undermine the usefulness of policy advice for policy-makers.

We cannot go into all the challenging aspects that are relevant in the context of value influences in scientific policy advice, but want to focus on two issues.[8] First, we explore the issue of value transparency to highlight certain shortcomings in current practice. Subsequently, we discuss the question of how we should select "the right values" to guide policymaking.

A major problem in dealing with value questions in predictive policy advice is that they often play an implicit role in the background in scientific concepts and decisions and are not openly addressed (Alexandrova, 2017). This makes navigating them an extremely difficult task. This is a very profound problem, because our value judgments are often opaque, even to ourselves. However, we want to emphasize that the intransparency of values in policy advice may (perhaps paradoxically) be caused, at least to some extent, by the value-free ideal—an ideal that is very much alive and kicking in the sciences and at the science policy-interface. Many scientists are adherents of this ideal and are keen to avoid non-epistemic value influences—and also: the *impression* of value influences—in their research and scientific predictions (see Brown, 2024) This could be observed time and again in the positioning of scientific experts in the public sphere during the coronavirus crisis. On the one hand, this is of course advantageous (and Max Weber would agree)—in particular, when scientists clearly distinguish between their own value judgments and the scientific basis of a prediction in public communication (Carrier, 2021). But it also harbors the danger that the actual influence of values is downplayed or ignored, for "that which must not, cannot be". In addition, there is hardly any suitable vocabulary available—at least to date—to communicate the role of

[8] Among other things, we do not discuss questions of deliberate value prioritization in the selection of scientific disciplines or individual scientists *by policy-makers*, for example in order to steer political decisions in a certain direction.

values in scientific policy advice in the public sphere. In fact, transparent communication is quite difficult since a simplistic ideal of science—as disinterested and purely objective—prevails in the public sphere, which leaves little room for a more refined discourse on the complexities of value influences in scientific research and policy advice (see Brown, 2024; Bleidorn in this volume discusses the intricacies of science communication in a post-truth society). In fact, too much openness and transparency may at times be detrimental in public communication as it can be misunderstood or exploited by malicious actors (John, 2017; Lewandowsky et al., 2015; for empirical results on ethical norms of science, see Hönnige and Nyhuis in this volume).

Even when scientists are transparent about the values that guided a specific piece of policy advice, this does not mean that they are *justified* in relying on those values, as opposed to others. Distinguishing legitimate from illegitimate value influences has been called the "new demarcation problem" (Holman & Wilholt, 2021) and is subject of a growing literature in the philosophy of science (e.g. Douglas, 2009; Intemann, 2015; Jebeile & Crucifix, 2021). A major problem for the purposes of predictive policy advice, however, is that no consensus has emerged from this literature as to what should be regarded as the proper demarcation criterion, and accordingly, what norms and duties this might imply for scientists as policy advisors. Should values only have *indirect* roles to play in internal stages of research (Douglas, 2009)? Are value judgments legitimate when they promote democratically endorsed aims of research (Intemann, 2015)? Or do they need to be in alignment with key principles of public institutions (Hilligardt, 2024)? Ironically, what might be to blame for this lack of consensus is, at least in part, again the value-free ideal—or, more precisely: a particular interpretation of it, because this has enticed theorists to conceptualize all contextual factors that may influence science as values, which may be too coarse-grained (see Hilligardt, 2022). To move forward, we need a richer conceptualization and evaluation of all the contextual factors that may influence scientific research and policy advice; but scientists cannot, at present, rely on helpful guidance from the philosophy of science concerning the question of which contextual influences are legitimate.

Summing up, values and other contextual influences inevitably influence predictive policy advice, but current practice exhibits considerable shortcomings concerning value transparency and the open communication of (frequently unavoidable) value influences. Furthermore, philosophers of science have had a hard time clearly articulating when these influences might be legitimate and when not, thus failing to provide guidance to stakeholders as to how value influences should be managed. In the absence of clear guidelines, there is a danger that some cases of predictive policy advice will rely on contentious values which may jeopardize its legitimacy. This is particularly problematic as predictive policy advice often shapes policy-making and individual behavior, as we will explore next.

4 Performativity Poses Epistemic and Ethical Problems for Predictive Policy Advice

There is broad consensus that models and theories, especially in the social sciences, often influence their own targets once they are disseminated or made public. This means that the very phenomena that we seek to understand by means of our models or theories are potentially changed through their deployment, sometimes in ways that may limit our understanding of the phenomena in question. For instance, polls, when made public, create expectations about who is likely to win or to lose an election, which may in various ways influence voters' behavior: some people may be more likely to vote for a candidate when they expect them to lose ("underdog effect", see Simon, 1954), while others may be more likely to vote for someone likely to win ("bandwagon effect", ibid.). Where there are strong bandwagon effects, a prediction of a candidate winning the election may thus become a "self-fulfilling prophecy", whereas strong underdog effects may generate "self-defeating" predictions. Polls are frequently plagued by these effects (e.g. Westwood et al., 2019 in the context of the 2016 US elections), yet performativity is a feature of almost any attempt at predicting social events. Performative effects seem to pose a severe problem for the credibility and epistemic import of predictions in the social sciences as we would expect them to be true or false *irrespective of their dissemination or deployment*, yet by affecting the expectations of the people whose actions they are supposed to describe, predictions often perform their target.

While performativity is a well-known feature of the social sciences, it also poses specific problems in the context of predictive policy advice, which are to date understudied. Here, the fundamental riddle is that predictions are made with the aim of informing important policy decisions, but these decisions threaten to impair the quality of the issued predictions. At the same time, however, it would seem that only high-quality predictions should form the basis of policy advice. Reflecting on this riddle is not a mere theoretical exercise in the philosophy of science, as the COVID-19 pandemic (among many other cases) forcefully shows. Here, dire projections of cases and deaths from the disease led to lockdown decisions and other non-pharmaceutical interventions, which in turn decreased case and death numbers, thus making some of the initial projections appear to be overblown, at least when interpreted as unconditional predictions (van Basshuysen et al., 2021). Since both the models and the decisions based on them have been subject of controversial debate, this brings to the fore the question of how, in general, we should assess the quality of performative models and use them to inform policy.

When seeking to make progress on this question, we are immediately faced with both epistemic and ethical obstacles. Among the epistemic problems is the fact that, ex ante, it is often difficult to predict what kinds of performative effects will materialize. Historically, this has generated skepticism about predictions or classifications in the social sciences, with theorists from Oskar Morgenstern (1928) to Ian Hacking (1995, 1999) viewing these effects as posing a severe epistemic threat. This pessimism may not be fully justified as social scientists have made progress on the possibility of incorporating performative effects in their models (Grunberg & Modigliani, 1954; Simon, 1954; in the context of epidemiological models, see Avery et al., 2020). Yet, this is a conceptual possibility that often evades us in practice, as there is typically significant uncertainty surrounding the correct response function for the deployment of a given model. It follows that performativity presents scientists and decision-makers with the difficult problem of gauging how reliable a piece of predictive policy advice will turn out to be, given that it may generate performative effects that are hard to foresee and to prevent. What is more, even from an ex post perspective, it is often difficult to evaluate the quality of a prediction in the presence of performative effects: when comparing the prediction with the actual outcomes, the prediction would appear to be flawed (in self-defeating cases), or "right for the wrong reasons" (in self-confirming cases). A better strategy might be to compare the prediction not with the actual outcomes but with the counterfactual outcomes that would result from suppressing the performative effects, which, however, adds a layer of complexity to the evaluation of the quality of predictions.

Apart from the evaluation of the epistemic performance of predictive policy advice, some have argued for a broader approach in which performative effects themselves may have a role to play in model evaluation. According to an adequacy-for-purpose view, the quality of a model should be assessed relative to a purpose, where this can be epistemic or practical (see Parker, 2020). Adopting such a view, some have argued that we may in some cases view the generation of performative effects as a possible practical model purpose, whose achievement should matter for model evaluation (van Basshuysen, 2023). How might this work in practice? For instance, van Basshuysen et al. investigate the possibility that early COVID-19 predictions led to increased social distancing and thus, fewer infections and deaths (2021). Assuming that this is a result that can be regarded as socially desirable, they argue that performativity here provided a way in which models were successful, even though, at the same time, it made the predictions less accurate. In this and in similar cases, performativity can thus drive a wedge between the epistemic purposes of science and social/ethical goals. This leaves open a range of questions that have only begun to be asked in the philosophy of science: How should we evaluate epistemic and performative model purposes, especially when they conflict with each other? Should epistemic purposes enjoy "lexical priority" over performative purposes, or should we grant science a more active role in society? Might it sometimes be justifiable for scientists to "steer" the course of events, or should this be rejected as undemocratic as scientists are not democratically elected to do so (Winsberg & Harvard, 2022)? Whom ought decision-makers trust when there are conflicting predictions pulling the ensuing effects "in different directions"? As with our discussion concerning values in the previous section, there is a dire need to learn to evaluate performative models and to demarcate legitimate from illegitimate instances of performativity, in order to be able to manage these effects (van Basshuysen, 2025). Yet there is no agreement among philosophers of science as to how to do this and current attempts at managing these effects fall short of providing well-developed strategies to manage the practical impacts of these effects (see Khosrowi, 2023; van Basshuysen, 2023).

5 Concluding Discussion

Each set of issues we have discussed so far individually poses formidable challenges for predictive policy advice. Yet in this final section, we wish to highlight important ways in which the issues discussed here can be inextricably interrelated, thus further complicating the task of providing and acting on this kind of policy advice. For instance, values may often be field-specific, as different fields tend to highlight different aspects of an issue, which may differ in their normative valence. Depending on which field "gets heard" and acted upon, different sets of values will then exert influence on policy and different performative effects may accordingly materialize. As we will argue, these links present us with a double-edged sword: while failing to properly handle one set of issues can exacerbate the problems posed by the others, sensible approaches to one set of issues can also pave the way for a more rational approach to the others (see Fig. 1). In the following, we discuss two of the key links in turn. For reasons of space, we will limit ourselves to a few aspects of the connections between pluralism/ value influences and pluralism/ performativity. Further aspects of these connections, as well as questions of value influences/ performativity, shall be addressed in future work.

First, the question of what fields should be drawn upon in predictive policy advice (i.e. the set of issues discussed in Sect. 2 on pluralism) can considerably complicate the issue of value-ladenness (i.e. the set of issues discussed in Sect. 3). For instance, the inclusion of social scientific perspectives during the COVID-19 pandemic presented policy-makers with *more* values to be considered, such as societal inequalities and intergenerational fairness, whereas the early tendency to focus primarily on epidemiological perspectives presented policy-makers with the relatively (!) simple task of considering the immediate health impacts of letting the virus spread vs. trying to contain it. While widening perspectives can in this way considerably complicate the task of reaching policy decisions, pluralism can also provide the basis for a more rational approach toward decision-making. For instance, diversity of perspectives can help to identify taken-for-granted but debatable value assumptions of certain models. This is precisely what happened with the above-mentioned epidemiological model projections in Switzerland whose underlying value-laden assumptions became only visible through a contrasting perspective.

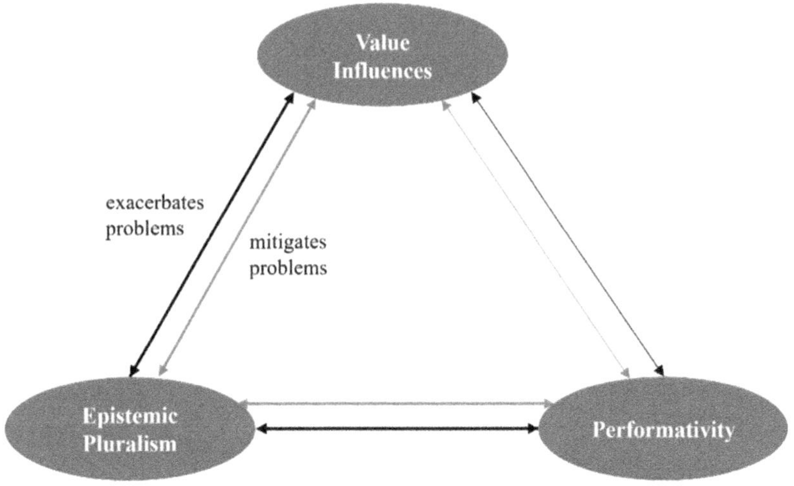

Fig. 1 Each set of issues can exacerbate the problems posed by the others (black arrows), yet sensible approaches to one set of issues may also mitigate the severity of the problems posed by the others (green arrows). Thick arrows are discussed in the main text

Unearthing hidden value judgments likes these via contrasting epistemic perspectives and approaches is a necessary step for discussing their legitimacy and prioritization (see de Melo-Martin, 2024 on this point).

The second link we wish to highlight is that, which fields and sources of evidence we draw upon in policy advice (as discussed in Sect. 2) can influence what performative effects will ensue (as discussed in Sect. 4), the anticipation of which might in turn influence policy-makers' selection of evidence they base their decisions on. Continuing the above example, we could imagine (whether or not this is true is irrelevant for our point) that the increasing inclusion of social scientific evidence, for instance on the unintended consequences of lockdown decisions, led to more skepticism about policy measures, and lower compliance with social distancing measures. It would seem to us to be illegitimate to exclude certain views from consideration in policy advice because they might arouse undesirable performative effects, yet there is a question of how to deal with such opposing views. More generally, when different disciplines, or individual scientists holding different views and values, may pull performative effects "in different directions", the question of which views to prioritize in policy and individual decisions comes to the fore. Pluralism in predictive policy advice might again provide a possible solution for this challenge. It can mitigate overly concentrated "performative power", that is, the ability of a single discipline, approach or research group to "steer" policy and individual behavior concerning important societal problems. It might be useful to have a plurality of approaches that highlight different aspects of a situation and are likely to generate different types of performative effects. This could establish a kind of democratic checks-and-balances mechanism at the level of predictive policy advice. How exactly such a mechanism can and should be designed is, of course, an open question and one that is extremely difficult to answer.

Our discussion shows that it will be important to promote a less idealized public understanding of predictive policy advice that pays more attention to the discussed conceptual problems and challenges. Further, it suggests that we will need to (a) pro-actively institutionalize more sophisticated ways of knowledge integration in predictive policy advice—as the dominant model of a mere juxtaposition of disciplines in policy advice is reaching its limits and may even be harmful—and (b) explore ways forward concerning the ethical and epistemic problems posed by performative and value-laden science. Rather than solving these complex issues, our goal was to increase awareness: there is a dire need to further explore and reflect on these issues and their interplay, in order to be able to meet the challenges of tomorrow's science and politics.

Acknowledgements The authors wish to thank Kilian Bizer, Karim Bschir, Christoph Hönnige, and the audience of a presentation of this paper at the colloquium of the Centre for Ethics and Law in the Life Sciences (Leibniz University Hannover) for constructive feedback and suggestions.

Competing Interests The authors have no conflicts of interest to declare that are relevant to the content of this chapter. Van Basshuysen's contribution is funded by the European Union (ERC, MAPS, 101115973). Views and opinions expressed are however those of the authors only and do not necessarily reflect those of the European Union or the European Research Council Executive Agency. Neither the European Union nor the granting authority can be held responsible for them.

References

Alexandrova, A. (2017). *A philosophy for the science of well-being*. Oxford University Press.

Avery, C., Bossert, W., Clark, A., Ellison, G., & Ellison, S. F. (2020). An economist's guide to epidemiology models of infectious disease. *The Journal of Economic Perspectives, 34*(4), 79–104.

Bellingeri, M., Bevacqua, D., Scotognella, F., & Cassi, D. (2024). The critical role of networks to describe disease spreading dynamics in social systems: A perspective. *Mathematics, 12*(6), Article 6.

Bogner, A. (2021). *Die Epistemisierung des Politischen: Wie die Macht des Wissens die Demokratie gefährdet*. Reclam.

Brown, M. J. (2024). For values in science: Assessing recent arguments for the ideal of value-free science. *Synthese, 204*(4), 112.

Bschir, K., & Lohse, S. (2022). Pandemics, policy, and pluralism: A Feyerabend-inspired perspective on Covid-19. *Synthese, 200*(6), 441.

Bschir, K., & Lohse, S. (2023). Taking pluralism seriously: A new perspective on evidence-based policy. *Science and Public Policy*, scad074.

Carrier, M. (2021). What does good science-based advice to politics look like? *Journal for General Philosophy of Science*.

Cartwright, N., & Stegenga, J. (2011). A theory of evidence for evidence-based policy. *Proceedings of the British Academy, 171*, 289–319.

Cartwright, N., & Hardie, J. (2012). *Evidence-based policy: A practical guide to doing it better*. Oxford University Press.

Dehning, J., Zierenberg, J., Spitzner, F. P., Wibral, M., Neto, J. P., Wilczek, M., & Priesemann, V. (2020). Inferring change points in the spread of COVID-19 reveals the effectiveness of interventions. *Science, 369*(6500).

Douglas, H. (2000). Inductive risk and values in science. *Philosophy of Science, 67*, 559–579.

Douglas, H. (2009). Science, policy, and the value-free ideal. *Choice Reviews Online, 47*(04), 47–1955.

Feyerabend, P. (1999[1968]). Outline of a pluralistic theory of knowledge and action. In J. Preston (Ed.), *Knowledge, science, and relativism. Philosophical Papers Vol. 3* (pp. 104–111). Cambridge University Press.

Grunberg, E., & Modigliani, F. (1954). The predictability of social events. *Journal of Political Economy, 62*(6), 465–478.

Hacking, I. (1995). The looping effects of human kinds. In D. Sperber, D. Premack, & A. J. Premack (Eds.), *Causal cognition: A multidisciplinary debate* (pp. 351–394). Clarendon Press/Oxford University Press.

Hacking, I. (1999). *The social construction of what?* Harvard University Press.

Herrmann, H. A., & Schwartz, J.-M. (2020). Why COVID-19 models should incorporate the network of social interactions. *Physical Biology, 17*(6), Article 065008.

Hilligardt, H. (2022). Looking beyond values: The legitimacy of social perspectives, opinions and interests in science. *European Journal for Philosophy of Science, 12*, 58.

Hilligardt, H. (2024). Science as public service. *European Journal for Philosophy of Science, 14*, 45.

Holman, B., & Wilholt, T. (2021). The new demarcation problem. *Studies in History and Philosophy of Science Part A, 91*, 211–220.

Hoyningen-Huene, P. (2013). *Systematicity: The nature of science*. Oxford University Press.

Intemann, K. (2015). Distinguishing between legitimate and illegitimate values in climate modeling. *European Journal for Philosophy of Science, 5*(2), 217–232.

Jebeile, J., & Crucifix, M. (2021). Value management and model pluralism in climate science. *Studies in History and Philosophy of Science Part A, 88*, 120–127.

John, S. (2017). Epistemic trust and the ethics of science communication: Against transparency, openness, sincerity and honesty. *Social Epistemology, 32*(2), 75–87.

Khosrowi, D. (2023). Managing performative models. *Philosophy of the Social Sciences, 53*(5), 371–395.

Lewandowsky, S., Oreskes, N., Risbey, J. S., Newell, B. R., & Smithson, M. (2015). Seepage: Climate change denial and its effect on the scientific community. *Global Environmental Change, 33*, 1–13.

Lohse, S., & Greve, J. (2017). Philosophie der Soziologie. In S. Lohse & T. A. C. Reydon (Eds.), *Grundriss Wissenschaftsphilosophie. Die Philosophien der Einzelwissenschaften* (pp. 543–582). Meiner.

Lohse, S., & Canali, S. (2021). Follow *the* science? On the marginal role of the social sciences in the COVID-19 pandemic. *European Journal for Philosophy of Science, 11*, Artikel 99.

Manzo, G. (2020). Complex social networks are missing in the dominant COVID-19 epidemic models. *Sociologica, 14*(1), Article 1.

De Melo-Martín, I. (2024). Concerns about contextual values in science and the legitimate/illegitimate distinction. *Philosophy of Science, 91*(4), 851–868.

Morgenstern, O. (1928). *Wirtschaftsprognose: Eine Untersuchung ihrer Voraussetzungen und Möglichkeiten*. Springer Vienna.

Mormina, M., Müller, B., Caniglia, G., Engebretsen, E., Löffler-Stastka, H., Marcum, J., Mercuri, M., Paul, E., Pfaff, H., Russo, F., Sturmberg, J., Tretter, F., & Weckwerth, W. (2024). Where to after COVID-19? Systems thinking for a human-centred approach to pandemics. *Humanities and Social Sciences Communications, 11*(1), 1–7.

Nehring, A., Reydon, T., & Bruckermann, T. (). Pluralism about the nature of science: Ethical challenges in science education and communication. In

J. Hinrichsen, K. Bizer, C. Hönnige, et al. (eds.) *Ethics of the sciences: New normative foundations for academic practice*. Palgrave Macmillan.

Noy, S., & Zhang, W. (2023). Experimental evidence on the productivity effects of generative artificial intelligence. *Science, 381*(6654), 187–192.

Parker, W. S. (2020). Model evaluation: An adequacy-for-purpose view. *Philosophy of Science, 87*(3), 457–477.

Popper, K. (1992 [1934]). *The logic of scientific discovery*. Routledge.

Rogelj, J., Popp, A., Calvin, K. V., Luderer, G., Emmerling, J., Gernaat, D., Fujimori, S., Strefler, J., Hasegawa, T., Marangoni, G., Krey, V., Kriegler, E., Riahi, K., Van Vuuren, D. P., Doelman, J., Drouet, L., Edmonds, J., Fricko, O., Harmsen, M.,... Tavoni, M. (2018). Scenarios towards limiting global mean temperature increase below 1.5 °C. *Nature Climate Change, 8*(4), 325–332.

Sanderson, I. (2002). Evaluation, policy learning and evidence-based policy making. *Public Administration, 80*(1), 1–22.

Simon, H. A. (1954). Bandwagon and underdog effects and the possibility of election predictions. *Public Opinion Quarterly, 18*(3), 245.

Steele, K. (2012). The scientist qua policy advisor makes value judgments. *Philosophy of Science, 79*(5), 893–904.

Strassheim, H., & Kettunen, P. (2014). When does evidence-based policy turn into policy-based evidence? Configurations, contexts and mechanisms. *Evidence & Policy: A Journal of Research, Debate and Practice, 10*(2), 259–277.

Van Basshuysen, P. (2023). Austinian model evaluation. *Philosophy of Science, 90*(5), 1459–1468.

Van Basshuysen, P. (2025). Performativity in science: Past and future. *Philosophy Compass*.

Van Basshuysen, P., White, L., Khosrowi, D. & Frisch, M. (2021). Three ways in which pandemic models may perform a pandemic. *Erasmus Journal for Philosophy And Economics, 14*(1).

Weingart, P., & Lentsch, J. (2008). *Wissen, Beraten, Entscheiden: Form und Funktion wissenschaftlicher Politikberatung in Deutschland*. Velbrück.

Westwood, S., Messing, S. & Lelkes, Y. (2019). Projecting confidence: how the probabilistic horse race confuses and demobilizes the public. *SSRN Electronic Journal*.

Winsberg, E., & Harvard, S. (2022). Purposes and duties in scientific modelling. *Journal of Epidemiology & Community Health, 76*(5), 512–517.

The Demand for Science in Politics: A Study of German Election Manifestos

Simon Fink

1 Introduction

Science is increasingly important for policy making (Weingart, 1999). In many policy fields, scientific knowledge is used to define problems, outline policy options, and legitimize chosen policies (Fink, 2025; Weible, 2008). While this "scientification of politics" has a long history in technology-intensive fields (Mintrom, 2013; Reber, 2007), or environmental politics (Garb et al., 2007), scientific assessments and predictions now affect virtually every policy field (van Baßhuysen and Lohse in this volume).

Much of the existing literature focuses on the *influence* of science on public policy (Christensen, 2021, 2023). Studies analyze the role of scientists as policy entrepreneurs (Mintrom, 2013), groups of scientists as parts of policy networks (Fourcade-Gourinchas & Babb, 2002; Haas, 1989; Heclo, 1978; Sabatier, 1988; Weible et al., 2009), or the institutional role of science in policy making (Gornitzka & Sverdrup, 2008; Timmermans & Scholten, 2006). In the larger picture, this development has led to the accusation of "expertocratism" (Frinken & Landwehr, 2022),

S. Fink (✉)
Institut Für Politikwissenschaft, Georg-August-Universität Göttingen, Göttingen, Germany
e-mail: simon.fink@sowi.uni-goettingen.de

J. Hinrichsen et al. (eds.), *Ethics of the Sciences*,
https://doi.org/10.1007/978-981-92-2821-8_10

which in turn may undermine trust in science (Hönnige and Nyhuis in this volume), and poses a thorny ethical problem for the relation between science and society (Hinrichsen et al. in this volume).

Nevertheless, the *demand* for scientific input across policy fields remains under-researched. The current narrative is one of scientification of politics as a broad and universal trend, driven by the growing complexity of societal and technological problems (Head, 2022; Weingart, 1999).

This contribution takes an empirical approach to the scientification of politics. The contribution argues that we should study the actor-centered bases of the demand for science in public policy. Specifically, I contend that *party politics* drive the demand for science in public policy. The argument is that parties use demands for science in their party manifestos to signal to voters that they pursue a responsible, fact-based, and to some extent de-politicized position, while their opponents' position—by way of reversing the argument—is ideology-driven (Cairney, 2021). Consequently, this chapter contributes to the debate on the interplay between societal values and science (Schicktanz and Reydon in this volume) by showing how parties translate societal demands into concrete expectations about the role of science.

As this topic has not yet been studied, I begin by mapping the field with a descriptive study of German electoral manifestos from 2002 to 2021. Based on the Manifesto Research on Political Representation (MARPOR) database (Lehmann et al., 2022), I assess: (1) *which parties* use references to science in their manifestos; (2) in *which policy fields* German parties refer to science; and (3) how these references have *changed over time*.

The results show no universal upward trend in scientific demand across parties or policy fields. The only consistent finding is that all parties regard science as a driver of economic growth. Apart from that, references to scientific studies or scientific advice are time-contingent and issue-specific. There are some patterns: The liberal FDP frequently calls for scientific evidence in environmental and climate-policy fields. The Green party uses references to science with regard to societal issues like discrimination and equality, as does the Left party. Parties of the socio-cultural left tend to demand science-oriented drug policy. But none of the parties appears to use references to science consistently over the years. The overall picture is one of contingent demands for science in some policy fields in some years by some parties that are not repeated in the next election. In essence, the political agenda is volatile and does not conform to a broad trend toward scientification.

This result suggests that the increasing demand for science in policy making seems to be a broad and universal trend on the surface (Head, 2022; Weingart, 1999), but might be driven by a series of highly contingent demands made by parties over time that privilege scientific evidence in one policy field or the other, depending on the circumstances (Weible, 2008). There might be other routes to increasing demand for science in public policy: Ministerial bureaucracies and government agencies may demand scientific input more regularly and over longer horizons (Veit et al., 2017). But the party-political demand for science is a short-term phenomenon and is not easily predicted. In terms of the overarching message of this volume, this means that the boundary between society and science is not only constantly renegotiated as a result of shifting societal demands (Hinrichs et al. in this volume), but also as a result of party-political strategies: Which policy fields are seen as being in need of scientific input may be a result of technical or societal developments, but also a result of vote-seeking parties that see electoral benefits in calling for scientification of some policy fields (but not others).

The chapter is structured as follows. Section 2 outlines the state of research on science and public policy and argues that we know little about the party-political bases of the demand for science. Section 3 discusses the literature on party manifestos and develops a theoretical argument about why parties use references to science in their manifestos. Section 4 discusses the data used. Section 5 contains the empirical analysis. Section 6 concludes.

2 The Context: The Scientification of Public Policy

A growing role for science in public policy has been documented (Christensen, 2021; Weingart, 1999). This diagnosis is prevalent in many different literatures: Political science, Science and Technology Studies (STS), and sociology describe how scientific knowledge is increasingly used in setting policy agendas, defining problems, and formulating and legitimizing policies (Meslin in this volume).

Part of this trend stems from the increasing institutionalization of scientific advice (Meslin in this volume). Earlier forms of including scientific advice into policy making were confined to high-profile cases and ad hoc advisory committees on special questions. Today, scientific fact-finding and prediction are part and parcel of any legislative process, for example in the German institution of Gesetzesfolgenabschätzung (Veit, 2014), or Regulatory Impact Assessments (RIAs) mandated by European law (Bunea & Chrisp, 2023; Commission of the European Communities, 2015; Dunlop et al., 2012; Staroňová, 2010). The idea is that any political measure must be scientifically evaluated ex ante, and its consequences studied using scientific methods (Radaelli, 2010). This is sometimes framed using the metaphor of "policy experiments" or "regulatory experiments" (Bizer in this volume), transferring the evidence-based approach of science to policy making.

A major research question that unifies nearly all studies of science and public policy is the question of the *influence* of science on public policies (Christensen, 2021, 2023). However, evidence for an influence of science on public policies is mixed.

Some scholars argue that institutionalized science exerts a substantial influence on public policy (Haas, 1989)—potentially at the cost of non-state actors (Hartley, 2016). The argument is that scientific knowledge reduces uncertainty about the consequences of public policies (Haas, 2004), and offers focal points to coordinate political action around shared ideas (van Nispen & Scholten, 2017). Especially if science is an institutionalized part of policy making, such as in the form of scientific advisory committees in regulatory agencies (Fink, 2024; Ruffing, 2022; Wonka & Rittberger, 2010, p. 736), science has a large impact in the early stages of policy formulation (Rimkutė & Haverland, 2015).

However, this influence is not only due to the power of facts in a simple model of evidence-based policy making (Christensen, 2021; Sanderson, 2006). Science is instrumentalized by political actors to advance their own goals (Dunlop et al., 2012; Jewell & Bero, 2007; van Nispen & Scholten, 2017). This corresponds to the experience of scientists who worked at the science-politics nexus and who emphasize the role of a plausible narrative that matches the political agenda (Cairney, 2021; Haas, 2004; Turner et al., 2020), underlining the important role of science communication (Bleidorn in this volume; Nehring et al. in this volume).

However, many of these studies of the influence of science on public policy have not adequately addressed the *demand* for scientific advice. To be sure, arguments about the *demand* for scientific advice are often implicit in theories of the *influence* of science on the policy process. Especially in functionalist arguments, the influence of science is derived from the demand for science. For example, if the function of science is to reduce uncertainty for policy makers, then the influence of science will be greatest if science serves that purpose. However, demand for science may stem not only from the functional demands of policy makers, but also from political demands of parties. The following section develops this theoretical argument.

3 The Theoretical Argument: Parties Formulate a "Demand for Science"

The central argument is that the scientification of politics is not merely a secular trend driven by the growing complexity of real-world problems. Many arguments about the scientification of public policy contain the argument that growing complexity necessitates scientification of policy, for example, arguing that "growing technical nature of problems [has] fostered an increase in the deference paid to technical expertise, and ... to that of scientists." (Adler & Haas, 1992, p. 381; Esmark, 2017; Haas, 1992, p. 11; Wonka & Rittberger, 2010). Although plausible as a broad condition, this argument lacks a micro-foundation explaining which actors anticipate that science can address complex problems and therefore design institutions that embed scientific advice. This mirrors the functional fallacy of explaining institutional design (Keohane, 1984, p. 80; Mayntz, 2008, p. 54), reasoning backwards from an effect—scientification of public policy addresses complexity—to the motives of the actors.

Science is always part of political conflicts, and is used as political ammunition, to legitimize or de-legitimize policies: "Science has become extremely politicized. The use of science is mediated and thus possibly distorted by the political goals of potential users. Truth claims are politically suspect because of potential funding bias or participation exclusion." (Haas, 2004, p. 572).

Thus, the demand for science in policy making may not only stem from science's contribution to problem-solving, but from its role as a weapon in political conflicts (Boswell, 2008; Rimkutė & Haverland, 2015; Weingart, 1999, p. 156). If we want to tackle the interplay of societal values and science (Schicktanz and Reydon in this volume), one major actor connecting society and science are *political parties*. Society can be characterized as large groups divided by major and long-lasting conflicts, such as labor vs. capital (Lipset & Rokkan, 1967), materialism vs. postmaterialism (Inglehart, 1977), or winners vs. losers of the liberal global order

(Hooghe & Marks, 2018). The values and preferences of these societal groups are then articulated by political parties that aggregate voter preferences, campaign for their preferred policies, and run for legislative and executive offices (Lipset & Rokkan, 1967; Stokes, 1999; Strøm, 1990).

Both electoral and policy-goal pursuits can benefit from scientific support and arguments. Empirical research shows that integrating scientific expertise into policy design and implementation is popular with voters (Bertsou, 2022), and that voters even favor decision making by scientific experts over decision making by elected officials, at least in some policy fields like environmental policy (Bertsou, 2022). This can be explained by voter's assumption that scientific experts make "better" policies that are more suitable for the problems at hand. From an instrumental perspective, voters delegate decisions to experts when they perceive expert preferences to align with their own (Beiser-McGrath et al., 2021). Moreover, delegating decision-making to experts can signal commitment to reforms (Alexiadou & Gunaydin, 2019), and depoliticize topics (Timmermans & Scholten, 2006).

Thus, vote-seeking and policy-seeking parties have an incentive to imbue their policy proposals with scientific authority. The empirical question is whether all parties do this to the same extent in all policy fields—which would in turn generate an overarching trend toward scientification of policies—or whether there is systematic variation in *which parties* demand science in *which policy fields* and *when*.

4 Research Design and Methods: Manifesto Data

To provide an initial description of parties' "demand for science," I analyze *electoral manifestos*.

Electoral manifestos are a suitable source for searching for parties' demand for science. Electoral manifestos have a pivotal role to play in defining and communicating parties' policy positions (Eder et al., 2017; Kercher & Brettschneider, 2013). Their purpose is manifold (Harmel et al., 2016; Laver & Garry, 2000, p. 620). *Internally*, manifestos resolve intra-party conflicts about policies (Bäck et al., 2016; Eder et al., 2017, pp. 76–77). Manifestos are approved by a party congress, and can thus be seen as the legitimate position of the "whole party" that can then mobilize its activists around shared aims (Harmel et al., 2016, p. 279). *Externally*, manifestos present parties as an organization with a homogenous position, recognizable by voters for their position, and distinct from other parties

(Snyder & Ting, 2002). Thus, if parties want to use references to science to appeal to voters, these references should be found in manifestos. Although many voters do not read manifestos (Kercher & Brettschneider, 2013), manifestos are picked up by the media, and streamline the electoral campaigns around key messages (Eder et al., 2017). If parties have the opportunity to enter government, manifestos form the basis for coalition negotiations (Bäck et al., 2011; Martin & Vanberg, 2020), and ultimately guide public policies. After the end of the electoral term, voters hold parties accountable for their electoral pledges (Matthieß, 2020).

Thus, electoral manifestos are vitally important documents for parties. If parties demand science to either attract voters or to justify their policies, these demands should be visible in manifestos.

I analyze the electoral manifestos of German parties from 2002 to 2021 as coded in the MARPOR database (Lehmann et al., 2022). The MARPOR data code party manifestos along a range of 56 categories that can be roughly thought of as policy fields or issues.[1] The manifestos are divided into quasi-sentences, and each quasi-sentence is assigned to a category like "Multiculturalism: Positive." In the original application, the frequencies of these codes are then used to scale party positions either on specific issues like multiculturalism (Abou-Chadi, 2014), or on a general left–right scale (Volkens, 2007). While these scaling techniques are methodologically contentious (Benoit et al., 2009; Mikhaylov et al., 2017), my application of the MARPOR data utilizes the underlying text corpus of manifestos (Merz et al., 2016).

I simply search for the occurrence of the German term "wissensch"[2] in the manifesto text corpus, with the main areas of interests being as follows: (1) which parties most often refer to science in their manifesto; (2) in which MARPOR categories these references occur; and (3) how the first two phenomena change over time.

[1] For example, Freedom and Human Rights, Decentralization, Market Regulation, Law and Order, Multiculturalism, and Agriculture and Farming.

[2] The German stem of "science," to capture terms like *wissenschaftlich*, *wissenschaftliche*, *Wissenschaftlerin*, etc. This search string probably underestimates the number of references to science, as it does not cover terms like "Studie" (study) that also may indicate references to science.

5 The Empirical Application: German Parties and Their "Demand for Science"

The first question is whether the references to science in German party manifestos over time suggest that there is an increasing demand for science over time. Figure 1 first suggests that the demand for science is a rare occurrence in party manifesto quasi-sentences. The 2021 election marks the peak: 1.17% of quasi-sentences (243 of 20.471) contain the term "Wissenschaft." There is hardly any trend of an increasing demand for science over time.

This uneven pattern is repeated if we look at the use of the term science party-by-party (Fig. 2 and Table 1). For all parties under consideration, the percentage of quasi-sentences containing science varies around 1%, with the Greens reaching nearly 2% in their 2021 manifesto, but a marked drop of references to science by the CDU/CSU and the FDP, and to some extent the Greens, in 2017. Thus, there is no significant difference in the demand for science by the German parties (Table 1).

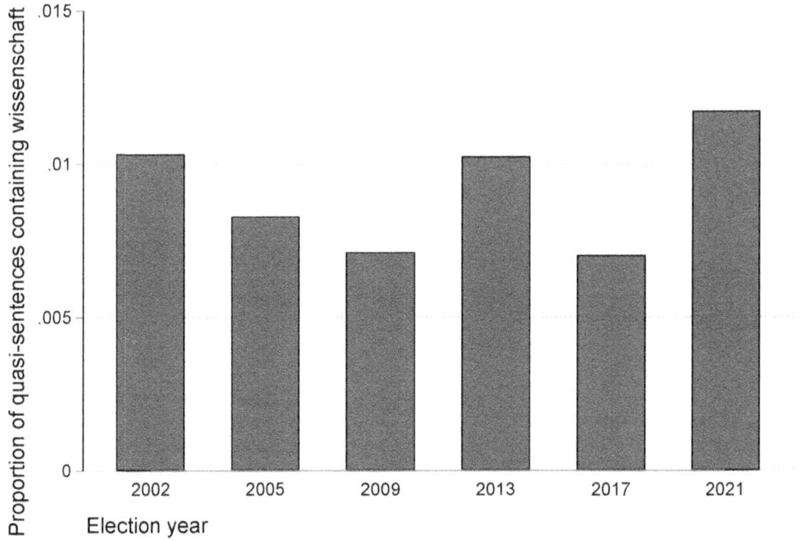

Fig. 1 Proportion of quasi-sentences mentioning science ("Wissenschaft") in German party manifestos

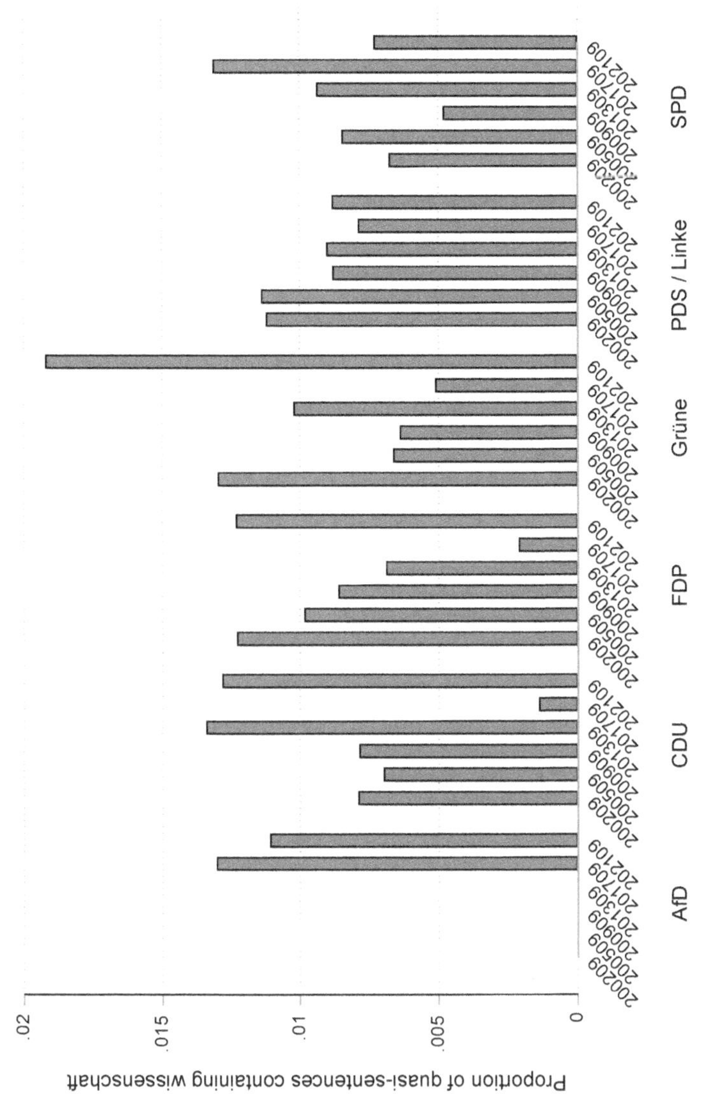

Fig. 2 Proportion of quasi-sentences mentioning science ("Wissenschaft") in German party manifestos, separated by parties

Table 1 Mean percentages of quasi-sentences containing science per party

Party	*Mean percentage of quasi-sentences referring to science*
AfD	1.1
CDU/CSU	0.9
FDP	0.8
Greens	1.0
PDS/Linke	0.8
SPD	0.8

If we disaggregate the data to see which parties use references to science in which policy fields, we first find several universals. All parties often refer to science in the category technology and infrastructure.[3] In other words, science is invoked as an essential infrastructure for economic growth. This science-growth nexus is used by nearly all parties throughout the period of study in statements such as "Science and industry must be more closely interlinked through competence centers and networks in order to accelerate the translation of research findings into new products and services"[4] (SPD, 2002), "the transfer of scientific research results into companies and thus into new products must be accelerated"[5] (FDP, 2002), "the innovative strength, especially of small and medium-sized enterprises, is to be promoted through increased cooperation between business and science"[6] (CDU, 2005), "basic research in all scientific disciplines is the central basis for innovation and progress"[7]

[3] The MARPOR code is 411, "Importance of modernization of industry and updated methods of transport and communication. May include: • Importance of science and technological developments in industry; • Need for training and research within the economy (This does not imply education in general (see category 506); • Calls for public spending on infrastructure such as roads and bridges; • Support for public spending on technological infrastructure (broadband internet, etc.)."

[4] "Wissenschaft und Wirtschaft müssen durch Kompetenzzentren und Netzwerke enger miteinander verzahnt werden, um die Umsetzung von Forschungsergebnissen in neue Produkte und Dienstleistungen zu beschleunigen."

[5] "Der Transfer von wissenschaftlichen Forschungsergebnissen in Unternehmen und damit in neue Produkte muss beschleunigt werden."

[6] "Die Innovationskraft, vor allem des Mittelstands, soll durch eine verstärkte Zusammenarbeit von Wirtschaft und Wissenschaft gefördert werden."

[7] "Grundlagenforschung in allen Wissenschaftsdisziplinen ist die zentrale Basis für Innovationen und Fortschritt."

(Bündnis 90/Die Grünen, 2009), and "more must be invested in science and research in order to strengthen innovation and universities, especially in Eastern Germany"[8] (Die Linke, 2017).

Another universal is reference to science under the heading of *education expansion*, that is, science policy in terms of universities as places of teaching.[9] This is also not surprising and can be found in nearly all party manifestos throughout the whole period of study. This includes questions of teacher training ("the FDP is in favour of carrying out model trials for tiered teacher training with bachelor's and master's degrees, as proposed by the science council"[10] (FDP, 2002)), scientific careers ("we create reliable career prospects for young scientists"[11] (CDU, 2021)), university financing ("we will maintain the temporary funding from the science and higher education pacts beyond 2019/20 and transfer it to reliable and permanent funding"[12] (SPD, 2017)), and student participation ("we want to involve students and their representatives more closely at federal level and expand their opportunities to have a say in higher education and science policy issues"[13] (Bündnis 90/Die Grünen, 2021)).

Both forms of mentioning science can be considered addressing science as a social system, that is either universally seen as beneficial, or in need of governance (e.g. to ensure funding, or viable career paths).[14] If we code

[8] "In Wissenschaft und Forschung muss umfangreicher investiert werden, um Innovationen und die Hochschulen besonders in Ostdeutschland zu stärken."

[9] The original MARPOR code is 506: "Education Expansion Need to expand and/or improve educational provision at all levels. Note: This excludes technical training which is coded under 411."

[10] "Die FDP befürwortet die Durchführung von Modellversuchen für eine gestufte Lehrerausbildung mit Bachelor- und Masterabschluss, wie es der Wissenschaftsrat vorgeschlagen hat."

[11] "Damit schaffen wir verlässliche Karriereperspektiven für den wissenschaftlichen Nachwuchs."

[12] "Die befristeten Mittel der Wissenschafts- und Hochschulpakte werden wir auch nach 2019/20 erhalten und in eine verlässliche und dauerhafte Finanzierung überführen."

[13] "Wir wollen Studierende und ihre Vertretungen auf Bundesebene stärker einbeziehen und ihre Mitsprachemöglichkeiten in hochschul- und wissenschaftspolitischen Fragen ausbauen."

[14] However, there are limits to this conclusion, as the MARPOR codes were not designed to distinguish between science-as-a-system and science-in-specific-policy-fields. The two MARPOR codes 411 and 506 contain general references to science as a system, but also statements like "Wir wollen eine naturwissenschaftlich fundierte Energiepolitik" (We want an energy policy based on science) or "...wollen wir die dauerhafte Verankerung

the policy fields in a more fine-grained manner (using context-sensitive hand coding that includes not only the quasi-sentence, but also the context in which science is mentioned), some policy fields are mentioned more often than others (see Fig. 3 and Table 2).

Contrary to expectations, the Greens are not the most frequent users of scientific references in *environmental and climate policy*; the liberal FDP leads in these domains. In all manifestos since 2002, the FDP flags areas of environmental and climate policy as being in need of scientific research or advice: Environmental protection measures,[15] waste disposal and recycling schemes,[16] biodiversity,[17] fishing quotas,[18] environmental performance evaluations,[19] genetically modified crops and

von Frauenstudien und Frauenforschung in den Hochschulen und der Wissenschaft" (We want to ensure that women's studies and women's research are firmly established in universities and the academic world), referring to energy policy and women's rights/equality policy.

[15] "Nature conservation measures must be scientifically substantiated and their effectiveness must be seriously assessed"/"Naturschutzmaßnahmen müssen naturwissenschaftlich begründet und seriös auf ihre Wirksamkeit hin überprüft werden" (FDP, 2005).

[16] "The fundamental ecological superiority of reusable packaging has been scientifically refuted"/"Die grundsätzliche ökologische Überlegenheit von Mehrwegverpackungen ist wissenschaftlich widerlegt" (FDP, 2002).

[17] "We need more funding for an internationally coordinated process for a scientifically sound inventory of biodiversity and the definition of clear targets"/"Wir brauchen mehr Mittel für einen international koordinierten Prozess zur wissenschaftlich fundierten Bestandsaufnahme der biologischen Vielfalt und zur Definition klarer Ziele" (FDP, 2009).

[18] "We also want to strengthen the competitiveness of our fishermen and aquaculture producers in international competition in the interests of consumers. we also rely on cooperative measures to monitor legal regulations, which are always based on scientific findings and must be further developed"/"Auch im Interesse der Verbraucherinnen und Verbraucher wollen wir die Konkurrenzfähigkeit unserer Fischerinnen sowie Fischer und Aquakulturproduzenten im internationalen Wettbewerb stärken. Wir setzen zudem auf kooperative Maßnahmen zur Überwachung gesetzlicher Regelungen, die immer auf Grundlage wissenschaftlicher Erkenntnisse basieren und weiterentwickelt werden müssen" (FDP, 2021).

[19] "Life cycle assessments make the environmental impact of products transparent and enable a science-based assessment on the basis of which necessary regulations can be made"/"Ökobilanzen machen die Umweltwirkung von Produkten transparent und ermöglichen eine wissenschaftsbasierte Bewertung, auf deren Basis notwendige Regulierungen erfolgen können" (FDP, 2021).

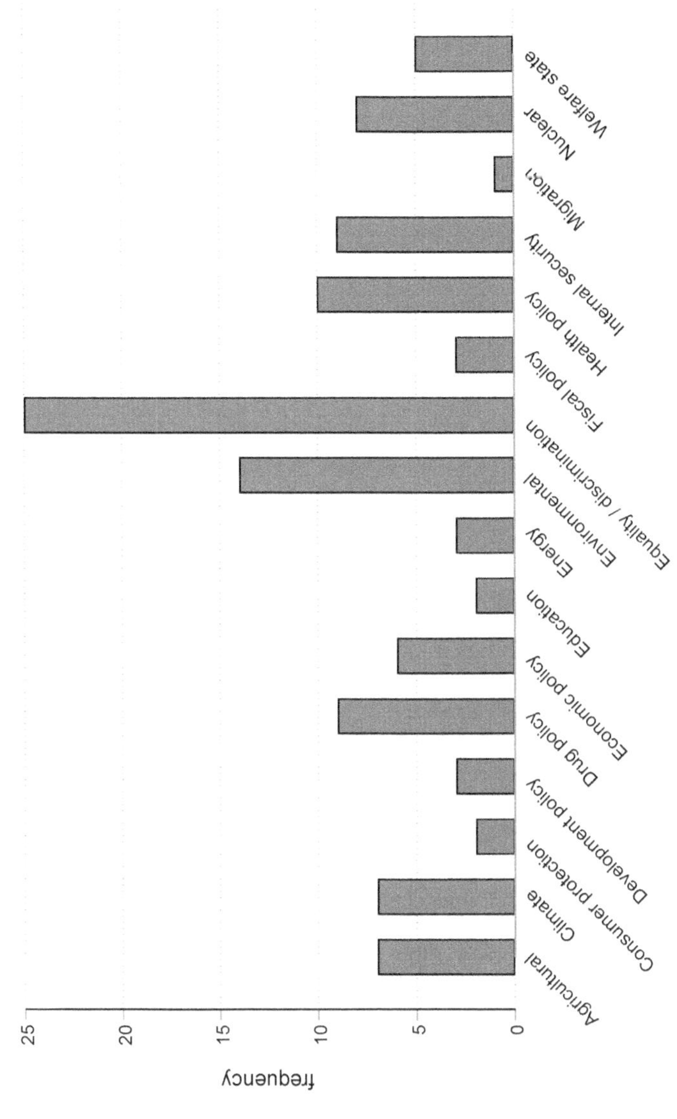

Fig. 3 Number of mentions of science by policy fields

Table 2 References to science by policy field and parties

	AfD	*CDU*	*FDP*	*Grüne*	*Linke*	*Piraten*	*SPD*	*SSW*	*Total*
Agricultural	0	0	5	1	1	0	0	0	7
Climate	3	0	3	1	0	0	0	0	7
Consumer protection	0	1	0	1	0	0	0	0	2
Development policy	0	0	1	2	0	0	0	0	3
Drug policy	0	0	2	2	4	1	0	0	9
Economic policy	0	0	1	3	2	0	0	0	6
Education	0	0	1	1	0	0	0	0	2
Energy	0	0	1	0	1	0	1	0	3
Environmental	1	1	6	0	3	2	1	0	14
Equality/discrimination	3	3	1	14	2	0	2	0	25
Fiscal policy	2	0	0	1	0	0	0	0	3
Health policy	0	2	0	1	4	2	1	0	10
Internal security	0	0	2	3	0	4	0	0	9
Migration	0	0	0	0	0	0	1	0	1
Nuclear	0	2	0	4	0	0	1	1	8
Welfare state	0	1	0	0	4	0	0	0	5
Total	9	10	23	34	21	9	7	1	114

food,[20] or ecological farming[21]—in all these areas, decisions should be made according to scientific advice. However, the variety of topics indicates that there is no one environmental issue that the FDP consistently highlights as being in need of scientific study. Instead, it seems that this reference to science is driven by current issue salience in the election years.

[20] "We therefore advocate the labelling of such foods so that this decision is possible in the first place. At the same time, we emphasize the importance of scientific and objective information and consumer education about the benefits and value of modern production methods"/"Wir setzen uns daher für eine Kennzeichnung solcher Lebensmittel ein, damit diese Entscheidung überhaupt erst möglich ist. Gleichzeitig legen wir Wert auf eine wissenschaftliche und objektive Information und Verbraucherbildung über den Nutzen und Wert moderner Herstellungsmethoden" (FDP, 2013).

[21] "There is no scientific evidence of differences in quality or flavour between products from conventional and organic farming"/"Qualitative, auch geschmackliche, Unterschiede zwischen Produkten aus konventionellem und ökologischem Landbau sind wissenschaftlich nicht belegbar" (FDP, 2002).

Another party that often refers to science in environmental policies is the Left, which calls for scientific evidence on fishing quotas[22] or animal welfare.[23]

A special case with regard to references to science in environmental policy is the AfD, which uses references to science in the negative to stigmatize policies as "nonscientific."[24]

The sole environmental issue on which the Greens consistently invoke scientific evidence is the siting of an atomic waste disposal facility. In four green manifestos from 2002 to 2017, "scientific criteria" are seen as important to determine the place for German radioactive waste disposal. The CDU and the SPD use similar rhetoric in their 2002 and 2013 manifestos.

All German parties under study that can be considered left in relation to socio-cultural dimensions use references to science to inform *drug policy*. This can be found in the manifestos of the FDP,[25] the Greens,[26]

[22] "And sustainable fishing with catch quotas based on scientific recommendations"/ "Und eine nachhaltige Fischerei mit Fangquoten, die auf wissenschaftlichen Empfehlungen basieren" (Dei Linke, 2021, similar in 2017).

[23] "Animals are subject to our actions, so we have a responsibility to minimise and avoid their suffering. Current policies and laws do not reflect the wishes of the public or scientific knowledge about the sentience of animals"/"Tiere sind unserem Handeln unterworfen, deshalb tragen wir die Verantwortung, ihr Leid zu vermindern und zu vermeiden. Die aktuelle Politik und die bestehenden Gesetze entsprechen weder den Wünschen der Bevölkerung noch den wissenschaftlichen Erkenntnissen über die Empfindungsfähigkeit von Tieren" (Dei Linke, 2021).

[24] "The statements of the Intergovernmental Panel on Climate Change (IPCC) that climate change is predominantly man-made are not scientifically proven"/"Die Aussagen des Weltklimarats (IPCC), dass Klimaänderungen vorwiegend menschengemacht seien, sind wissenschaftlich nicht gesichert, AfD 2017, Scientifically inadequately proven measures, such as diesel driving bans or environmental lanes, do not achieve this./ Wissenschaftlich mangelhaft belegte Maßnahmen, wie z. B. Dieselfahrverbote oder Umweltspuren, erreichen das nicht" (AfD, 2021).

[25] "Once the results are available, it will be possible to decide on a scientific basis whether and to what extent heroin substitution is superior to methadone substitution"/ "Nach Vorliegen der Ergebnisse wird auf wissenschaftlicher Grundlage entschieden werden können, ob und inwieweit die Heroinvergabe der Methadonsubstitution überlegen ist" (FDP, 2002).

[26] "We call for an evaluation of the current drug policy with the involvement of science, drug help and health policy"/"Wir fordern unter Einbeziehung von Wissenschaft, Drogenhilfe und Gesundheitspolitik eine Evaluierung der aktuellen Drogenpolitik" (Bündnis 90/Die Grünen, 2013).

and the Left,[27] but not consistently over all years. Instead, 2013 and 2017 seem to be the two years in which scientific drug policy was a major issue, but not afterwards.

Another issue on which the Greens and the FDP seem to agree on more science-driven policy is *internal security*. Both parties want more scientific evidence to inform internal security legislation[28] and demand more scientific competence in the working of the security apparatus.[29]

Nearly all parties look to science for guidance on *societal policy, equality, discrimination*. Here, all parties seem to favor scientific studies of the societal fields in which they are most invested ideologically. The conservative CDU demands scientific study of displaced persons' cultural heritage (meaning displaced persons from the former German territories after 1945)[30] and of wrongdoings by former GDR

[27] "DIE LINKE is calling for an independent scientific review of whether the repressive drug policy to date has achieved its objectives and what side effects it has had"/"DIE LINKE fordert eine unabhängige wissenschaftliche Überprüfung, ob die bisherige repressive Drogenpolitik ihre Ziele erreicht und welche Nebenwirkungen sie entfaltet hat" (De Linke, 2017).

[28] "Many of the current measures result in less security, less freedom and a less livable society. They should be scrutinized according to scientific criteria and, in case of doubt, corrected"/"Viele der aktuellen Maßnahmen sorgen für weniger Sicherheit, weniger Freiheit und eine weniger lebenswerte Gesellschaft. Sie gehören nach wissenschaftlichen Kriterien auf den Prüfstand und im Zweifel korrigiert" (Bündnis 90/Die Grünen, 2021); "We are in favour of an ongoing, scientifically supported review of current firearms legislation"/"Wir setzen uns für eine fortlaufende, wissenschaftlich begleitete Überprüfung des geltenden Waffenrechts ein" (FDP, 2013).

[29] "The scientific expertise of the security authorities in combating various forms of extremism must be expanded"/"Die wissenschaftliche Expertise in den Sicherheitsbehörden bei der Bekämpfung verschiedener Extremismusformen muss ausgebaut werden" (FDP, 2021); "We want to reorganize the structure of the Office for the Protection of the Constitution: on the one hand, with an independent institute working scientifically from public sources to protect the constitution"/"Wir wollen den Verfassungsschutz strukturell neu aufstellen: zum einen mit einem unabhängigen, wissenschaftlich aus öffentlichen Quellen arbeitenden Institut zum Schutz der Verfassung…" (Bündnis 90/Die Grünen, 2021).

[30] "We want to strengthen federal funding for the preservation of the cultural heritage of displaced persons and refugees and for the promotion of scientific research in accordance with Section 96 of the Federal Expellees Act as an anchor point of policy on displaced persons and ethnic German repatriates"/"Die Bundesförderung zur Pflege des Kulturgutes der Vertriebenen und Flüchtlinge und zur Förderung der wissenschaftlichen Forschung nach § 96 Bundesvertriebenengesetz wollen wir als einen Ankerpunkt der Vertriebenen- und Aussiedlerpolitik stärken" (CDU, 2021).

governments.[31] The SPD favors scientific study of the women's movement,[32] while the Left favors the study of colonialism and right-wing tendencies.[33] The Greens are a major player in the demand for science in societal policies. In earlier manifestos, they called for scientific study of feminism[34] and democracy.[35] However, the 2021 manifesto of the Greens contains a plethora of issues in which scientific evidence is either directly called for, or scientific experts are called on

[31] "The CDU and the CSU are in favour of establishing a permanent chair at a Berlin university for research into and academic analysis of the GDR injustice regime"/"CDU und CSU setzen sich dafür ein, dass ein Lehrstuhl zur Erforschung und wissenschaftlichen Aufarbeitung des DDR-Unrechtsregimes an einer Berliner Universität dauerhaft etabliert wird" (CDU, 2009).

[32] "We want to further advance the academic reappraisal of the women's movement with the involvement of women's archives, digitally secure their holdings and make them accessible to the public"/"Wir wollen die wissenschaftliche Aufarbeitung der Frauenbewegung unter Einbeziehung der Frauenarchive weiter vorantreiben, ihre Bestände digital sichern und sie der Öffentlichkeit zugänglich machen" (SPD, 2017).

[33] "The federal government should set up an institution on the history of colonialism with the involvement of academics from the former colonies. A scientific investigation into extreme right-wing attitudes and racist practices in the police and the Bundeswehr is needed"/"Der Bund soll unter Einbeziehung von Wissenschaftlerinnen aus den ehemaligen Kolonien eine Institution zur Geschichte des Kolonialismus aufbauen. Es braucht eine wissenschaftliche Untersuchung extrem rechter Einstellungen und rassistischer Praktiken bei Polizei und Bundeswehr" (Die Linke, 2021).

[34] "Feminist critique of science makes the category of gender the subject of scientific work and thus makes an important contribution to the reflection of social conditions"/"Feministische Wissenschaftskritik macht die Kategorie Geschlecht zum Gegenstand wissenschaftlicher Arbeit und leistet damit einen wichtigen Beitrag zur Reflexion der gesellschaftlichen Verhältnisse" (Bündnis 90/Die Grünen, 2002); "A nationwide research institution on gender equality should bundle scientific expertise, initiate and support discussion processes"/"Eine bundesweite Forschungseinrichtung zur Gleichstellung soll wissenschaftliche Expertise bündeln, Diskussionsprozesse initiieren und unterstützen" (Bündnis 90/Die Grünen, 2013).

[35] "In future, an independent institute for the analysis of anti-democratic and anti-human endeavors is to monitor and analyse anti-democratic and anti-human endeavours on the basis of public sources using scientific means"/"Die Beobachtung und Analyse demokratie- und menschenfeindlicher Bestrebungen anhand öffentlicher Quellen soll künftig ein unabhängiges Institut zur Analyse demokratie- und menschenfeindlicher Bestrebungen mit wissenschaftlichen Mitteln leisten" (Bündnis 90/Die Grünen, 2002, similar in 2017).

to be part of policy-making committees: Equal opportunities policies,[36] discrimination,[37] religion,[38] colonialism,[39] racism,[40] and antisemitism[41] are all areas in which the Greens demand more scientific knowledge or

[36] "We will develop the newly created Federal Foundation for Gender Equality into an effective, reliably funded and independent institution that provides reliable knowledge on the living conditions of all genders and develops and bundles effective measures for equality and makes them accessible to business, politics and the public. The social sciences and gender studies make an indispensable contribution to this"/"Die neu geschaffene Bundesstiftung Gleichstellung werden wir zu einer effektiven, verlässlich finanzierten und unabhängigen Institution ausbauen, die gesichertes Wissen zu den Lebenslagen aller Geschlechter bereitstellt und wirksame Maßnahmen für Gleichberechtigung entwickelt, bündelt und für Wirtschaft, Politik und Öffentlichkeit zugänglich macht. Hierfür leisten die Sozialwissenschaften und die Genderstudies einen unverzichtbaren Beitrag" (Bündnis 90/Die Grünen, 2021).

[37] "To ensure that the perspectives and expertise of those affected by discrimination and structural disadvantage are heard and that they are given the opportunity to fully participate as equals, we want to introduce a Participation Council, similar to the German Ethics Council, as a legally anchored and independent body with representatives from (post-)migrant civil society, academia and research who represent the different dimensions of diversity"/"Damit die Perspektive und Expertise derjenigen, die von Diskriminierung und struktureller Benachteiligung betroffen sind, gehört werden, sie als Gleichberechtigte die Möglichkeit zur vollen Teilhabe erhalten, wollen wir einen Partizipationsrat, ähnlich dem Deutschen Ethikrat, als ein gesetzlich verankertes und unabhängiges Gremium einführen, mit Vertreter*innen aus der (post-)migrantischen Zivilgesellschaft, Wissenschaft und Forschung, die die unterschiedlichen Dimensionen von Vielfalt abbilden" (Bündnis 90/Die Grünen, 2021).

[38] "Advocate for an independent scientific institution to research the religious and ideological landscape"/"... uns für eine unabhängige wissenschaftliche Einrichtung zur Erforschung der religiösen und weltanschaulichen Landschaft einsetzen" (Bündnis 90/Die Grünen, 2021).

[39] "We will expand independent research on post-colonialism, discrimination and racism, regularly collect anti-discrimination and equality data and conduct scientific studies on state institutions and the effectiveness of anti-discrimination measures"/"Wir werden die unabhängige Forschung zu Postkolonialismus, Diskriminierung und Rassismus ausbauen, regelmäßig Antidiskriminierungs- und Gleichstellungsdaten erheben und wissenschaftliche Studien in Bezug auf staatliche Institutionen und Wirksamkeit von Antidiskriminierungsmaßnahmen durchführen" (Bündnis 90/Die Grünen, 2021).

[40] "We will initiate independent scientific studies on racism and right-wing extremism in the various security authorities, record hate violence, and pursue it consistently"/"Wir werden unabhängige wissenschaftliche Studien zu Rassismus und Rechtsextremismus in den verschiedenen Sicherheitsbehörden initiieren, Hassgewalt erfassen und konsequent verfolgen" (Bündnis 90/Die Grünen, 2021).

[41] "Independent scientific studies on right-wing extremism, antisemitism and racism in the security and law enforcement agencies are long overdue"/"Längst überfällig sind unabhängige wissenschaftliche Studien zu Rechtsextremismus, Antisemitismus und

advice. On the other side of the political spectrum, the AfD tries to discredit gender studies and gender policies as "unscientific."[42]

Economic policies are a major policy field both in the view of voters and of parties. However, references to science are only occasionally found in economic fields—apart from the general science–innovation–growth nexus outlined above. The Left and the Greens mobilize science for a minimum wage and alternative economic models,[43] while the AfD wants scientific input into tax policies.[44]

Given that this is a relatively obvious field close to science, one would have expected many references to science in *health policy* (Hoppe and Stauch in this volume). However, the references to science in health policy are few in number, mostly related to the call that health insurance benefits

Rassismus in den Sicherheits- und Strafverfolgungsbehörden" (Bündnis 90/Die Grünen, 2021).

42 "Gender ideology contradicts both the scientific findings of biology and developmental psychology as well as the practical everyday experience of many generations"/"Die Gender-Ideologie widerspricht sowohl den wissenschaftlichen Erkenntnissen der Biologie und der Entwicklungspsychologie als auch der lebenspraktischen Alltagserfahrung vieler Generationen" (AfD, 2017).

43 "We want to promote pluralistic approaches in research and teaching at colleges, universities and in policy advice in order to counter the neoliberal mainstream in economics with critical thinking"/"Um dem neoliberalen Mainstream in den Wirtschaftswissenschaften kritisches Denken zur Seite zu stellen, wollen wir plurale Ansätze in Forschung und Lehre an Hochschulen, Universitäten und in der Politikberatung fördern" (Dei Linke, 2021); "Minimum wage commission based on the British model is responsible for the general wage floor.

and should also decide if the collective bargaining parties in the sectors are not strong enough to enforce fair minimum wages. The commission—consisting of employers, trade unions and academics—draws up proposals that are then implemented by the federal government"/"Eine Mindestlohn-Kommission nach britischem Vorbild ist für die generelle Lohnuntergrenze verantwortlich und soll auch dann entscheiden, wenn die Tarifparteien in den Branchen nicht stark genug sind, um faire Mindestlöhne durchzusetzen. Die Kommission – bestehend aus Arbeitgebern, Gewerkschaften und Wissenschaft – erarbeitet Vorschläge, die anschließend von der Bundesregierung umgesetzt werden" (Bündnis 90/Die Grünen, 2009).

44 "The tax system's reason for levying them is unclear, which is why their right to exist and their constitutionality are highly controversial in legal scholarship"/"Ihr steuersystematischer Erhebungsgrund ist unklar, weshalb ihre Existenzberechtigung und ihre Verfassungsmäßigkeit in der Rechtswissenschaft hochumstritten ist" (AfD, 2021).

have to take into account scientific evidence and scientific progress,[45] and that authorization of new medical products is based on scientific criteria.[46]

Besides these major fields in which at least several parties call for science in multiple manifestos, there are several policy fields in which only one party in one manifesto calls for science: Education policy,[47] migration,[48] or development aid.[49]

6 Conclusion

All in all, it seems that there is no universal increase of the demand for science in party manifestos. Neither on the aggregate level nor on the level of single parties do we see a steady increase of demand for science in manifestos.

A similar conclusion applies to the policy fields. The only universal pattern is that all parties—from the left to the right—view science as an engine of innovations and economic growth. Moreover, there are a few

[45] "[The benefits catalog of statutory health insurance] is adapted to medical progress on the basis of sound scientific knowledge"/"[Der Leistungskatalog der gesetzlichen Krankenversicherung] wird auf der Basis gesicherter wissenschaftlicher Erkenntnisse an den medizinischen Fortschritt angepasst" (SPD, 2002); "Standard dental care in the health care system must also be regularly adapted to the current state of scientific knowledge"/"Auch die zahnmedizinische Regelversorgung in der GKV muss regelmäßig an den aktuellen Stand der Wissenschaft angepasst werden" (Bündnis 90/Die Grünen, 2021).

[46] "All medicines should only be authorized on the basis of scientific studies. Specialized routes without a scientific basis must be abolished"/"Alle Medikamente dürfen nur durch wissenschaftliche Studien ihre Zulassungen bekommen. Sonderwege ohne wissenschaftliche Grundlagen müssen abgeschafft werden" ((Dei Linke, 2021).

[47] "Which measures for educational equity work well, also in international comparison, are to be investigated more intensively scientifically and provided with recommendations for action"/"Welche Maßnahmen für Bildungsgerechtigkeit, auch im internationalen Vergleich, gut funktionieren, soll wissenschaftlich intensiver eruiert und mit Handlungsempfehlungen versehen werden" (Bündnis 90/Die Grünen, 2021).

[48] "We should learn more about the effects of immigration and emigration on social change. This requires science-based analyses"/"Wir sollen mehr über die Auswirkungen von Ein- und Auswanderungen auf den gesellschaftlichen Wandel erfahren. dafür sind wissenschaftsbasierte Analysen notwendig" (SPD, 2017).

[49] "From now on, German development policy will for the first time be subject to a scientifically sound, independent evaluation of its activities in order to learn from the experience of its work in the long term"/"Von nun an stellt sich die deutsche Entwicklungspolitik erstmals einer wissenschaftlich fundierten, unabhängigen Begutachtung ihrer Aktivitäten, um aus den Erfahrungen ihrer Arbeit nachhaltig zu lernen" (FDP, 2013).

noteworthy patterns: The FDP often tries to de-politicize environmental policies by appealing to science, socio-culturally left-of-center parties try to de-politicize drug policy, and the Greens in 2021 started a major attempt to infuse science into all kinds of policies concerning discrimination and equality. The AfD—in contrast to other parties that mostly use science as a positive attribute—uses the term "science" mostly in a negative sense, discrediting other worldviews as unscientific.

However, in the long run, there are no stable patterns. This casts doubt on the policy-seeking logic in which parties have stable policy preferences and use science to support their preferred policies. For example, a pattern in which the Green party consistently uses appeals to science to support environmental policies would be more consistent with this view. Instead, the pattern of which parties demand scientific evidence at which time and in which policy field seems highly contingent on external factors. This pattern is consistent with a perspective on parties that emphasizes short-term calculations on how to generate votes by appeals to science in which policy fields.

This result suggests that the increased demand for science in policy making is not due to an underlying stable increase in demand for science by party-political actors. This also suggests that how societal values are transformed into political guidance for the conduct of science—the interplay of societal values and scientific practice (Schicktanz and Reydon in this volume)—is constantly renegotiated not only as a result of changing societal values, but also as a result of the shifting agendas of political parties. While the conclusion that science constantly navigates an ever-changing societal environment is not new (see Hinrichs et al. in this volume), the volatility of party competition adds another layer of volatility and uncertainty in the chain between societal values and scientific practice.

Competing Interests The author has no conflicts of interest to declare that are relevant to the content of this chapter.

References

Abou-Chadi, T. (2014). Niche party success and mainstream party policy shifts—How green and radical right parties differ in their impact. *British Journal of Political Science, 46*(2), 417–436.

Adler, E., & Haas, P. M. (1992). Conclusion: Epistemic communities, world order, and the creation of a reflective research program. *International Organization, 46*(1), 367–390. https://www.jstor.org/stable/2706960

AfD (Alternative für Deutschland). (2017). *Programm für Deutschland. Wahlprogramm der Alternative für Deutschland für die Wahl zum Deutschen Bundestag am 24. September 2017*. AfD. https://www.afd.de/wp-content/uploads/2017/06/2017-06-01_AfD-Bundestagswahlprogramm_Onlinefassung.pdf

AfD (Alternative für Deutschland). (2021). *Deutschland. Aber normal. Programm der Alternative für Deutschland für die Wahl zum 20. Deutschen Bundestag*. AfD. https://www.afd.de/wp-content/uploads/2021/06/20210611_AfD_Programm_2021.pdf

Alexiadou, D., & Gunaydin, H. (2019). Commitment or expertise? Technocratic appointments as political responses to economic crises. *European Journal of Political Research, 58*(3), 845–865.

Bäck, H., Debus, M., & Dumont, P. (2011). Who gets what in coalition governments? Predictors of portfolio allocation in parliamentary democracies. *European Journal of Political Research, 50*(4), 441–478.

Bäck, H., Debus, M., & Klüver, H. (2016). Bicameralism, intra-party bargaining, and the formation of party policy positions: Evidence from the German federal system. *Party Politics, 22*(3), 405–417.

Beiser-McGrath, L. F., Huber, R. A., Bernauer, T., & Koubi, V. (2021). Parliament, people or technocrats? Explaining mass public preferences on delegation of policymaking authority. *Comparative Political Studies, 55*(4), 527–554.

Benoit, K., Laver, M., & Mikhaylov, S. (2009). Treating words as data with error: Uncertainty in text statements of policy positions. *American Journal of Political Science, 53*(2), 495–513.

Bertsou, E. R. I. (2022). Bring in the experts? Citizen preferences for independent experts in political decision-making processes. *European Journal of Political Research, 61*(1), 255–267.

Boswell, C. (2008). The political functions of expert knowledge: Knowledge and legitimation in European Union immigration policy. *Journal of European Public Policy, 15*(4), 471–488.

Bündnis 90/Die Grünen. (2002). *Grün wirkt! Unser Wahlprogramm 2002—2006*. Bündnis 90/Die Grünen. https://www.boell.de/sites/default/files/assets/boell.de/images/download_de/publikationen/2002_Wahlprogramm_Bundestagswahl.pdf

Bündnis 90/Die Grünen. (2009). *Der Grüne Neue Gesellschaftsvertrag. Klima—Arbeit—Gerechtigkeit—Freiheit. Bundestagswahlprogramm 2009*. Bündnis 90/Die Grünen. http://www.gruene.de/fileadmin/user_upload/Dokumente/Wahlprogramm/BTW_Wahlprogramm_2009_final_screen_060709.pdf

Bündnis 90/Die Grünen. (2013). *Zeit für den grünen Wandel. Teilhaben. Einmischen. Zukunft schaffen. Bundestagswahlprogramm 2013.* Bündnis 90/ Die Grünen. http://www.gruene.de/fileadmin/user_upload/Dokumente/Gruenes-Bundestagswahlprogramm-2013.pdf

Bündnis 90/Die Grünen. (2021). *Deutschland. Alles ist drin. Bundestagswahlprogramm 2021.* Bündnis 90/Die Grünen. https://cms.gruene.de/uploads/assets/Wahlprogramm-DIE-GRUENEN-Bundestagswahl-2021_barrierefrei.pdf

Bunea, A., & Chrisp, J. (2023). Reconciling participatory and evidence-based policymaking in the EU Better Regulation policy: Mission (im)possible? *Journal of European Integration, 45*(5), 729–750.

Cairney, P. (2021). *The politics of policy analysis.* Springer.

CDU (Christlich Demokratische Union Deutschlands) & CSU (Christlich-Soziale Union in Bayern). (2005). *Deutschlands Chancen nutzen. Wachstum. Arbeit. Sicherheit. Regierungsprogramm 2005—2009.* CDU/CSU. http://www.regierungsprogramm.cdu.de/download/regierungsprogramm-05-09-cducsu.pdf

CDU (Christlich Demokratische Union Deutschlands) & CSU (Christlich-Soziale Union in Bayern). (2009). *Wir haben die Kraft—Gemeinsam für unser Land. Regierungsprogramm 2009—2013.* CDU/CSU. https://www.hss.de/fileadmin/user_upload/HSS/Dokumente/ACSP/Bundestagswahlen/BTW-2009.pdf

CDU (Christlich Demokratische Union Deutschlands) & CSU (Christlich-Soziale Union in Bayern). (2021). *Das Programm für Stabilität und Erneuerung. Gemeinsam für ein modernes Deutschland.* CDU/CSU. https://www.csu.de/common/download/Regierungsprogramm.pdf

Christensen, J. (2021). Expert knowledge and policymaking: A multi-disciplinary research agenda. *Policy & Politics, 49*(3), 455–471.

Christensen, J. (2023). Studying expert influence: A methodological agenda. *West European Politics, 46*(3), 600–613.

Commission of the European Communities. (2015). Commission staff working document: Better regulation guidelines. https://eur-lex.europa.eu/legal-content/EN/TXT/?uri=swd:SWD_2015_0111

Die Linke. (2017). *Sozial. Gerecht. Frieden. Für alle. Die Zukunft, für die wir kämpfen! Wahlprogramm zur Bundestagswahl 2017.* Die Linke. https://www.die-linke.de/fileadmin/download/wahlen2017/wahlprogramm2017/die_linke_wahlprogramm_2017.pdf

Die Linke. (2021). *Zeit zu handeln! Für soziale Sicherheit, Frieden und Klimagerechtigkeit. Wahlprogramm zur Bundestagswahl 2021.* Die Linke. https://www.die-linke.de/fileadmin/download/wahlen2021/Wahlprogramm/DIE_LINKE_Wahlprogramm_zur_Bundestagswahl_2021.pdf

Dunlop, C. A., Maggetti, M., Radaelli, C. M., & Russel, D. (2012). The many uses of regulatory impact assessment: A meta-analysis of EU and UK cases. *Regulation & Governance, 6*(1), 23–45.

Eder, N., Jenny, M., & Müller, W. C. (2017). Manifesto functions: How party candidates view and use their party's central policy document. *Electoral Studies, 45*, 75–87.

Esmark, A. (2017). Maybe it is time to rediscover technocracy? An old framework for a new analysis of administrative reforms in the governance era. *Journal of Public Administration Research and Theory, 27*(3), 501–516.

FDP (Freie Demokratische Partei). (2002). *Bürgerprogramm 2002*. FDP. http://www.documentarchiv.de/brd/2002/wahlprogramm_fdp_2002.html

FDP (Freie Demokratische Partei). (2005). *Arbeit hat Vorfahrt. Deutschlandprogramm 2005*. FDP. http://files.liberale.de/fdp-wahlprogramm.pdf

FDP (Freie Demokratische Partei). (2009). *Deutschlandprogramm 2009*. FDP. http://www.liberale.de/files/653/Deutschlandprogramm09_Endfassung.PDF

FDP (Freie Demokratische Partei). (2013). *Bürgerprogramm 2013*. FDP. https://fdp-hessen.de/wp-content/uploads/2015/08/2013-BTW-Programm.pdf

FDP (Freie Demokratische Partei). (2021). *Nie gab es mehr zu tun. Das Programm der Freien Demokraten zur Bundestagswahl 2021*. FDP. https://www.fdp.de/sites/default/files/2021-06/FDP_Programm_Bundestagswahl2021_1.pdf

Fink, S. (2024). Balancing reputational strategies in the European administrative space: How private actors and agencies talk about regulation. *Journal of European Public Policy*, 1–26.

Fink, S. (2025). Regulatory talk about climate policies: National plans or commission streamlining. *JCMS: Journal of Common Market Studies*.

Fourcade-Gourinchas, M., & Babb, S. L. (2002). The rebirth of the liberal creed: Paths to neoliberalism in four countries. *American Journal of Sociology, 108*(3), 533–579.

Frinken, J., & Landwehr, C. (2022). Information and deliberation in the Covid-19 crisis and in the climate crisis: How expertocratic practices undermine self-government and compliance. *Acta Polit*, 1–19.

Garb, Y., Manon, M., & Peters, D. (2007). Environmental impact assessment: Between bureaucratic process and social learning. In F. Fischer, G. J. Miller, & M. S. Sidney (Eds.), *Handbook of public policy analysis* (pp. 481–492). Taylor & Francis.

Gornitzka, Å., & Sverdrup, U. (2008). Who consults? The configuration of expert groups in the European Union. *West European Politics, 31*(4), 725–750.

Haas, P. (2004). When does power listen to truth? A constructivist approach to the policy process. *Journal of European Public Policy, 11*(4), 569–592.
Haas, P. M. (1989). Do regimes matter? Epistemic communities and Mediterranean pollution control. *International Organization, 43*(3), 377–403.
Haas, P. M. (1992). Introduction: Epistemic communities and international policy coordination. *International Organization, 46*(1), 1–35.
Harmel, R., Tan, A. C., Janda, K., & Smith, J. M. (2016). Manifestos and the "two faces" of parties: Addressing both members and voters with one document. *Party Politics, 24*(3), 278–288.
Hartley, S. (2016). Policy masquerading as science: An examination of non-state actor involvement in European risk assessment policy for genetically modified animals. *Journal of European Public Policy, 23*(2), 276–295.
Head, B. W. (2022). *Wicked problems in public policy: Understanding and responding to complex challenges*. Springer Nature.
Heclo, H. (1978). Issue networks and the executive establishment. In A. King (Ed.), *The New American political system* (pp. 87–124). American Enterprise Institute.
Hooghe, L., & Marks, G. (2018). Cleavage theory meets Europe's crises: Lipset, Rokkan, and the transnational cleavage. *Journal of European Public Policy, 25*(1), 109–135.
Inglehart, R. (1977). *The silent revolution*. Princeton University Press.
Jewell, C., & Bero, L. (2007). Public participation and claimsmaking: Evidence utilization and divergent policy frames in California's ergonomics rulemaking. *Journal of Public Administration Research and Theory, 17*(4), 625–650.
Keohane, R. O. (1984). *After Hegemony: Cooperation and discord in the World political economy*. Princeton University Press.
Kercher, J., & Brettschneider, F. (2013). Wahlprogramme als Pflichtübung? Typen, Funktionen und Verständlichkeit der Bundestagswahlprogramme 1994–2009. In B. Weßels, H. Schoen, & O. W. Gabriel (Eds.), *Wahlen und Wähler* (pp. 269–290). Springer.
Laver, M., & Garry, J. (2000). Estimating policy positions from political Texts. *American Journal of Political Science, 44*(3), 619–634.
Lehmann, P., Burst, T., Matthieß, T., Volkens, A., Weßels, B., & Zehnter, L. (2022). The Manifesto data collection. Manifesto Project (MRG/CMP/MARPOR). Version 2022a: Wissenschaftszentrum Berlin für Sozialforschung (WZB).
Lipset, S. M., & Rokkan, S. (1967). Cleavage structures, party systems and voter alignments: An introduction. In S. Rokkan & S. M. Lipset (Eds.), *Party systems and voter alignments: Cross-national perspectives* (pp. 1–64). Free Press.
Martin, L. W., & Vanberg, G. (2020). Coalition government, legislative institutions, and public policy in parliamentary democracies. *American Journal of Political Science, 64*(2), 325–340.

Matthieß, T. (2020). Retrospective pledge voting: A comparative study of the electoral consequences of government parties' pledge fulfilment. *European Journal of Political Research, 59*(4), 774–796.

Mayntz, R. (2008). Von der Steuerungstheorie zu global Governance. In G. F. Schuppert & M. Zürn (Eds.), *Governance in einer sich wandelnden Welt* (pp. 43–60). Springer.

Merz, N., Regel, S., & Lewandowski, J. (2016). The manifesto corpus: A new resource for research on political parties and quantitative text analysis. *Research & Politics, 3*(2).

Mikhaylov, S., Laver, M., & Benoit, K. R. (2017). Coder reliability and misclassification in the human coding of party manifestos. *Political Analysis, 20*(1), 78–91.

Mintrom, M. (2013). Policy entrepreneurs and controversial science: Governing human embryonic stem cell research. *Journal of European Public Policy, 20*(3), 442–457.

Radaelli, C. M. (2010). Regulating rule-making via impact assessment. *Governance, 23*(1), 89–108.

Reber, B. (2007). 33 Technology assessment as policy analysis: From expert advice to participatory approaches. In F. Fischer, G. J. Miller, & M. S. Sidney (Eds.), *Handbook of public policy analysis* (pp. 493–511). Routledge.

Rimkutė, D., & Haverland, M. (2015). How does the European Commission use scientific expertise? Results from a survey of scientific members of the Commission's expert committees. *Comparative European Politics, 13*(4), 430–449.

Ruffing, E. (2022). European (networked) agencies between independence and influence. *Journal of European Public Policy, 29*(10), 1546–1567.

Sabatier, P. A. (1988). An advocacy coalition framework of policy change and the role of policy-oriented learning therein. *Policy Sciences, 21*(2), 129–168.

Sanderson, I. (2006). Complexity, 'practical rationality' and evidence-based policy making. *Policy & Politics, 34*(1), 115–132.

SPD (Sozialdemokratische Partei Deutschlands). (2002). *Erneuerung und Zusammenhalt. Wir in Deutschland. Regierungsprogramm 2002—2006*. SPD. https://www.spd.de/fileadmin/Dokumente/Beschluesse/Bundesparteitag/regierungsprogramm_bundesparteitag_berlin_2002.pdf

SPD (Sozialdemokratische Partei Deutschlands). (2017). *Es ist Zeit für mehr Gerechtigkeit: Zukunft sichern, Europa stärken. Das Regierungsprogramm 2017 bis 2021*. SPD. https://www.spd.de/fileadmin/Dokumente/Regierungsprogramm/SPD_Regierungsprogramm_BTW_2017_A5_RZ_WEB.pdf

Snyder, J. M., & Ting, M. M. (2002). An informational rationale for political parties. *American Journal of Political Science, 46*(1), 90–110.

Staroňová, K. (2010). Regulatory impact assessment: Formal institutionalization and practice. *Journal of Public Policy, 30*(Special Issue 01), 117–136.

Stokes, S. C. (1999). Political parties and democracy. *Annual Review of Political Science, 2*(1), 243–267.
Strøm, K. (1990). A behavioral theory of competitive political parties. *American Journal of Political Science, 34*(2), 565–598.
Timmermans, A., & Scholten, P. (2006). The political flow of wisdom: Science institutions as policy venues in The Netherlands. *Journal of European Public Policy, 13*(7), 1104–1118.
Turner, K., Alabi, O., & Race, J. (2020). Nudging policymakers: A case study of the role and influence of academic policy analysis. *Journal of European Public Policy, 27*(8), 1270–1286.
Van Nispen, F. K. M., & Scholten, P. W. A. (2017). The utilization of expert knowledge in times of crisis: Budgetary and migration policies in the Netherlands. *European Policy Analysis, 3*(1), 81–100.
Veit, S. (2014). Evidenzbasierte Politik durch Gesetzesfolgenabschätzungen? Erfahrungen aus Deutschland und Schweden. In J. Ziekow (Ed.), *Bewerten und Bewertet-Werden* (pp. 37–56). Nomos Verlagsgesellschaft mbH & Co. KG.
Veit, S., Hustedt, T., & Bach, T. (2017). Dynamics of change in internal policy advisory systems: The hybridization of advisory capacities in Germany. *Policy Sciences, 50*(1), 85–103.
Volkens, A. (2007). Strengths and weaknesses of approaches to measuring policy positions of parties. *Electoral Studies, 26*(1), 108–120.
Weible, C. M. (2008). Expert-based information and policy subsystems: A review and synthesis. *Policy Studies Journal, 36*(4), 615–635.
Weible, C. M., Sabatier, P. A., & McQueen, K. (2009). Themes and variations: Taking stock of the advocacy coalition framework. *Policy Studies Journal, 37*(1), 121–140.
Weingart, P. (1999). Scientific expertise and political accountability: Paradoxes of science in politics. *Science and Public Policy, 26*(3), 151–161.
Wonka, A., & Rittberger, B. (2010). Credibility, complexity and uncertainty: Explaining the institutional independence of 29 EU agencies. *West European Politics, 33*(4), 730–752.

Trust in Science in Modern Democracies: Determinants and Empirical Findings

Christoph Hönnige and Dominic Nyhuis

1 Introduction

Standard political science models of politics do not include the role of science in politics. While we find chapters on parties, interest groups, social movements, and even religion in textbooks on comparative politics, there are no chapters on science or higher education policy. The most we can find are chapters on education policy in handbooks on policy research. Science and scientists are not integrated in common models of comparative politics, e.g. those explaining the legitimacy of political systems (Easton, 1965) or political change and decision making (Tsebelis, 2002).

This gap is surprising as scientists play a direct a role in political decision making on several levels. Firstly, legislative processes typically involve hearings of interest groups (Schmid, 2007), but also experts in the respective fields, which are often university professors (Lehmann et al., 2017). Secondly, expert institutions, which play an important role in

C. Hönnige · D. Nyhuis (✉)
Institut für Politikwissenschaft, Leibniz Universität Hannover, Hannover, Germany
e-mail: d.nyhuis@ipw.uni-hannover.de

C. Hönnige
e-mail: c.hoennige@ipw.uni-hannover.de

J. Hinrichsen et al. (eds.), *Ethics of the Sciences*,
https://doi.org/10.1007/978-981-92-2821-8_11

political decision making (Kaiser, 1997)—like central banks or constitutional courts—are traditionally composed to a large extent by professors of law or economics. Such institutions spread worldwide in the 1990s in the so-called "rise of the unelected" (Vibert, 2007).

The best option for integrating science into models of politics is to treat them as intermediate actors (Jarren & Donges, 2011). To this group of actors belong political parties, interest groups, social movements and the media. They serve to aggregate and communicate citizen preferences to political institutions, and to communicate political decisions to citizens. The aim of the scientific system is not to communicate and aggregate societal preferences, but to explain how the natural and social world works.

A free and functioning science system is a precondition for modern democracies. Science is an instrument to de-personalize and objectivize power, similar to bureaucracies and the legal system. While science is thus an instrument to constrain majority rule, it does so by applying standards which are external to political decision making. If applied well, scientists can provide politicians with expertise which is free from political bias (Ezrahi, 1990).

Prompted by the COVID-19 pandemic, research on trust and the acceptance of science has recently exploded (Flores et al., 2022; Mihelj et al., 2022; Taraktas et al., 2024). It has become clear that scientists are involved in political decision making, and trust in science and trust in political institutions has become linked.

The aim of this contribution is therefore to analyze the current state of research on the relationship between politics and the role of science in these systems. Our central argument is that trust in science is a prerequisite for modern democracies (Ezrahi, 1990), but its function and mechanisms are not well understood.

This chapter proceeds as follows. Firstly, we discuss trust in science as a pre-requisite for the place of science in modern societies and elaborate on the status of research on science in modern democracies. Secondly, we look at the empirical research on trust in science in order to understand which conceptual aspects of trust are actually covered. We provide an overview of the content and weaknesses of existing survey data dealing with questions on science. We place a focus on trust in science here and outline the commonalities and differences between the surveys and their major gaps. Thirdly, we look at findings from the major datasets, especially the Eurobarometer (EB), the World Values Survey (WVS), Varieties of Democracy (V-Dem), and the *Wissenschaftsbarometer* (WB), while also showing the research potential of existing data. Lastly, we develop a research agenda for the theoretical and empirical analysis of the ethical norms of science. By distinguishing between endogenous and exogenous norms with regard to scientific ethics, we highlight key factors that have contributed to the increasing skepticism toward science.

2 Trust in Science and Academic Freedom in Modern Democracies

With few exceptions, scientific policy advice is non-binding. Therefore, to have any impact at all, politicians and citizens need to trust in science, as the advice of scientists runs the risk of being ignored otherwise. Indeed, science is increasingly coming under fire when scientific advice collides with policy preferences—similar to how the rule of law has been threatened in recent years. Examples are the debates around Brexit, where Minister Michael Gove has said that "We had enough of experts" (Nature, 2020) or Trump's "War on Science" (Lin, 2018).

In his seminal book *The Descent of Icarus*, Yaron Ezrahi argues that science has a similar function as law and bureaucracies: it de-personalizes political power in liberal democracies and makes decisions more objective and rational. Compared to autocratic political systems, power is shifted away from the heads of government to these rather rational institutional functions (Ezrahi, 1990). In this sense, science places limits on power. While not formally included in the consensus democracy concept by Lijphart, which describes political systems with a high degree of power-sharing (Lijphart, 2012), we can infer that a strong science system has the same effect as strong bureaucracies, judiciaries, second chambers, or interest groups, namely adding checks and balances.

This also explains why autocratic leaders and democratic leaders who aim to reduce restrictions on their power often attack and try to de-legitimize science and scientists who deliver advice which does not fit with the leaders' worldviews (Bellolio, 2024; Meyer, 2024). The aim of such leaders is to reduce academic freedom directly or indirectly by institutional or financial means or de-legitimization (Druckman, 2017; Jaspal et al., 2012; Mede & Schäfer, 2020; Oliveira et al., 2022; Taraktas et al., 2024).

Only recently has academic freedom come under systematic scrutiny. The concept has been defined by Karran who analyzed the constitutionally guaranteed freedoms of science in 23 EU Member States (Karran, 2007). Academic freedom has become far more popular and better analyzed after the integration into the multinational research project V-Dem (https://www.v-dem.net) through the Academic Freedom Index (AFI). The index aims to assess de facto academic freedom across the world. Academic freedom has a clear relationship with democracy. A higher level of democracy is linked to a higher level of academic freedom in the AFI data. A positive relationship is also evident with judicial independence and the presence of second chambers (Berggren & Bjørnskov, 2022). This means that in consensus-oriented democracies, academic freedom seems better protected.

If we follow the argument by Ezrahi (1990), science is in a similar position as courts. This is also true for the mechanisms behind the acceptance of scientific knowledge by politicians and citizens. Montesquieu (1748) has argued that the power of the judiciary is almost non-existent as it relies on the executive—and sometimes the legislative—to implement its decisions. This is similar to science. No citizen or politician is obliged to follow scientific advice. They only do so if they trust scientists and the scientific system. Trust in science is thus an important pre-requisite for the acceptance and implementation of scientific knowledge. Measuring trust in science (Algan et al., 2021; Bromme et al., 2022; Cologna et al., 2025; Gauchat, 2012; Krause et al., 2019; Schäfer, 2016) and appropriate measures of trust (Besley, 2023; Nadelson et al., 2014; Reif & Guenther, 2021) are thus essential for understanding the role of science in politics and society. In the following section, we will thus discuss different measures.

Research on trust in science has grown significantly during the COVID-19 pandemic. Studies on trust in science also moved away from specialized higher education journals to generalist journals such as *Nature* and *Science* (Algan et al., 2021; Cologna et al., 2025; Flores et al., 2022; Merkley & Loewen, 2021). Here, we find large cross-country studies (Cologna et al., 2025) and many national studies (Groeniger et al., 2021; Heinzel & Liese, 2021; Luna et al., 2021; Mede & Schäfer, 2022).

As the central normative question of the COVID-19 pandemic was how much the state should restrict personal freedoms to ensure collective public health, many of the independent variables being tested were related to political attitudes. We thus find a lot of research focusing on right-wing political attitudes, populism, and anti-intellectualism (Lockhart et al., 2020; Mede & Schäfer, 2022; Merkley & Loewen, 2021; Taraktas et al., 2024). Other studies focused on trust in other institutions, beliefs about misinformation, or other individual attitudes (Daoust et al., 2021; Santirocchi et al., 2023), as well as attitudes toward scientists and experts (Evans & Hargittai, 2020; Flores et al., 2022; Heinzel & Liese, 2021; Mihelj et al., 2022; Scheitle & Corcoran, 2021).

The COVID-19 pandemic was a big step forward for research into trust in science and the test of possible explanatory variables. However, the COVID-19 studies also suffer from shortcomings. As they were conducted during the COVID-19 pandemic, they focus on a very specific time period which greatly impacted respondents' lives. Moreover, COVID-19 was a health issue and it is not obvious whether this issue can be generalized to other science policy areas. Lastly, it was a singular event producing a straw fire. It is still an open question as to what lessons we can draw from COVID-centric research for research on the science system beyond this special case.

3 Trust in Science and Academic Freedom: Comparative and German Datasets

With regard to internationally comparative surveys, we focus on the World Values Survey WVS (https://www.worldvaluessurvey.org/wvs.jsp), the Eurobarometer EB (https://europa.eu/eurobarometer/), and the Varieties of Democracy V-Dem project (https://www.v-dem.net). They all cover attitudes in multi-annual waves or expert estimations over time, but all three remain on the surface regarding trust in science. The available survey questions can be divided into three subsets: The evaluation of science and the scientific process; the evaluation of the content of science; and the evaluation of scientists as actors.

At the time of writing, the Eurobarometer (see Table 1) contains seven survey waves covering trust in science in various sub-questions: on the benefits of science, the life-changing quality, the relation to faith, improving lives, and the dangers of science. In the older waves, the World Values Survey only covers the question whether science will help or harm humanity; in the last two waves (2010, 2017), it contains four questions on the relationship with religion, on the dangers of scientists, and whether science makes the world a better place.

Table 1 Eurobarometer and European Values Study/World Values Survey waves containing questions about science

	Dataset													
	Eurobarometer							*European Values Study/ World Values Survey*						
Item	*'89*	*'92*	*'01*	*'05*	*'10*	*'13*	*'21*	*'81*	*'90*	*'95*	*'99*	*'05*	*'10*	*'17*
The benefits of science are greater than any harmful effects it may have	✓	✓	✓	✓	✓									
Science makes our way of life change too fast	✓	✓	✓	✓	✓	✓	✓							
We depend too much on science and not enough on faith	✓	✓	✓	✓	✓	✓	✓					✓	✓	✓
Science and technology are making our lives healthier, easier, and more comfortable	✓	✓	✓	✓	✓	✓	✓					✓	✓	✓
Because of their knowledge, scientific researchers have a power that makes them dangerous		✓	✓	✓	✓									
We can no longer trust scientists to tell the truth about controversial scientific issues					✓		✓							
Is the overall influence of science and technology on society positive or negative?						✓								

(continued)

Table 1 (continued)

	Dataset													
	Eurobarometer							*European Values Study/ World Values Survey*						
Item	*'89*	*'92*	*'01*	*'05*	*'10*	*'13*	*'21*	*'81*	*'90*	*'95*	*'99*	*'05*	*'10*	*'17*
Scientists should intervene in public debate to ensure that decisions take into account scientific evidence							✓							
Do you think the scientific advances we are making will help or harm mankind?								✓	✓	✓	✓	✓		
Would you say that the world is better off, or worse off, because of science and technology?												✓	✓	✓
Whenever science and religion conflict, religion is always right													✓	✓

Sources Eurobarometer 31, Eurobarometer 38.1, Eurobarometer 55.2, Eurobarometer 63.1, Eurobarometer 73.1, Eurobarometer 79.2, Eurobarometer 95.2; European Values Study Wave 1/World Values Survey Wave 1, European Values Study Wave 2/World Values Survey Wave 2, World Values Survey Wave 3, European Values Study Wave 3 / World Values Survey Wave 4, World Values Survey Wave 5, World Values Survey Wave 6, World Values Survey Wave 7

Within the V-Dem set of indicators, there is the AFI, which aims to assess de facto academic freedom across the world (https://academic-freedom-index.net). The AFI is an expert survey and does not cover institutional rules protecting the scientific system. It contains five indicators: freedom to research and teach, freedom of academic exchange and dissemination, institutional autonomy, campus integrity, and freedom of academic and cultural expression (Lott and Spannagel forthcoming; Spannagel & Kinzelbach, 2023). It covers 179 countries over time based on more than 2,000 estimations by country experts. While the advantage of the AFI is the long-time horizon and the large number of countries covered, it also has disadvantages: Expert assessments make estimation of historical values difficult. While the expert data are on a micro level, they are aggregated to the macro level. However, the questions are disconnected from common variables like "trust in science" which are often asked in citizen surveys. While it is good to understand relationships at the macro level, such as between regime types and academic freedom, it is difficult to understand the mechanisms which increase or decrease academic freedom.

In Germany, the main data source is the *Wissenschaftsbarometer* (https://wissenschaft-im-dialog.de/projekte/wissenschaftsbarometer), which asks a set of varying science-related questions of citizens (see Table 2). The *Wissenschaftsbarometer* has been published annually since 2014. The questionnaire can be divided into three subsets: The evaluation of science and the scientific process; the evaluation of the content of science; and the evaluation of scientists as actors.

Unfortunately, the set of questions in the *Wissenschaftsbarometer* is not stable. Questions which are contained in almost each round are "How much do you trust in science and research?," and the two general evaluation statements "People trust science too much" and "Science harms more than it is useful." These three items can be used to study citizen attitudes toward science. The other items in the survey mainly focus on scientists and aim to identify reasons as to why citizens hold positive or negative attitudes toward science. Here, the focus lies on rules and standards, expertise, research in the public interest as drivers of trust and errors, manipulated research, and the dependency on funders as drivers of mistrust.

A benefit of the *Wissenschaftsbarometer* is the potential for understanding endogenous drivers of trust in science. It focuses mainly on drivers of trust based on the behavior of individual researchers. A not fully realized potential is studying the effect of scientific processes on trust. In the *Wissenschaftsbarometer* 2018, the replication crisis is analyzed and funding mechanisms are indirectly analyzed by asking about the dependence on third-party funding. Other important aspects are left aside, like plagiarism, data manipulation, or the employment situation of younger researchers. Teaching is also not part of this set of questions, even it is

Table 2 *Wissenschaftsbarometer* waves and questions about science, translation by the authors

	Wissenschaftsbarometer										
Item	*'14*	*'15*	*'16*	*'17*	*'18*	*'19*	*'20*	*'20 (Apr)*	*'20 (May)*	*'21*	*'22*
People trust science too much and enough their feelings and belief	✓	✓	✓	✓		✓	✓			✓	
Science harms more than it is useful	✓	✓	✓	✓		✓	✓			✓	
Scientists work for the good of society				✓	✓	✓				✓	
How much do you trust in science and research?				✓	✓	✓	✓	✓	✓	✓	✓
Reasons for trust: Because scientists are experts in their field				✓	✓	✓	✓			✓	✓
Reasons for trust: Because scientists follow rules and standards				✓	✓	✓	✓			✓	✓
Reasons for trust: Because scientists work in the interests of the public				✓	✓	✓	✓			✓	✓
Reasons for mistrust: Because scientists frequently make mistakes				✓	✓	✓	✓			✓	✓

(continued)

Table 2 (continued)

		Wissenschaftsbarometer									
Item	*'14*	*'15*	*'16*	*'17*	*'18*	*'19*	*'20*	*'20 (Apr)*	*'20 (May)*	*'21*	*'22*
Reasons for mistrust: Because scientists frequently make their results fit their expectations				✓	✓	✓	✓			✓	✓
Reasons for mistrust: Because scientists are dependent on their funders				✓	✓	✓	✓			✓	✓
Science and research changes our way of life too fast				✓	✓	✓					
Replication crisis: You cannot trust science and research					✓						
Replication crisis: The review shows that science and research conducts quality assurance					✓						
Replication crisis: Errors and their correction are part of science and research					✓						

(continued)

Table 2 (continued)

	Wissenschaftsbarometer										
Item	*'14*	*'15*	*'16*	*'17*	*'18*	*'19*	*'20*	*'20 (Apr)*	*'20 (May)*	*'21*	*'22*
How much do you trust scientists at universities and public research institutions?						✓					
Scientists should make public states when political decisions do not incorporate research results						✓				✓	✓
It is not the job of scientists to engage in politics						✓	✓	✓		✓	✓
We should rely more on intuition and less on scientific studies						✓	✓	✓		✓	
Political decisions should be based on scientific results						✓	✓	✓	✓	✓	✓

Sources GESIS ZA6944 (2014), ZA6945 (2015), ZA6946 (2016), ZA6947 (2017), ZA6965 (2018), ZA7634 (2019), ZA6939 (2020), ZA7635 (April 2020), ZA7636 (May 2020), ZA7640 (2021), ZA7641 (2022)

increasingly part of the media debate, especially because of enrolment fees or public debates about post-colonialism in academia and gender and diversity studies at universities.

The *Wissenschaftsbarometer* also covers the exogenous perspective. Repeated questions are: "Political decisions should be based on scientific results," i.e. how much politicians should rely on science, but also on how much scientists should interfere in politics: "It is not the job of scientists to engage in politics." Also, the relationship with faith and common sense is covered. Issues which are not covered include, for example, media appearances of scientists, participation in parliamentary hearings or expert committees, or general transfer activities of universities and scientists.

Most surprisingly, standard surveys in the social sciences largely ignore the subject. For example, the German Longitudinal Elections Study (GLES) (https://www.gesis.org/gles) does not include questions about trust in science. The German Internet Panel (GIP) (https://www.uni-mannheim.de/gip) only contains questions about trust in science since the COVID-19 pandemic. The GESIS Panel (https://www.gesis.org/gesis-panel) and the bi-annual ALLBUS (https://www.gesis.org/allbus) cover trust in universities at least in a few survey questions, but again, more frequently so since the pandemic. But all these questions mainly focus on trust as the dependent variable and not on reasons for varying degrees of trust or changes over time.

In summary, current surveys suffer from several shortcomings. Firstly, standard surveys in the social sciences largely ignore the scientific system and its interaction with politics and society. Questions are not asked consistently. What we can observe is that the number of questions has increased during and after COVID-19. Secondly, because of the lack of a panel structure, we cannot observe any systematic in-person changes over time. Thirdly, there is no consistent measure for trust in science or acceptance of scientific findings by the public. The questions vary considerably between surveys and over time, and often relate to trust or acceptance in relation to other sources of authority, like faith. Fourthly, while the major surveys focus on the relationship between citizens and science—the exogenous perspective—the *Wissenschaftsbarometer* also includes the rules and actors' behavior within science—the endogenous perspective. In the next section, we will move on to the analysis of the available comparative data.

4 Trust in Science and Academic Freedom: Some Empirical Findings

To provide some empirical insights, we first consider Eurobarometer data, which are collected by the European Commission to survey the population in the EU Member States. The Eurobarometer does not directly ask questions about trust in science, but only in relation to faith ("We depend too much on science and not enough on faith"). Other questions in the Eurobarometer are about possible positive effects of science or the dangers of science.

In Fig. 1, we present data from the Eurobarometer on the relationship between science and faith. There is a tendency for an increase in the belief in science over time. However, we also observe strong differences between the European regions. In Northern and Western Europe, we find not only the strongest shifts from faith to science, but also the highest trust in science in the 2000s. Central and Eastern Europe along with Southern Europe exhibit a far higher preference for faith.

A second perspective is possible with data from the V-Dem project (https://www.v-dem.net). Figure 2 presents data on academic freedom in Europe since 1945. The estimates are based on expert judgments. We can observe stark movements over time in the different European regions. Northern and Western Europe exhibit a high level of academic freedom. There are low levels of academic freedom in Eastern Europe until the collapse of communism in 1990. After that, we see improvements in academic freedom, especially for the Central and Eastern European Member States, but a slow decline for countries like Ukraine or Hungary, and especially for Russia. On the contrary, we observe improvements in Southern Europe during the 1980s. These developments suggest that academic freedom correlates with the degree of democratization (Lott, 2024).

For both data sources, we observe differences not only between countries, but also within countries over time. Neither how citizens view science nor how experts perceive academic freedom is stable. This variaion needs to be explained. In the following section, we will develop a heuristic about what influences (changes in) the perception of science.

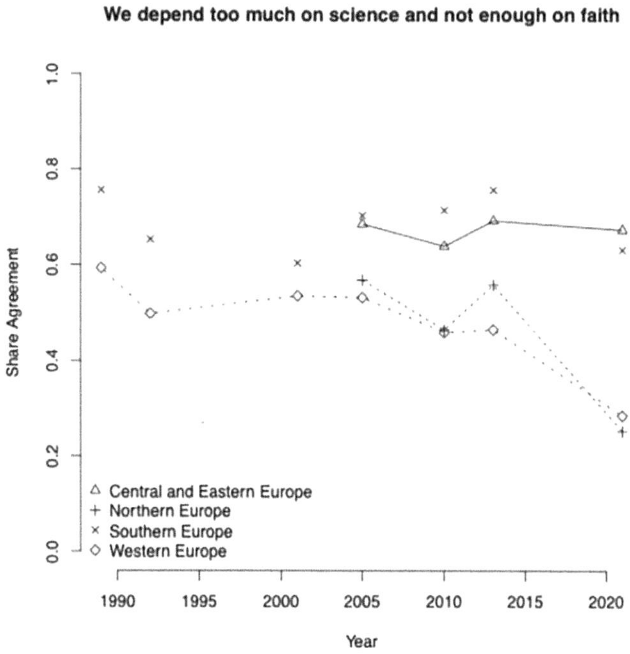

Fig. 1 Eurobarometer, science-faith-relationship. *Sources* Eurobarometer 31, Eurobarometer 38.1, Eurobarometer 55.2, Eurobarometer 63.1, Eurobarometer 73.1, Eurobarometer 79.2, Eurobarometer 95.2

5 Trust in Science: Ethical Norms of Science as a Future Research Agenda

In this section we develop a heuristic model to understand the relationship between internal and external ethical norms, specific support for decisions, trust in science, and the stability of democratic systems (see Fig. 3).

In light of the observed changes in trust in science, the question arises which factors increase or decrease trust in science. The political science and sociology research focuses on variables exogenous to the science system to explain trust. It identifies factors like political preferences (Krange et al., 2021; McCright et al., 2013) and standard demographic

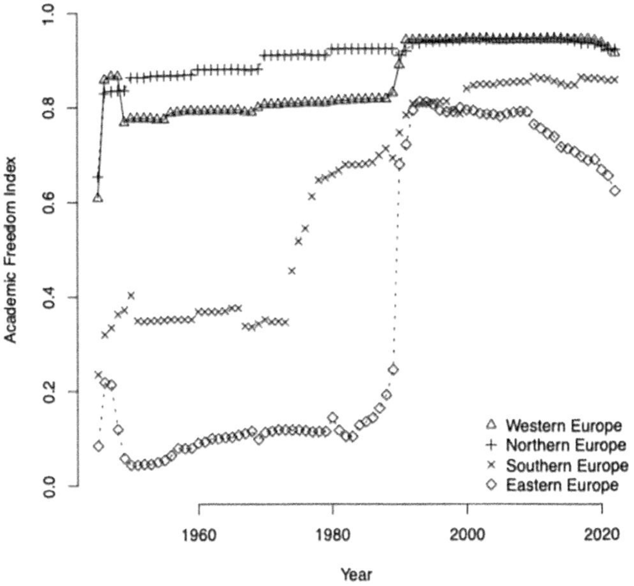

Fig. 2 V-Dem 1945 to 2025, Academic Freedom Index. *Source* https://www.v-dem.net

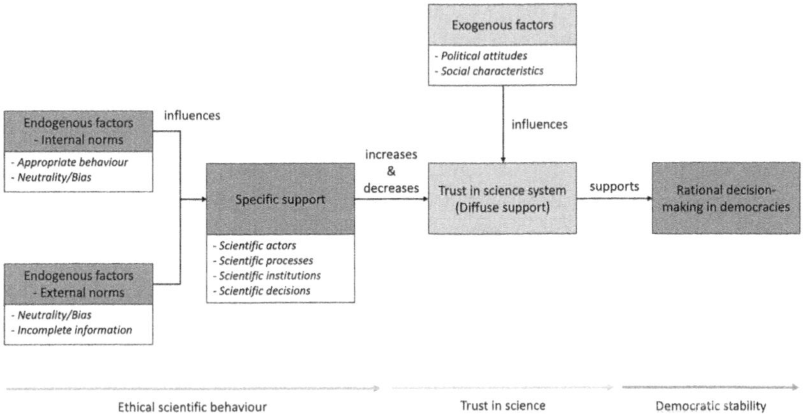

Fig. 3 A heuristic model of endogenous and exogenous factors influencing trust in science

and sociological variables like education (Achterberg et al., 2015) or institutional trust (Huber et al., 2022).

However, some determinants of trust are also endogenous to the scientific system, especially the norms within the scientific system and the perception of these norms as appropriate ethical behaviors. Ethical scientific behavior is an important factor which can be separated into internal and external norms (see Schicktanz and Reydon in this volume):

(a) Internal norms refer to rules of behavior for scientists when conducting research and teaching. This may shape the perception of citizens and politicians about the scientific process from personal experience or what is being reported in the media about the "inside" of science.
(b) External norms cover the relationship between scientists and societal actors, politics, and the media. They determine the relationship of scientists as well as scientific institutions with respect to serving the interests of society.

To understand how trust increases or decreases from a political science perspective, the model by David Easton (1975) is typically used. It was originally developed to understand trust in democracy. The basic idea is that in political systems, we can observe diffuse support, the general support for the system or an institution. Diffuse support is often measured as "trust in the system" or "trust in an institution." It is often equated with the legitimacy of a system or of an institution. Diffuse support builds on the many outputs of a system. Such outputs receive specific support, but evaluations of specific outputs can be positive or negative. Repeated positive support increases overall levels of diffuse support. On the other hand, many individual decisions which are disliked may lower the level of diffuse support. Thus, trust can increase or decrease. In this model, it is important to identify and understand the mechanisms which receive high or low specific support, as they can change generalized systemic trust in the long run.

Sticking to internal and external norms is viewed positively by citizens, violating norms is seen as negative. We can analyze four components of the science system with regard to specific support: (1) scientific actors; (2) the processes linking these actors; (3) scientific institutions; and (4) the decisions taken in the system. In any of these four components, support can be gained or lost through various mechanisms supporting or violating internal or external norms. The sum of the various specific behaviors leads to an in- or decrease of diffuse support of the science system, which can be measured as trust in the system.

Science is typically a self-referential system in free societies. It largely sets its own norms of what is right and what is wrong. The first important issue with regard to internal ethical norms is transparency and honesty in the research process (Merton, 1973). Typical forms of violations are plagiarism or data manipulation, or the impossibility of replicating results (Xie et al., 2021). A violation of such norms is typically sanctioned through formalized processes by universities or research organizations. If cases are prominent, media report about them and politicians and citizens may take notice of such cases. The reporting can influence trust in science in the long run (Zillich et al., 2024). Beyond elements of the research process, teaching and management issues might also play a role. Power abuses or inappropriate behaviors toward students or staff members might influence the level of trust when they are reported by the media (Egelhofer et al., 2024).

Another aspect are institutions of academic self-governance in research, teaching, and internal management (see Schicktanz and Reydon in this volume). Important institutions which could help foster trust are neutral funding institutions, course accreditation, ethics committees, or equal opportunities officers.

Scientists are increasingly expected to communicate their findings to society and to translate their findings into practical applications. With regard to external ethical norms, the focus is on the ways in which scientists communicate with citizens, the media, and the public. While there are strong and widely accepted internal norms, the picture is less clear with regard to external norms. There are typically no guidelines for scientists on how to communicate research results, and media training is often not available. Open questions are, for example, how to communicate the uncertainty of scientific results, how to translate complex analyses into simple messages, or how to deal with the question of political neutrality (Hilligardt, 2023). Not only are the norms unclear, but also how public-facing communication by scientists influences trust in the science system. While even high-profile researchers do not always believe in the neutrality of science (Ambrosj, Desmond, and Dierickx forthcoming), for citizens it seems important that scientists are not linked to political elites (Mihelj et al., 2022) and the trustworthiness of experts has a strong effect (Wintterlin et al., 2021).

Consequently, there are major research gaps regarding the relationship between ethics of science and trust in the science system. Research needs to be strengthened on how internal and external norms in the science system are perceived by citizens, as well as which norms receive specific support and thus increase trust in the system. Especially unclear are effects related to the internal norms which have been identified in similar research areas: the social composition, training, and selection processes of scientists, as well as decision-making rules within the scientific community. Other aspects relate to the external norms, such as how scientific communication and transfer activities are perceived by citizens. Such factors should be incorporated into future surveys.

6 Conclusion

Ezrahi (1990) views a well-functioning science system as a pre-requisite of modern democracies. Research shows that democratic systems exhibit a higher degree of academic freedom (Lott, 2024), while populist and autocratic leaders try to curb academic freedom and try to reduce trust in the science system (Druckman, 2017; Jaspal et al., 2012; Mede & Schäfer, 2020; Oliveira et al., 2022; Taraktas et al., 2024).

The science system is not systematically integrated into common models of politics like the Easton model of political systems (Easton, 1965) and is not usually discussed in political science textbooks. We argued that the scientific system could be seen as part of the intermediate system in political systems, like interest groups, social movements, or parties.

With regard to research, we find that in the major political science surveys, questions on the science system occur only sporadically. The best source for European cases is arguably the Eurobarometer and the Academic Freedom Index of the V-Dem project for a comparative overview. The surveys suffer from the fact that their variables—broadly covering trust in science—are inconsistent or are asked in relation to faith as another source of legitimate decision making. There is also no data source that covers trust in the science system over time in a consistent way.

Most research on the relationship between the political system and the science systems stems from climate change and COVID-19 research, and research on this has exploded since the pandemic (Flores et al., 2022; Mede & Schäfer, 2022; Mihelj et al., 2022; Scheitle & Corcoran, 2021). During the COVID-19 crisis, one of the main normative political questions was how far the state should reduce personal freedoms of citizens to guarantee public health as a collective good. Research thus focused on common factors in political science which might explain the attitudes of citizens toward government measures. The central finding was that support for right-wing and populist parties reduces trust in science (Lockhart et al., 2020; Mede & Schäfer, 2022; Merkley & Loewen, 2021; Taraktas et al., 2024). Overall, the existing research has focused on explanatory factors which are exogenous to the science system.

Far less researched are factors endogenous to the science system which lead to a build-up of trust in science. So far, the focus has been on experts/scientists and institutions (Berggren & Bjørnskov, 2022; Heinzel & Liese, 2021; Mihelj et al., 2022). We developed a future research agenda for the theoretical and empirical analysis of the ethical norms of science. By distinguishing between endogenous and exogenous norms with regard to scientific ethics, we highlight key factors that may have contributed to the increasing skepticism toward science.

From the perspective of political science, it seems important to find a systematic place for the science system in systemic models of political science and to systematically assess not only factors that decrease trust, but also those that increase trust. If higher education is considered a policy field in political science, the understanding of these mechanisms is important in a time of democratic backsliding (Bermeo, 2016).

Competing Interests The authors have no conflicts of interest to declare that are relevant to the content of this chapter.

References

Achterberg, P., de Koster, W., & van der Waal, J. (2015). A science confidence gap: Education, trust in scientific methods, and trust in scientific institutions in the United States, 2014. *Public Understanding of Science, 26*(6), 704–720.

Algan, Y., Cohen, D., Davoine, E., Foucault, M., & Stantcheva, S. (2021). Trust in scientists in times of pandemic: Panel evidence from 12 countries. *Proceedings of the National Academy of Sciences, 118*(40).

Ambrosj, J., Desmond, H., & Dierickx, K. (Forthcoming). 'I don't believe in the neutrality of research. OK?' Mapping researchers' attitudes toward values in science. *Accountability in Research.*
Bellolio, C. (2024). An inquiry into populism's relation to science. *Politics, 44*(3), 486–500.
Berggren, N., & Bjørnskov, C. (2022). Political institutions and academic freedom: Evidence from across the world. *Public Choice, 190*, 205–228.
Bermeo, N. (2016). On democratic backsliding. *Journal of Democracy, 27*(1), 5–19.
Besley, J. C. (2023). What are you assessing when you measure, trust in scientists with a direct measure? *Public Understanding of Science, 32*(6), 709–726.
Bromme, R., Mede N. G., Thomm E., Kremer, B., & Ziegler, R. (2022). An anchor in troubled times: Trust in science before and within the COVID-19 pandemic. *PLoS ONE, 17*(2).
Cologna, V., Mede, N. G., Berger, S., Besley, J., Brick, C., Joubert, M., Maibach, E. W., et al. (2025). Trust in scientists and their role in society across 68 countries. *Nature Human Behaviour, 9*(4), 713–730.
Daoust, J.-F., Bélanger, É., Dassonneville, R., Lachapelle, E., Nadeau, R., Becher, M., Brouard, S. et al. (2021). A guilt-free strategy increases self-reported non-compliance with COVID-19 preventive measures: Experimental evidence from 12 countries. *PLoS ONE, 16*(4).
Druckman, J. N. (2017). The crisis of politicization within and beyond science. *Nature Human Behavior, 1*, 615–617.
Easton, D. (1965). *A systems analysis of political life.* John Wiley & Sons.
Easton, D. (1975). A re-assessment of the concept of political support. *British Journal of Political Science, 5*(4), 435–457.
Egelhofer, J. L., Seeger, C., & Binder, A. (2024). The effects of witnessing harassment of scientists on public perceptions of science. *Journal of Science Communication, 23*(9).
Evans, J. H., & Hargittai, E. (2020). Who doesn't trust Fauci? The public's belief in the expertise and shared values of scientists in the COVID-19 pandemic. *Socius, 6.*
Ezrahi, Y. (1990). *The descent of Icarus: Science and the transformation of contemporary democracy.* Harvard University Press.
Flores, A., Cole, J. C., Dickert, S., Eom, K., Jiga-Boy, G. M., Kogut, T., Loria, R., et al. (2022). Politicians polarize and experts depolarize public support for COVID-19 management policies across countries. *Proceedings of the National Academy of Sciences, 119*(3).
Gauchat, G. (2012). Politicization of science in the public sphere: A study of public trust in the United States, 1974 to 2010. *American Sociological Review, 77*(2), 167–187.

Groeniger, J. O., Noordzij, K., van der Waal, J., & de Koster, W. (2021). Dutch COVID-19 lockdown measures increased trust in government and trust in science: A difference-in-difference analysis. *Social Science and Medicine, 275.*

Heinzel, M., & Liese, A. (2021). Expert authority and support for COVID-19 measures in Germany and the UK: A survey experiment. *West European Politics, 44*(5–6), 1258–1282.

Hilligardt, H. (2023). Partisan science and the democratic legitimacy ideal. *Synthese, 202*(5).

Huber, R. A., Greussing, E., & Eberl, J.-M. (2022). From populism to climate scepticism: The role of institutional trust and attitudes toward science. *Environmental Politics, 31*(7), 1115–1138.

Jarren, O., & Donges, P. (2011). Strukturen des intermediären Systems der Interessenvermittlung. In O. Jarren & P. Donges (Eds.), *Politische Kommunikation in der Mediengesellschaft* (pp. 119–28). VS Verlag für Sozialwissenschaften.

Jaspal, R., Nerlich, R., & Koteyko, N. (2012). Contesting science by appealing to its norms: Readers discuss climate science in the Daily Mail. *Science Communication, 35*(3), 383–410.

Kaiser, A. (1997). Types of democracy: From classical to new institutionalism. *Journal of Theoretical Politics, 9*(4), 419–444.

Karran, T. (2007). Academic freedom in Europe: A preliminary comparative analysis. *Higher Education Policy, 20*, 289–313.

Krange, O., Kaltenborn, B. P., & Hultman, M. (2021). ‚Don't confuse me with facts': How right wing populism affects trust in agencies advocating anthropogenic climate change as a reality. *Humanities and Social Sciences Communications, 8.*

Krause, N. M., Brossard, D., Scheufele, D. A., Xenos, M. A., & Franke, K. (2019). Americans' trust in science and scientists. *Public Opinion Quarterly, 83*(4), 817–836.

Lehmann, H., Mortsiefer, H., Müller-Neuhof, J., Woratschka, R., Maicher, L., Bartsch, F., Grega, K. et al. (2017). Gesetze der Macht: Diese Verbände nehmen am meisten Einfluss auf Gesetze. *Der Tagesspiegel.*

Lijphart, A. (2012). *Patterns of democracy: Government forms and performance in thirty-six countries* (2nd ed.). Yale University Press.

Lin, A. C. (2018). President Trump's war on regulatory science. *Harvard Environmental Law Review, 43*, 247–306.

Lockhart, M., Hill, S. J., Merolla, J., Romero, M., & Kousser, T. (2020). America's elecorate is increasingly polarized along partisan lines about voting by mail during the COVID-19 crisis. *Proceedings of the National Academy of Sciences, 117*(40), 24640–24642.

Lott, L. (2024). Academic freedom growth and decline episodes. *Higher Education, 88*(3), 999–1017.

Lott, L., & Spannagel, J. (Forthcoming). *Quality assessment of the academic freedom index: Strengths, weaknesses, and how best to use it. Perspectives on Politics.*

Luna, D. S., Bering, J. M., & Halberstadt, J. B. (2021). Public faith in science in the United States through the early months of the COVID-19 pandemic. *Public Health in Pratice, 2.*

McCright, A. M., Dentzman, K., Charters, M., & Dietz, T. (2013). The influence of political ideology on trust in science. *Environmental Research Letters, 8*, 1–9.

Mede, N. G., & Schäfer, M. S. (2020). Science-related populism: Conceptualizing populist demands toward science. *Public Understanding of Science, 29*(5), 473–491.

Mede, N. G., & Schäfer, M. S. (2022). Science-related populism declining during the COVID-19 pandemic: A panel survey of the Swiss population before and after the Coronavirus outbreak. *Public Understanding of Science, 31*(2), 211–222.

Merkley, E., & Loewen, P. (2021). Anti-intellectualism and the mass public's response to the COVID-19 pandemic. *Nature Human Behavior, 5*, 706–715.

Merton, R. K. (1973). *The sociology of science theoretical and empirical investigations.* The University of Chicago Press.

Meyer, J. M. (2024). Power and truth in science-related populism: Rethinking the role of knowledge and expertise in climate politics. *Political Studies, 27*(3), 520–537.

Mihelj, S., Kondor, K., & Stetka, V. (2022). Establishing trust in experts during a crisis: Expert trustworthiness and media use during the COVID-19 pandemic. *Science Communication, 44*(3), 292–319.

Montesquieu, C.-L. de. (1748). *De l'Esprit des Loix.* Garnier-Flammarion.

Nadelson, L., Jorcyk, J., Yang, D., Smith, M. J., Matson, S., Cornell, K., & Husting, V. (2014). I just don't trust them: The development and validation of an assessment instrument to measure trust in science and scientists. *School Science and Mathematics, 114*(2), 76–86.

Nature, Editor. 2020. Editorial. *Nature, 588*, 370.

Oliveira, T., Wang, Z., & Jingxin, X. (2022). Scientific disinformation in times of epistemic crisis: Circulation of conspiracy theories on social media platforms. *Online Media and Global Communication, 1*(1), 164–186.

Pelke, L. (2023). Academic freedom and the onset of autocratization. *Democratization, 30*(6), 1015–1039.

Reif, A., & Guenther, L. (2021). How representative surveys measure public (dis)trust in science: A systematisation and analysis of survey items and open-ended questions. *Journal of Trust Research, 11*(2), 94–118.

Santirocchi, A., Spataro, P., Alessi, F., Rossi-Arnaud, C., & Cestari, V. (2023). Trust in science and belief in misinformation mediate the effects of political orientation on vaccine hesitancy and intention to be vaccinated. *Acta Psychologica*, 237.

Schäfer, M. S. (2016). Mediated trust in science: Concept, measurement and perspectives for the ‚science of science communication'. *Journal of Science Communication, 15*(5).

Scheitle, C. P., & Corcoran, K. E. (2021) COVID-19 skepticism in relation to other forms of science skepticism. *Socius, 7*, 1–12.

Schmid, J. (2007). „Interessenvermittlung und Policy-Making imeuropäischen Mehrebenensystem: Vom Korporatismus zum Pluralismus zur organisierten Anarchie?" In N. C. Bandelow & W. Bleek (Ed.), *Einzelinteressen und kollektives Handeln in modernen Demokratien* (pp. 43–57). VS Verlag für Sozialwissenschaften.

Spannagel, J., & Kinzelbach, K. (2023). The Academic Freedom Index and its indicators: Introduction to new global time-series V-Dem data. *Quality and Quantity, 57*, 3969–3989.

Taraktas, B., Esen, B., & Uskudarli, S. (2024). Tweeting through a public health crisis: Communication strategies of right-wing populist leaders during the COVID-19 pandemic. *Government and Opposition, 59*(1), 207–228.

Tsebelis, G. (2002). Legislation. In *Veto players: How political institutions work* (pp. 165–186). Princeton University Press.

Vibert, F. (2007). *The rise of the unelected: Democracy and the new separation of powers*. Cambridge University Press.

Wintterlin, F., Hendriks, F., Mede, N. G., Bromme, R., Metag, J., & Schäfer, M. S. (2021). Predicting public trust in science: The role of basic orientations toward science, perceived trustworthiness of scientists, and experience with science. *Frontiers in Communication, 6*.

Xie, Y., Wang, K., & Kong, Y. (2021). Prevalence of research misconduct and questionable research practice: A systematic review and meta-analysis. *Science and Engineering Ethics, 27*(4).

Zillich, A. F., Schlütz, D., Roehse E.-M., Möhring W., & Link, E. (2024). Populism, research integrity, and trust: How populist beliefs shape the relationship between ethical conduct and trust in scientists. *International Journal of Public Opinion Research, 36*(3).

The Idea of Regulatory Experiments as an Instrument of Learning

Kilian Bizer and Daniel Feser

1 Introduction

How do we ensure that regulation improves over time? In economics, regulators such as legislative assemblies are often portrayed as actors who maximize voters' perceived utility in order to secure reelection (Frey & Kirchgässner, 1994). Failing to do so, the argument goes, they will quickly lose their seats due to voter dissatisfaction and competition from other politicians, whether from their own party or another. If vote maximization is the core interest of members of parliament, their learning is directed toward understanding the electorate's preferences, anticipating what voters will demand in the future, and meeting these demands as

K. Bizer (✉)
Faculty of Business and Economics, Georg-August-Universität Göttingen, Göttingen, Germany
e-mail: bizer@wiwi.uni-goettingen.de

D. Feser
Society for Institutional Analysis, Hochschule Darmstadt University of Applied Sciences, Darmstadt, Germany

European University of Technology (EUt+), Darmstadt, Germany

D. Feser
e-mail: daniel.feser@h-da.de

J. Hinrichsen et al. (eds.), *Ethics of the Sciences*,
https://doi.org/10.1007/978-981-92-2821-8_12

effectively as possible, either through legislative action or, at minimum, through symbolic gestures.

This is far from trivial. Voters' preferences may be hidden, unstable over time, or internally inconsistent; moreover, different combinations of political messages and policy actions may be perceived differently by heterogeneous groups of voters. Consequently, this line of thinking explains regulatory differences primarily as a function of differing political preferences across countries. For example, voters in the Nordic countries have shown persistent support for higher taxes and publicly funded education, healthcare, and other social protections, whereas voters in Anglo-Saxon countries tend, broadly speaking, to favor lower taxes and a more limited welfare state. These varieties of capitalism also tend to change only slowly over time (Hall & Soskice, 2013).

If representatives in legislative assemblies act accordingly, choosing one regulatory instrument over another may appear relatively straightforward: they assess which groups in the electorate are likely to be harmed and which are likely to benefit, and then select the option that is politically advantageous. In practice, however, matters are rarely that simple. The more complex the situation underlying a regulatory choice, the more difficult it becomes for regulators to identify effective responses (Hensel et al., 2010). Consider the seemingly simple case of setting a carbon-dioxide emissions target such as climate neutrality by 2045. While the goal is easy to state, achieving it is ambitious and requires reshaping consumption and production across large parts of society. Because nearly everyone is involved, whether through heating homes or traveling by car, plane, train, or bus, policy instruments affect many constituencies in many different contexts. This complexity should be reflected both in the formulation of normative goals, such as climate neutrality by a specified date, and in the regulatory choice problem of which instruments, or which mix of instruments, should be implemented to achieve the goal on time. And while carbon neutrality is already complex, the challenge of closing material cycles appears even more difficult. When regulators face such a "wicked problem," regulatory choices must be broader in scope and take into account dynamics that extend far beyond their own constituency, potentially affecting global value chains.

Many challenges related to climate change, demographic change, and digitalization can be described as "wicked" because of their inherent complexity, the need for collaboration among many actors, and pervasive uncertainty (Incropera, 2016). The search for new answers pushes regulators into uncharted territory. Moreover, interdependencies among climate change, sustainability, and digitalization often turn regulatory choice problems into highly complex tasks with significant ethical implications. For instance, should economic growth be prioritized, or should poverty be addressed more directly? Should policy promote technological progress in digitalization, or should it focus on empowering individuals through education to make their own market choices? And, in the context of regulatory experimentation, is epistemic learning about societal dynamics more important than political learning about voter preferences?

At the global level, the Sustainable Development Goals (SDGs), with 17 goals and 169 underlying targets, provide broad normative guidance for addressing these challenges (United Nations, 2015). At the same time, each legislative assembly must develop a concrete roadmap for its own jurisdiction. Such a roadmap should advance normative goals, enable reasonable economic growth, and avoid alienating voters. In addition, uncertainty extends beyond electoral reactions: regulators may also question whether incremental reforms to existing rules are adequate given the scale of transformation required. In such situations, it is often unclear how regulatory bodies should establish new procedures and policy instruments capable of driving change. If regulatory bodies are also confronted with multiple overlapping crises, learning becomes an urgent challenge in developing plausible responses.

In this paper, we reflect on regulatory learning by legislative assemblies in a broad sense. If a regulatory body adopts specific and ambitious objectives, such as carbon-dioxide neutrality by 2045, appropriate policy instruments are required to achieve them. We understand learning in this context as "a process of updating beliefs about key components of policy (such as problem definition, results achieved at home or abroad, goals, but also actors' strategies and paradigms)" (Radaelli, 2009, pp. 1046–1047). We view learning not only by scientists, but also by politicians, as imperative if wicked problems are to be addressed with the help of regulatory experiments.

Our research question is: What implications do regulatory experiments have for learning by politicians and (social) scientists in the context of wicked problems? We draw on an empirical study of diverse cases of regulatory experiments (Bauknecht et al., 2020) and examine one case in particular that we co-designed with municipalities in Nienburg County in northern Germany to illustrate how wicked problems can be addressed by both politicians and scientists (Bizer et al., 2019a, 2019b). Our working hypothesis, both for this case study and for this contribution more generally, is that learning must be enabled through deliberate design of the interaction process before, during, and after a regulatory experiment.

The paper addresses the volume's theme of "futurization" by understanding regulatory experiments as a way to manage uncertainty. When a legislative body knows the general direction of change but cannot specify concrete goals, or when goals conflict, regulatory experiments can demonstrate, in a relatively controlled setting, what a specific regulatory intervention contributes. Learning from regulatory experiments may occur in different arenas. In the political arena, political learning takes place; in the scientific arena, epistemic learning is more prevalent. These forms of learning are guided by different internal and external ethical norms, each relevant within its own arena. The paper is conceptual in that it depicts and emphasizes learning processes in the context of regulatory experiments. It does so by drawing on a survey of empirical case studies (Bauknecht et al., 2020) as well as a case study we conducted ourselves (Bizer et al., 2019a, 2019b).

The paper also connects to several contributions in this volume that examine relationships between the political arena and the academic arena. Simon Fink's chapter on "The Demand for Science Politics" addresses tensions between politics and academia. Christoph Bleidorn's contribution, "Communicating Science in a Post-Truth Society," examines linkages among science, policy, and society at large. Finally, the paper by Christoph Hönnige and Dominic Nyhuis, which analyzes the relationship between politics and the role of science in democratic systems, is highly relevant because experimentation by regulatory bodies can affect trust in science.

Along these lines the paper emphasizes the importance of regulatory learning through regulatory experiments that integrate the roles of researchers and regulators. We begin by describing the range of existing regulatory experiments and presenting specific cases (Sect. 2). We treat regulatory experiments as a means of empirically examining how actors respond to policy instruments and of drawing conclusions for improving regulation. We then distinguish between the political learning mode that politicians may prioritize and the epistemic learning mode that social scientists may prefer (Sect. 3). Next, we discuss regulatory experiments and their potential effectiveness in addressing complex ("wicked") problems (Sect. 4), including a municipal-county case in northern Germany involving a collective fund. Finally, we draw conclusions about the potential of regulatory experiments to generate learning (Sect. 5).

2 Varieties of Regulatory Experiments

This section introduces regulatory experiments. In the first part, we lay out the origins, definition and goals of regulatory experiments, as well as the types of regulatory experiments. In the second part, we present three empirical cases of regulatory experiments.

Governance research has identified two main drivers behind the rise of experimentation. The first is the need to develop new governance instruments in response to increasingly complex problems. The difficulty of governing environmental challenges, where established problem-solving mechanisms prove ineffective or deliver insufficient results, has been a particularly strong catalyst and has inspired a growing body of work in the sustainability transitions literature (Berkhout et al., 2010; Sengers et al., 2019). This research often examines grassroots, bottom-up projects that test new governance models, such as initiatives promoting small-scale renewable energy. As a result, experimentation has become a tool through which actors from research, public administration, politics, and civil society collectively address difficult regulatory choice problems with the aim of improving regulation (Bizer et al., 2010).

The second driver draws on experimentation as a scientific method. Experiments are a well-established approach in the (social) sciences for generating new knowledge and are especially valuable for assessing causal claims. In economics, experimental methods range from laboratory studies to field experiments conducted in real-world settings. A prominent example, often considered the gold standard, is the randomized controlled trial (RCT), which uses random assignment to treatment and control groups to isolate the effects of a specific intervention. The provision of school materials to students in developing countries, for instance, has been studied using RCT designs (Banerjee & Duflo, 2009). This line of work underscores the importance of rigorous methodological foundations as a prerequisite for experiments intended to renew governance mechanisms.

Building on these two drivers, regulatory experimentation refers to experiments that involve regulatory bodies, state actors, or parliaments. Regulatory experiments can be defined as "a means to deliberately deviate from the current regulatory framework to try out new or different rules in a real-world setting" (Bauknecht et al., 2021). The central idea is that such experiments facilitate two-way communication between regulators and stakeholders. This interaction supports the development of effective governance mechanisms through reflexive regulatory processes (Feindt & Weiland, 2018). Regulatory experiments may aim to generate scientific information, test policy options and pilot programs, or enable innovation (Bauknecht et al., 2020). They may also facilitate market entry for new actors by temporarily relaxing or adapting regulatory barriers (Schittekatte et al., 2021). Overall, regulatory experiments hold promise for reflexive governance by enabling policy tools to be developed, tested, disseminated, and implemented with observable real-world effects (Bos & Brown, 2012; Ceschin, 2014).

Two fundamental types of regulatory experiments can be distinguished. First, regulatory sandboxes create a legal framework for socio-technical experimentation that would otherwise conflict with existing law. Sandboxes typically rely on exemptions or experimental clauses that allow innovators to test new approaches under specified conditions. Second, regulatory innovation trials treat regulation itself as the object of experimentation. In this case, the experiment is designed to generate evidence about how a specific regulatory change affects a population and how particular reforms might be configured.

To date, regulatory experiments remain the exception rather than standard practice. While the literature documents many early "lighthouse" projects, large-scale diffusion and systematic upscaling have not yet occurred. In the following section, we present three empirical cases of regulatory experiments.

2.1 *Regulatory Sandbox in the Netherlands*

In 2015, the Dutch Ministry of Economic Affairs and Climate Policy established a regulatory sandbox to foster innovation in the electricity sector. The initiative responded to the need to integrate renewable energy into the power system, improve efficiency through new services, and enable "prosumer" models in the market. The sandbox was open to locally focused projects involving renewable generation and combined heat and power. Beyond exploring novel approaches to advancing renewable energy systems, its primary purpose was to test solutions at a small scale before considering broader implementation.

2.2 Finland's Basic Income Experiment: A Regulatory Innovation Trial

Public debate about reforming Finland's social insurance system led to the launch of a basic income experiment. In response to changing labor-market conditions, the proposal aimed to adjust existing incentive structures, streamline the tax system, and reduce administrative burdens. While the theoretical arguments were compelling, empirical evidence was limited. For this reason, the Finnish government initiated a regulatory experiment administered by the Finnish Social Insurance Institution (Kela). The experiment included 2,000 randomly selected participants who received €560 per month, along with a control group of 175,000 individuals. The two-year pilot ran from January 1, 2017, to December 31, 2018.

2.3 Municipal Development of Villages and Town Centers: Nienburg County, Germany

In Germany, demographic change and population shifts have left many villages and small towns with vacant older buildings in their historic centers while newer housing is built on the outskirts. As a result, these places may appear less attractive and fear losing residents, small businesses, and investors. In Nienburg County, northern Germany, nearly all municipalities jointly established a fund to support the revitalization of village and small-town centers (Kommunaler Innenentwicklungsfonds, KIF). Only participating municipalities are eligible to apply for funding. The fund is financed through annual contributions from all participating municipalities, including the county government.

Municipalities participate because they value structured competition for strong ideas to make town and village centers more attractive again. Funding is awarded to the most convincing proposals, depending on the resources available in a given year. No municipality can be certain it will receive support for its projects. By 2025, the KIF had been operating for five years, and participants intended to extend it for another five. In this case, the "regulatory" change was initiated by the municipalities themselves: they voluntarily created and financed a mechanism that, in turn, provided resources to improve village and town centers.

3 Learning by Regulatory Experiments

Regulatory learning can take various forms. We focus on two types, political learning and epistemic learning, to contrast the perspectives of politicians and social scientists. Political learning in regulatory experiments involves shifts in stakeholders' preferences as well as changes in how stakeholders collaborate, commit to political goals, and benefit from governance arrangements. Epistemic learning, by contrast, concerns building and refining the scientific knowledge base at a more general level, for example, by examining how stakeholder groups respond to incentives and disincentives, or how price signals (such as surcharges) affect consumer and producer behavior. In the case of Finland's basic income experiment, epistemic learning would include evidence on rates of entry into and exit from the labor market in response to a guaranteed income. Political and epistemic learning are often intertwined in regulatory experiments, particularly when epistemic learning is generated by observing political learning processes among participants.

From a scientific perspective, regulatory experiments primarily support epistemic learning because (social) scientists seek insights that can be reproduced in controlled settings or are at least accepted as valid by scholarly peers. Researchers often aim to identify causal effects between a specific instrument and actors' behavior, not least because causal inference increases the likelihood of publication in high-ranking journals and may attract broader public attention. Scientific reputation is also built through conducting experiments within and across disciplinary boundaries.

Policy makers, however, often approach regulatory experiments differently. Their primary concern is how stakeholders will respond in the next election cycle. If a problem is widely recognized and within the authority of the legislature, policy makers may enhance their standing by solving it or, at minimum, by visibly addressing it. In some situations, symbolic action may be sufficient to secure voter support. From this viewpoint, a regulatory experiment can also function as a way to buy time, postponing a politically costly decision until after an election. At the same time, experiments may also be framed as an active effort to identify an effective solution, in which case politicians may benefit from initiating them.

These heterogeneous perspectives make learning through regulatory experiments challenging. For researchers, the primary concern is typically internal validity: whether procedures were implemented correctly, how participants were selected, whether treatment and control groups were clearly separated (Morton & Williams, 2009), and whether participants understood the experimental tasks. Regulators, by contrast, are often most concerned with external validity: what can be inferred from a small-scale experiment for the broader constituency and how voters may be affected. If scientists and politicians attempted to learn separately, their learning objectives and outcomes would likely diverge. In practice, neither group learns exclusively in one mode, and additional learning modes exist. Still, political learning and epistemic learning capture two central motivations and objectives for learning under the respective incentive structures faced by politicians and scientists.

3.1 Advantages and Disadvantages of Regulatory Experiments

Regulatory experiments have several advantages and disadvantages as a vehicle for learning. One key advantage is that scientists generally cannot conduct regulatory experiments without politicians, and politicians cannot implement them without scientific support. Politicians are needed to create temporary exemptions from existing rules or to establish alternative rules for the duration of the experiment. Scientists, in turn, are needed to anticipate likely effects using qualitative and quantitative methods and to specify plausible outcomes and impacts. Ideally, expected results should not be assessed through a single method but should be triangulated through multiple approaches, which often requires broader, interdisciplinary teams.

Designing a regulatory experiment therefore requires accommodating both scientific and political perspectives and their respective learning priorities. It is an overly idealistic assumption that politicians could engage exclusively in epistemic learning while ignoring political learning. It is equally unrealistic to assume that social scientists would engage only in political learning. While both groups may accept hybrids of learning modes, neither can abandon its primary orientation given its motives, incentives, professional practices, and shared mental models. In this sense, regulatory experiments almost inevitably combine political and epistemic learning and therefore depend on cooperative learning processes among participants.

Another advantage is that regulatory experiments are inherently transdisciplinary. Transdisciplinarity begins from the recognition that behavioral change is required and asks why such change can be expected under different institutional conditions. This involves modeling behavior and predicting responses to institutional change. Relevant evidence can be generated through a wide range of empirical methods, including qualitative and quantitative approaches as well as laboratory experiments, field experiments, simulation games, and related designs.

A further advantage is that regulatory experiments can reduce costs by testing an alternative regulatory approach on a limited scale before extending it to the full population of regulated actors. Trialing a regulation in a bounded setting may reduce costs for citizens, firms, and public agencies if the experiment reveals how the instrument can be better aligned with its objectives and with observed behavioral responses.

At the same time, regulatory experiments have notable disadvantages. First, they require substantial resources to prepare. Regulatory experiments are among the most demanding ways to test regulatory change because they do not merely simulate an alternative framework but implement real regulatory change for a limited group of actors and a limited period. For that reason, they typically require far more preparation than many field experiments, where the regulatory framework remains constant and only the incentive structure is adjusted temporarily.

Second, regulatory experiments require political majorities at two points in time: when the experiment is enacted and when results are available and can be translated into broader regulatory reform. If parliamentary majorities change in the interim, the new majority may have little interest in following up.

Third, considerable time often elapses between preparation, implementation, evaluation, and the translation of findings into general law. For scientists accustomed to three-to-five-year research horizons, this may be acceptable. For politicians, the time lag may even be attractive in some contexts, for example when they want to demonstrate action but are uncertain about the best course of action and prefer to postpone a potentially costly decision until after an election. In other contexts, however, the delay may be politically unattractive, especially when leaders seek to appear decisive and proactive.

Finally, it is important to note that regulatory experiments can counteract political disengagement. Scientific policy advice often fails to reach the political arena. By contrast, when a regulatory experiment is initiated, political actors have already signaled that they expect relevant learning; otherwise, they would not have invested in establishing the experiment in the first place.

3.2 Difficulties Implementing Regulatory Experiments

In this section, we analyze the preconditions for establishing regulatory experiments and integrating them into routine legislative practice. This is not a trivial task, given that experimentation in the context of sustainability challenges often fails to move beyond initial project phases. We use three criteria, scalability, transferability, and accounting for unintended consequences, to characterize challenges and interdependencies in learning through regulatory experiments, especially during the early stages of preparation (Feser et al., 2024; Luederitz et al., 2017). Table 1 summarizes the findings. It shows that researchers and regulators often approach these dimensions from different perspectives that must be integrated. Put differently, their motives, incentives, and learning modes must be aligned within a given regulatory experiment along these three criteria.

The three categories, scalability, transferability, and accounting for unintended consequences, are shaped by the sometimes conflicting perspectives of researchers and regulators when implementing regulatory experiments. This tension reflects their different operational logics. Researchers are incentivized to produce publishable findings that will withstand peer review, whereas regulators are constrained by election cycles, political campaigns, and public scrutiny. This fundamental difference helps explain the divergent approaches summarized in Table 1.

With respect to scalability, researchers prioritize the robustness of the intervention and the quality of the data, while political support matters primarily insofar as it enables continued funding. Regulators, by contrast, are closely tied to electoral timelines and may need to demonstrate visible success to voters within a limited period. Regarding transferability, researchers focus on comparing the appropriate contextual elements, such as regulatory settings, geographic conditions, and time horizons. They also rely on standardized procedures for participant selection and evaluation in order to protect methodological integrity. Regulators, however, tend to place greater emphasis on public opinion and on managing the political and reputational risks associated with experimentation. Finally, in accounting for unintended consequences, Table 1 points to a key asymmetry: regulators are often directly affected by outcomes and potential backlash, whereas researchers can remain more detached and analytical.

Overall, the table suggests that aligning these perspectives is critical for the effective implementation and uptake of regulatory experiments.

Table 1 Own elaborations on the basis of Feser et al. (2024)

Category	*Topic*	*Researcher*	*Regulator*
Scalability	Timeframe	Collecting enough empirical data	Restricted by parliamentary timeframe
	Timing	Depending on planning of intervention	Restricted by legislative term
	Political support	Only interesting for funding	Binds continuation
Transferability	Regulatory context	Contextual factor	Focus on change
	Geographical particularities	Point of comparison	Focus on electoral level
	Timeframe	Adjusted behavior of participants desire for long-term experiments	Risk management with shorter periods
	Selection process for participants	Randomization for making sure that sample is representative	Targeted for supporters of regulators to win next election
	Evaluation procedures	Rigid to compare results	Avoidance of negative publicity/ accountability for regulatory experiment
Accounting for unintended consequences	Variety of design options	Variation increases the robustness and validity of the results	Limiting options due to budget restrictions
	Heterogeneity of participants	No specific focus	Risk of creating negative publicity
	Communication	Focus on learning for scientific progress, e.g. interviews with respondents	Communication toward voters

Understanding the underlying differences is therefore essential for integrating experimentation into legislative practice.

Regulatory experiments also entail significant risks. An experiment may fail in the sense that it produces unintended effects or does not achieve its initial objectives. This helps explain why regulators and parliaments are not automatically inclined to adopt experimental instruments. Put bluntly, elected officials often secure reelection more reliably by taking limited risks than by taking substantial ones. Experiments can also provoke resistance if voters feel treated as "laboratory subjects." If an experiment does not lead to adoption of the tested instrument, it may be labeled a "failed experiment," reinforcing the perception that the burdens of participation were incurred without benefit.

This risk is real, but equating non-adoption with failure is misleading. When conducted rigorously, experiments generate empirical knowledge, including knowledge about feasibility, side effects, and design limits. Such evidence can improve regulatory capacity even when a specific instrument is not implemented in full. Regulatory experiments remain far from standard practice in legislative procedures, and researchers and policy makers are still developing diverse experimental formats and pushing the field forward. As a result, the landscape of regulatory experimentation is evolving rapidly. Tools such as guidelines for regulatory experiments represent an important first step, but they also leave substantial room for further development (Bauknecht et al., 2021).

4 Learning by Regulatory Experiments for "Wicked Problems"

The previous section examined the advantages and disadvantages of learning through regulatory experiments from the perspectives of regulators and researchers. We acknowledge that addressing the broader challenge of sustainable development, especially in the case of "wicked problems," would require engaging a much wider set of stakeholders. In this paper, however, we limit our analysis to the learning opportunities and requirements of politicians and social scientists, as these two groups are central to guiding the design of regulatory experiments.

4.1 Wicked Problems

"Wicked problems" are complex. They require actors to reconcile competing objectives and diverse interests, which may also shift over time, and they demand substantial resources. Moreover, wicked problems are not confined to sustainable development: pension reform in aging societies, corruption among politicians and administrative staff, and bureaucratic procedures and red tape are just a few examples from a long list of issues that are widely perceived as difficult to solve even outside the sustainability context.

The term "wicked" refers to problems that are "highly resistant to solution" (Australian Government, 2007; Rittel & Webber, 1973). Wicked problems cannot be solved through a traditional linear, analytical approach; instead, they require cooperation and collaboration to reconcile conflicting interests. In some cases, the parties involved do not even agree on what the problem is, nor do they fully understand the complex and indirect effects that a given solution may have on third parties. First introduced by Rittel and Webber (1973) in the context of social planning, wicked problems arise at multiple levels of governance, including the global level. For Rittel and Webber (1973), wicked problems are not merely complicated; they call for fundamentally different approaches. They contrast wicked problems with "tame" problems, which may be challenging but for which established routines exist to produce solutions.

Because such routines are largely absent for wicked problems, they require learning among the actors involved. Learning, however, does not occur automatically. It depends on processes that bring relevant parties together, create time and space for them to articulate how they understand the problem, expose them to expert input, and enable them to begin imagining what plausible solutions might look like.

One such alternative approach is to experiment even with regulatory norms. Laws are typically designed to apply uniformly within a jurisdiction, which makes "experimental" regulation seem counterintuitive at first: it implies treating some people differently than others. Yet regulatory experiments require precisely this temporary differentiation in order to generate insights into how and why actors behave differently under alternative institutional arrangements. If differentiated treatment produces better outcomes with respect to regulatory objectives, the experiment may provide evidence, other things equal, for adjusting regulation accordingly.

Regulatory experiments are only one component of learning-oriented approaches to wicked problems. Their distinctive advantage, as outlined above, is that they can facilitate learning by politicians and social scientists as well as by other stakeholders. By design, regulatory experiments test hypotheses and generate empirical results. While such results remain open to interpretation, they nevertheless expand the available knowledge base.

At the same time, regulatory experiments are demanding because they require preparatory research to define one or more treatments that can be tested empirically. Learning processes aimed at addressing wicked problems typically do not begin with a regulatory experiment. Rather, experiments tend to occur at later stages, when alternative regulatory options are already on the table and additional empirical evidence is needed before scaling a particular solution more broadly (Bauknecht et al., 2021).

4.2 *Case Study: Addressing a Wicked Problem in Nienburg County, Germany*

Compared with challenges such as greenhouse gas mitigation or closing material cycles, this section examines a smaller-scale "wicked problem": the increasing number of vacant buildings in the centers of villages and small towns in northern and eastern Germany. While the scope may be more limited, the problem is no less wicked. It is driven by fundamental demographic change, and any solution requires contributions from many actors. It is difficult to resolve and calls for new approaches.

At the core of the problem is a spatial and generational shift. Older residents tend to live in village and town centers, while younger households often build new homes on the outskirts rather than moving into existing housing stock. As the older generation passes away, more central houses become vacant, creating a stark contrast with expanding developments at the periphery. Incentives to refurbish central homes are often weak: living standards and housing preferences have changed, and building a new home on undeveloped land is frequently cheaper than renovating an older building in the center.

In 2016, officials in Nienburg County in northern Germany approached us to explore whether a research design could support local mayors in addressing this challenge (Bizer et al., 2019a, 2019b). After extensive consultations, we developed a simulation game (see other such games Dopfer et al., 2018) to support an ideation and evaluation process for municipalities applying for funding to develop and implement projects in their centers. Financial support for the research and implementation project, the Kommunaler Innenentwicklungsfonds (KIF), was provided by the Federal Ministry of Research, Technology and Space. Our design combined an ongoing process of idea development with an initial funding mechanism financed collectively by participating municipalities. Municipalities would compete with one another, and the best projects as rated by an internal evaluation committee would be selected, depending on available funds.

We conducted in-depth interviews with interested mayors and prepared them through multiple meetings for the fund's simulation game, encouraging them to develop concrete project ideas for their own village or town centers. Following the interviews, participants took part in a one-and-a-half-day simulation in which each person played their actual role as mayor of their municipality. In the simulation, participants worked in groups, proposed measures to counteract vacancy in central areas, drafted simplified grant applications, agreed on a jury procedure, and simulated the selection of winning proposals. They then reflected on the process, proposed revisions, and ultimately agreed to implement such a fund for a five-year period. Implementation began in 2021. In 2025, all participating municipalities agreed to extend the arrangement for an additional five years starting in 2026 in order to sustain the learning process.

Mayors emphasized that each municipality, given its specific demographic situation, required its own tailored response. At the same time, many reported lacking actionable ideas and sufficient funding to initiate planning processes, obtain expertise on building-related challenges, address infrastructure gaps, or mobilize citizen initiatives. As local politicians, mayors clearly have incentives for political learning. Simultaneously, they also sought epistemic learning: while they could anticipate negative trends they wished to prevent, they lacked reliable knowledge about which interventions would work and how to implement them.

The subsequent shift in attitudes was notable. When we first proposed the KIF, many mayors and officials, including members of the project's advisory board and economists we consulted, assumed the fund was unlikely to succeed. Economists argued that municipalities, acting in their self-interest, would not contribute without guaranteed returns. Mayors voiced similar concerns, emphasizing that voluntary payments into a county-level fund could jeopardize reelection.

As discussions progressed, however, mayors increasingly recognized the potential value of the KIF. They highlighted the importance of structured competition, anticipated spillover effects across the county if some projects succeeded, and reported that collective discussion of the problem had already altered their perspectives. After the simulation game, there was near-uniform agreement to proceed with implementation.

What changed? The simulation game and its preparatory stages created structured opportunities for learning (Dopfer et al., 2018; Führ et al., 2018a, 2018b; Scheier et al., 2013), and participants actively used them. Mayors overcame skepticism about whether other municipalities would defect. During the simulation, they shared ideas, assembled them into short stylized grant applications, compared proposals, evaluated them, and made collective decisions. Over the course of the one-and-a-half days, participants engaged in both epistemic and political learning.

Epistemic learning emerged because many participants initially could not envision how such a fund would function in practice: who would carry which workload, how application and selection procedures would operate, and what conflicts might arise. The simulation made these issues tangible and enabled open discussion of the advantages and disadvantages of alternative design options. This epistemic learning process helped participants form realistic expectations about implementation. At the same time, exchanges among mayors, county officials, and representatives from the county assembly enabled political learning across party lines. Political learning was evident in formal sessions when participants discussed electoral constraints and differences among municipalities. It was also visible in informal conversations during breaks, when participants debated specific design elements and articulated their support or opposition more directly.

The KIF addresses a wicked problem at the local level. Challenges such as closing material cycles in the chemical industry, achieving carbon neutrality by 2045, or establishing effective water resource management in Europe operate at very different scales. Nevertheless, the KIF case illustrates that seemingly intractable problems can be addressed when actors converge on a shared problem definition and create conditions for epistemic and political learning. The mayors chose to undertake a regulatory experiment by implementing the KIF. The experiment was successful in demonstrating that municipalities can agree to contribute to a shared fund even without guaranteed individual returns. It also created a cooperation mechanism for a problem that municipalities could not solve independently. Finally, this ongoing process of generating ideas and motivating follow-through provided a foundation for extending the KIF for an additional eight years until 2034.

5 Conclusion

This paper examines how regulation can improve over time under conditions in which legislators are often modeled as vote-maximizers. While political learning about voter preferences, coalition dynamics, and electoral incentives is rational and necessary, the paper argues that wicked problems associated with normatively defined sustainability goals require an additional learning mode: epistemic learning. Epistemic learning produces more reliable knowledge about the intended and unintended effects of regulatory options. Regulatory experiments are proposed as a mechanism for regulatory learning, but only if they are deliberately designed to enable both political and epistemic learning and to manage the friction between the political and scientific arenas. Against this background, the paper investigates what regulatory experiments imply for how politicians and (social) scientists learn when addressing wicked problems.

Conceptually, the analysis links regulatory experimentation to two drivers identified in governance research: the need for new governance instruments in response to increasingly complex sustainability challenges, where established mechanisms often underperform; and experimentation as a scientific method for testing causal claims, ranging from laboratory studies to field experiments and randomized controlled trials. Building on this foundation, regulatory experiments are defined as deliberate deviations from existing regulatory frameworks to test alternative rules in real-world settings. Two broad types are distinguished: regulatory sandboxes, which create temporary legal exemptions for socio-technical innovations that would otherwise conflict with existing law; and regulatory innovation trials, in which the regulation itself becomes the object of testing in order to generate evidence for reform.

The paper illustrates the variety of regulatory experimentation through three empirical cases: a Dutch electricity sandbox (2015) aimed at integrating renewable energy, improving efficiency via new services, and enabling prosumer models; Finland's basic income experiment (2017–2018), administered by Kela with 2,000 randomly selected recipients of €560 per month and a large control group; and a locally initiated German case in Nienburg County, where municipalities created a collective funding instrument to address vacancy and decline in village and small-town centers. Across these cases, experiments are framed as tools for managing uncertainty when policy goals are ambitious, contested, or difficult to operationalize.

A central analytical result is that implementing regulatory experiments is challenging because researchers and regulators follow different operational logics. Scientists tend to prioritize internal validity (rigorous design, clear separation of treatment and control groups, and standardized evaluation), while regulators are often driven by external validity, reputational concerns, and election cycles. These differences shape implementation prospects along three criteria: scalability, transferability, and accounting for unintended consequences. Researchers prioritize robust data and methodological consistency, whereas regulators prioritize politically feasible timeframes, public acceptance, and risk management. The paper argues that successful experimentation therefore depends on aligning motives and learning objectives across arenas rather than treating experiments as purely technical evaluations.

The Nienburg County case (Kommunaler Innenentwicklungsfonds, KIF) provides detailed evidence on how such alignment can be achieved. Nienburg faced a wicked problem driven by demographic change: older residents remained in historic centers while younger households built on the outskirts, leading to increasing vacancy in central housing stock. Renovation incentives were weak because greenfield construction often appeared cheaper than refurbishing older buildings, and municipalities lacked ideas, resources, and expertise to respond. In 2016, county officials sought a policy instrument that could support local mayors. After extensive consultations, we developed a structured, participatory process centered on a simulation game that allowed mayors to test an institutional solution before committing to it: a collectively financed, competitive fund through which participating municipalities and the county contributed annually and selected the best projects for funding.

In Nienburg, initial reactions were strongly skeptical: economists predicted municipalities would not contribute without guaranteed payback, and mayors feared electoral punishment for transferring local funds to a county-level instrument. During preparatory interviews, meetings, and the one-and-a-half-day simulation, participants collaboratively explored design choices, anticipated conflicts, clarified workloads, and rehearsed decision rules (applications, jury procedures, selection). This produced epistemic learning about how the mechanism would work in practice and what trade-offs different design options entailed. At the same time, it generated political learning as mayors exchanged views about constituency constraints, distributional concerns across municipalities, and the political acceptability of various rules. Importantly, the process reduced expectations of defection and increased trust that other municipalities would contribute even without guaranteed returns, partly because participants recognized spillover benefits from successful projects elsewhere in the county.

The result was a notable shift from widespread doubt to near-unanimous commitment. Municipalities agreed to implement the KIF for five years starting in 2021 and, after operating it through 2025, collectively decided to extend it for another year, preparing at the beginning of 2026 another eight years until 2034. The paper interprets this as evidence that well-designed interaction processes can translate experimentation into sustained institutional practice, even under conditions where standard economic predictions of narrow self-interest would suggest non-cooperation. More broadly, the KIF case supports the paper's core claim: regulatory experiments are most effective when they are designed not only to test instruments, but also to structure learning, coordination, and commitment across political and scientific actors, thereby enabling regulatory improvement in the face of wicked problems.

Competing Interests We are both grateful for funding by the Federal Ministry of Research, Technology and Space, specifically the projects "Regulatory Experiments for reflexive and adaptive Governance of Innovations (REraGI—grant no. 16ITA213)" and "Kommunaler Innenentwicklungsfonds (KIF—grant no. 033L171)." Both projects contributed greatly to this paper. The authors have no conflicts of interest to declare that are relevant to the content of this chapter.

References

Australian Government. (2007). *Tackling wicked problems. A Public Policy Approach*. Australian Public Service Commission. https://www.enablingchange.com.au/wickedproblems.pdf

Banerjee, A. V., & Duflo, E. (2009). The experimental approach to development economics. *Annual Review of Economics, 1*(1), 151–178.

Bauknecht, D., Heyen, D. A., Gailhofer, P., Bizer, K., Feser, D., Führ, M., Winkler-Portmann, S., Bischoff, T. S., & Proeger, T. (2021). *How to design and evaluate a regulatory experiment? A guide for public administrations*. Oeko-Institut e.V; Georg-August-University Goettingen; University of Applied Sciences Darmstadt; Institute for Small Business Economics, 26 pp. Retrieved January 14, 2025, from https://www.oeko.de/fileadmin/oekodoc/Regulatory_Experiments-Guide_for_Public_Administrations.pdf

Bauknecht, D., Bischoff, T. S., Bizer, K., Führ, M., Gailhofer, P., Heyen, D. A., Proeger, T., & von der Leyen, K. (2020). Exploring the pathways: Regulatory experiments for sustainable development—An interdisciplinary approach. *Journal of Governance and Regulation, 9*(3), 49–71.

Berkhout, F., Verbong, G., Wieczorek, A. J., Raven, R., Lebel, L., & Bai, X. (2010). Sustainability experiments in Asia: Innovations shaping alternative development pathways? *Environmental Science & Policy, 13*(4), 261–271.

Bizer, K., Lechner, S., & Führ, M. (2010). The European impact assessment and the enviroment.

Bizer, K., Gödecke, D., Klein, T., Jacuniak-Suda, M., Sack, A., & Sommer, A. (2019a). Kommunaler Innenentwicklungsfonds. Ein Instrument zur Stabilisierung und Weiterentwicklung der Ortskerne im ländlichen Raum. Planerin: Innen entwickeln. *Kommunikativ und Kooperativ Lebenswerte Orte Gestalten, 6*, 38–40.

Bizer, K., Gödecke, D., Jacuniak-Suda, M., Klein, T., Sack, A., & Sommer, A. (2019b). Kommunaler Innenentwicklungsfonds. *Ein Instrument Zur Stabilisierung und Weiterentwicklung der Ortskerne Im Ländlichen Raum, Planerin, 6*(2019), 38–40.

Bos, J. J., & Brown, R. R. (2012). Governance experimentation and factors of success in socio-technical transitions in the urban water sector. *Technological Forecasting and Social Change, 79*(7), 1340–1353.

Ceschin, F. (2014). How the design of socio-technical experiments can enable radical changes for sustainability. *International Journal of Design, 8*(3), 1–21.

Dopfer, J., Bizer, K., Schentenh, J., Balla, S., Schicketanz, S., & Bunge, T. (2018). Planspiele in der Gesetzesfolgenabschätzung—Teil 2—Empfehlungen anhand ausgewählter Planspiele. *UVP-Report, 32*(2), 79–86.

Feindt, P. H., & Weiland, S. (2018). Reflexive governance: Exploring the concept and assessing its critical potential for sustainable development. Introduction to the special issue. *Journal of Environmental Policy & Planning, 20*(6, SI), 661–674.

Feser, D., Winkler-Portmann, S., Bischoff, T. S., Bauknecht, D., Bizer, K., Führ, M., Heyen, D. A., Proeger, T., von der Leyen, K., & Vogel, M. (2024). Institutional rules for the up-take of regulatory experiments: A comparative case study. *Futures, 156*(28), Article 103318.

Führ, M., Dopfer, J., Bizer, K., Balla, S., Bunge, T., Schenten, J., Schicketanz, S. (2018a). Simulation games in the Regulatory Impact Assessment—Simulation of the implementation of the EIA Amending Directive 2014/52/EU. elni Review, 17–24.

Führ, M., Dopfer, J., Bizer, K., Schenten, J., Balla, S., Schicketanz, S., & Bunge, T. (2018b). Planspiele in der Gesetzesfolgenabschätzung—Teil 1: Planspiel zur Umsetzung der UVP-Änderungsrichtlinie 2014/52/EU. *UVP-Report, 32*(2), 68–78.

Frey, B. S., & Kirchgässner, G. (1994). *Demokratische Wirtschaftspolitik, 2.* Verlag Franz Vahlen.

Hall, P. A., & Soskice, D. W. (Eds.). (2013). *Varieties of capitalism: The institutional foundations of comparative advantage* (Reprint). Oxford University Press.

Hensel, S., Bizer, K., Führ, M., & Lange, J. (2010). Gesetzesfolgenabschätzung in der Anwendung - Perspektiven und Entwicklungstendenzen. *Interdisziplinäre Studien zu Recht und Staat*. Bd. 48. Nomos.

Incropera, F. P. (2016). *Climate change: A wicked problem*. Cambridge University Press.

Luederitz, C., Schäpke, N., Wiek, A., Lang, D. J., Bergmann, M., Bos, J. J., Burch, S., Davies, A., Evans, J., König, A., Farrelly, M. A., Forrest, N., Frantzeskaki, N., Gibson, R. B., Kay, B., Loorbach, D., McCormick, K., Parodi, O., Rauschmayer, F., ... Westley, F. R. (2017). Learning through evaluation—A tentative evaluative scheme for sustainability transition experiments. *Journal of Cleaner Production, 169*, 61–76.

Morton, R. B., & Williams, K. C. (2009). Experimentation in political science. In J. M. Box-Steffensmeier, H. E. Brady, & D. Collier (Eds.), *The Oxford handbook of political methodology*. Oxford University Press.

Radaelli, C. M. (2009). Measuring policy learning: Regulatory impact assessment in Europe. *Journal of European Public Policy, 16*(8), 1145–1164.

Rittel, H. W. J., & Webber, M. M. (1973). Dilemmas in a general theory of planning. *Policy Sciences, 4*, 155–169.

Scheier, J., Spiwoks, M., & Bizer, K. (2013). Planspiel Kapitalmarktprognose—Ein empirischer Vergleich der Prognosekompetenz von Amateuren und Experten. Darmstadt: 13–2, 2013. *sofia-Studien zur Institutionenanalyse, 13*(2).

Schittekatte, T., Meeus, L., Jamasb, T., & Llorca, M. (2021). Regulatory experimentation in energy: Three pioneer countries and lessons for the green transition. *Energy Policy, 156*, Article 112382.

Sengers, F., Wieczorek, A. J., & Raven, R. (2019). Experimenting for sustainability transitions: A systematic literature review. *Technological Forecasting and Social Change, 145*, 153–164.

United Nations (UN). (2015). Transforming our world: The 2030 agenda for sustainable development. A/RES/70/1. United Nations (UN). Retrieved February 19, 2020, from https://sustainabledevelopment.un.org/post2015/transformingourworld/publication

Part 4: Ethics in Science Communication and Education

The Dual Lens: Connecting AI Ethics in Education with Science Ethics

Andreas Nehring and Elisabeth Hildt

1 Introduction

Artificial intelligence (AI) continues to influence educational systems. In particular, the advent of large language models (LLMs) such as ChatGPT presents notable opportunities to transform educational environments by enhancing both teaching and learning experiences. Integrating reflection on ethical aspects and societal implications of AI into education and educational environments becomes increasingly crucial. This is true in various regards: to provide students and educational environments with AI technologies that are ethical and trustworthy; and to educate students about the responsible use of AI in the various contexts of their current

A. Nehring (✉)
Institute of Science Education, Leibiz University Hannover, Hannover, Germany
e-mail: nehring@idn.uni-hannover.de

E. Hildt
Center for the Study of Ethics in the Professions, Illinois Institute of Technology, Chicago, USA

Institute of Data Science, Leibniz University Hannover, Hannover, Germany

E. Hildt
e-mail: ehildt@illinoistech.edu

J. Hinrichsen et al. (eds.), *Ethics of the Sciences*,
https://doi.org/10.1007/978-981-92-2821-8_13

situation (at school, at home, during leisure time) and future roles as citizens and professionals.

Recent reviews of AI ethics education, such as Wiese et al. (2025), show that while the field has rapidly expanded, much of the discussion remains limited to individual moral awareness. This chapter extends that perspective by situating ethical reflection within academic and scientific practice.

Beyond providing an overview of ethical challenges of AI in education, this chapter contributes to the broader discourse of the ethics of science by exploring how academic research and educational practice mutually shape the ethical landscape of AI. From this perspective, ethical reflection is not limited to general users, students, or teachers but also extends to scientists, developers, and institutions designing and evaluating AI systems. Following the principle of continuous procedural reflection central to the ethics of science, we sketch how responsible academic practice—aligned with frameworks such as the Responsible Conduct of Research (see below)—can foster trustworthy and educationally sound AI.

Discussing the responsible and ethical use of AI in education involves a number of disciplines and topic areas. Especially when AI tools are used in science, technology, engineering, and mathematics (STEM) education, questions about responsible uses of AI in these fields are closely connected to research ethics and ethics of science.

The interdisciplinary debate of ethical aspects of AI in education can build on two related fields: On the one hand, the broad field of ethics of science (also called "research ethics") and on the other hand, the relatively new field of AI ethics with its direct focus on ethical and societal aspects of AI.

In the United States, research ethics has a clear focus on what is called Responsible Conduct of Research (RCR), the standards and rules of responsible research practice. The US National Institutes of Health (NIH) define responsible conduct of research as "the practice of scientific investigation with integrity. It involves the awareness and application of established professional norms and ethical principles in the performance of all activities related to scientific research". (National Institutes of Health [NIH], 2009). Over the years, the subject matter relevant to education in RCR has been expanded. In the most recent guidance document "Requirement for Instruction in the Responsible Conduct of Research" (National Institutes of Health [NIH], 2022), the NIH provides an updated list of topics that can subject to higher or graduate education as part of formation of training junior or established researchers. Some of the topics on this list are primarily of relevance for lab-based research, such as policies regarding human subjects, animal research, and safe laboratory practices; mentor/mentee responsibilities and relationships; safe research environments; collaborative research; peer review and research misconduct. Other topics on this list, while certainly relevant for lab-based research, address norms and ethical principles that are also of central importance when using AI-based tools in research and education. These include conflict of interest and conflict of commitment, in allocating time, effort, or other research resources; data acquisition and analysis; laboratory tools (e.g., tools for analyzing data and creating or working with digital images); secure and ethical data use; data confidentiality, management, sharing, and ownership; responsible authorship; and societal impacts of scientific research.

When it comes to education about or education involving the ethical and responsible use of AI, these norms and ethical principles are worth considering. What is needed is a transfer of the rather general standards listed in the RCR document to the specific field of education about AI or education involving AI tools.

In contrast, the field of AI ethics, with its focus on ethical and societal aspects of AI, provides a relatively specific debate on ethical aspects such as privacy, bias, or informed consent in the specific context of AI-based tools and human-AI interaction. Education involving AI tools and education on ethical and responsible uses of AI can directly profit from the discussion and reflection in this interdisciplinary field. What is needed is the identification of the topics and aspects considered relevant in education and the transfer to educational contexts.

In this chapter, we focus on the latter, more direct approach, and build on the field of AI ethics to discuss ethical aspects of AI in education. Given the high potential AI has for supporting education, this chapter provides an overview of ethical aspects of AI tools in educational contexts, organized around five areas of concern: bias, AI-related disparities, data privacy, agency and autonomy, as well as opacity and transparency. In addition, we present and discuss approaches for integrating AI ethics into education, including frameworks and concrete examples. We address the importance of fostering AI literacies among students and educators, emphasizing the need for educational frameworks that equip users with the skills necessary to navigate AI technologies responsibly. Toward the end of the chapter, we then compare the AI ethics-based approach with the AI-related topics mentioned in the NIH guidance on RCR instruction and identify future directions for research on ethical aspects of AI in education. By providing not only a condensed overview of the current debate but also seeking to develop the debate further by integrating a broader research ethics perspective, this chapter aims to contribute to the ongoing discourse on ethical AI implementation in education.

2 AI-based Systems in Education: Potentials and Promises

AI-based systems have the capability to provide personalized learning opportunities at a large and economically feasible scale, tailoring educational content to fit individual learners‘needs and pacing, thus promoting inclusivity and accessibility in education (e.g., Holmes et al., 2022; Kasneci et al., 2023; Paulsen & Lindsay, 2024). By enabling real-time feedback and interactive conversations, large language models and cognitive tutors can individually support students in a range of disciplines, offering explanations, elaborations, and feedback that enrich traditional learning methods (Hwang et al., 2020; Shin & Shim, 2021). Furthermore, these tools can aid educators by automating administrative tasks, such as grading and responding to common student inquiries, allowing teachers to allocate more time to personalized instruction and mentorship (e.g., Adler & Shani, 2026). Additionally, AI has even the potential to foster critical thinking and creativity by engaging students in simulated dialogues and problem-based learning scenarios, preparing them for complex, real-world challenges.

An AI-based personalization of learning is especially valuable in regions with a shortage of qualified teachers, such as rural or socioeconomically disadvantaged areas, where AI tools can fill instructional gaps. Furthermore, AI-based tools can have a supportive role for students with disabilities by adapting educational content to their specific requirements, thereby fostering inclusivity (Roski et al., 2024). The ability of AI tools to provide individualized support according to learners‘achievement levels can help bridge performance gaps between low and high achievers. By facilitating differentiated learning, AI technologies can also enable teachers to effectively manage mixed-level classrooms and support remote learning environments.

When it comes to combining AI and education, a distinction can be made between four different areas (Holmes et al., 2022): (a) "Learning with AI", the use of AI-based tools in teaching and learning; (b) "Using AI to learn about learning", i.e. the field of learning analytics or data mining that seeks to learn how students learn and develop effective learning approaches; (c) "Learning about AI", i.e. an area that seeks to increase the AI knowledge and skills to increase people's technical understanding of AI technology; and (d) "Preparing for AI", the goal of which is to prepare citizens for the social and ethical aspects of interacting with AI technology. Learning about AI and Preparing for AI closely go together, as learning about the technological dimension of AI technology should always include learning about its ethical and societal implications (see Holmes et al., 2022).

Here, we focus on the various ways in which AI technologies can be used to support learning (a combination of (a) and (c)), and the concomitant fourth area, i.e. educating students and teachers about the chances and risks of AI application in educational contexts.

3 AI Ethics

As with any transformative technology, AI applications may raise new ethical questions and reinforce existing issues. In a generic perspective, Mittelstadt et al. (2016) delineate several specific challenges, each carrying significant implications for education and beyond. One significant concern is the reliance on inconclusive evidence due to the inherent limitations in data quality and availability (see also Messeri et al., 2024). Algorithms may utilize incomplete or uncertain datasets, resulting in decision-making processes that lack a sound evidential basis, ultimately leading to actions that are unjustified or unsupported by robust data. This challenge is particularly salient in fields where high-stakes decisions require rigorous justification. Ultimately, integrating AI ethics thoughtfully into educational research and practice offers the potential for equitable and sustainable progress in the field.

Given such challenges, AI ethics encompasses various aspects, including bias, data privacy, student autonomy in learning processes, and transparency. Effectively incorporating AI ethics into educational frameworks necessitates innovative approaches that address these critical issues (Akgun & Greenhow, 2022; Holmes et al., 2022; Krasneci et al., 2023; UNESCO, 2022; U.S. Department of Education, 2024).

3.1 Bias

Bias is a central critical concern, as highlighted by the potential for algorithms to generate misguided evidence (Li et al., 2023). When algorithms are developed or trained using biased datasets, they can perpetuate or magnify these biases, leading to decisions that are systematically unfair. Such biases may disadvantage particular groups, raising significant ethical and social justice issues across various domains (Cheuk, 2021).

These inequities manifest as unfair outcomes, which can lead to discrimination. Discriminatory impacts may emerge when algorithmic decisions favor certain populations over others, often reinforcing existing societal biases and inequalities (Ifenthaler et al., 2024). As AI becomes increasingly pervasive, it has the potential to alter traditional power dynamics and influence personal decision-making processes, challenging individual autonomy. However, the opaque nature of many AI systems complicates efforts to trace outcomes back to their origins, obscuring who is responsible for algorithmic decisions. This lack of interpretability poses ethical challenges, as it can prevent appropriate attribution of responsibility and hinder efforts to address potential harms caused by possibly biased AI systems (Rudin, 2019).

The use of AI tools in education presents significant concerns regarding bias and fairness, with the potential to perpetuate existing biases and reinforce disadvantages and forms of discrimination (Akgun & Greenhow, 2022). The datasets used to develop these technologies often reflect the social and cultural contexts from which they originate, posing risks of embedding systemic biases and perpetuating historical sociocultural norms. For instance, the seemingly mundane choices involved in selecting tasks or names in educational content can inadvertently reflect and reinforce societal stereotypes, such as the selection of gender-specific or descriptive language that associates certain behaviors with specific demographics.

A critical issue is that most AI-based educational tools are developed by companies located in Western, English-speaking countries. Consequently, the underlying data and algorithms are often biased toward English-speaking contexts, putting at disadvantage non-English-speaking learners by not adequately representing their linguistic and cultural nuances.

3.2 AI-Related Disparities

While AI technologies used in classrooms offer each student access to the same resources, promoting a semblance of equal opportunity, the differential access to technology outside of school settings can exacerbate disparities. Not all students have equal access to updated technology or reliable internet at home, which is essential for utilizing AI tools for individualized learning. Additionally, some educational tools are proprietary, requiring purchase, which can be a barrier for socioeconomically disadvantaged students.

Acquiring the skills necessary to use these technologies effectively, such as prompt engineering, may also be easier for some students than others, potentially widening the gap between those who can and cannot leverage these technologies to their full potential. These factors collectively contribute to increased unfairness and disadvantage certain student groups, particularly those who are socioeconomically or otherwise marginalized.

The reliance on AI-based technology for educational purposes becomes particularly problematic if it cannot be accessed equally by all students, leading to systemic disadvantages in learning opportunities and assessments. Automated grading tools, which are often rooted in datasets that favor English-language expressions and conventions, tend to impose further disadvantages on non-English-speaking learners.

The deployment of AI-based educational tools may inadvertently exacerbate educational disparities, given the diverse linguistic capabilities and needs of students. AI tools, often developed within English-speaking Western contexts, may inherently reflect biases that hinder their effectiveness in different cultural settings or non-English-speaking regions.

What is more, although the positive aspects of AI-based tools have been acknowledged, the evidence supporting their efficacy in improving student learning in the long run remains insufficient. Robust scientific studies are lacking, partly because many are conducted by companies promoting their own tools, thus raising concerns about independence. The rapid pace of technological development further complicates objective assessment, as new educational tools continuously evolve (e.g., Holmes et al., 2022; Valle et al., 2021).

3.3 Data Privacy

The integration of AI-based systems in educational settings brings significant privacy concerns related to the collection and use of personal information. These tools often require tracking students'and teachers'digital activities to gain insights into learning behaviors, improve educational experiences, and individualize learning paths. The AI-based personalization of learning necessitates the collection of detailed data about student interactions with the tools, including their responses, performance metrics, preferences, and usage patterns. This data collection allows the system to tailor instructional content and suggest future learning activities based on students'past interactions.

Despite the requirements under law (such as the General Data Protection Regulation of the EU; European Union, 2016) for obtaining consent from students or their parents for data collection, there can still remain a lack of transparency regarding the specifics of how and where this data is subsequently shared if it is further processed through AI-based systems. In those cases where there is a clear expectation within classroom settings to use such tools, this further complicates matters, as opting out for any reason, including privacy concerns, might place students at a disadvantage compared to their peers. This pressure can effectively limit the genuine consent and agency of students or their guardians regarding data privacy.

Moreover, if teachers can access and monitor this data, additional privacy considerations are introduced. While having access to detailed information about student learning can aid instructional decision-making, it also borders on excessive surveillance that risks infringing upon students'privacy. Continuous monitoring of student behavior may shift classroom dynamics, engendering a sense of constant surveillance among learners, which could have negative implications for student autonomy and trust.

3.4 Agency and Autonomy

AI-based learning tools are designed with the intent to enhance the learning experience by offering personalized tutorials to meet the individual needs of students. When effectively implemented, these tools can support student autonomy by allowing learners to tailor their educational paths and improve their skills (e.g., Oliveira et al., 2025). Such systems can empower both teachers and students to select tutorials and resources they find most beneficial, fostering a more individualized and self-directed learning environment.

There is a lack of research on how the use of AI tools, including LLMs like ChatGPT, influence learning, however. Concerning the use of GPT-based tutors in educational settings, a recent study by Hamsa Bastani and colleagues (Bastani et al., 2024) provides a cautionary tale. Their study results indicate that while these tools may be useful in providing immediate support in problem-solving, they may negatively affect learning in that their use prevents students from learning critical skills.

Moreover, predictive systems, which generate learning pathway recommendations based on historical data, may inadvertently limit user autonomy (Akgun & Greenhow, 2022; Holmes et al., 2022). As students are guided through predetermined pathways, they need to follow prescribed sequences and complete various levels, potentially hindering the development of independent learning strategies. While these structured approaches might benefit some students, they may constrain others, posing a risk of students becoming "stuck" within the automated system without perceiving it as beneficial to their learning journey.

The automated pathways suggested by AI tools heavily rely on the system's perceived assessment of a student‘s skill level. Issues such as language barriers or sociocultural differences can lead to inaccuracies in this assessment, resulting in tutorials that may misalign with a student‘s actual learning requirements. What is intended as a supportive, individualized learning approach may, therefore, inadvertently disrupt or derail a student‘s learning efforts.

In addition, the focus of AI-based tools on technology-mediated learning may inadvertently overlook the development of social skills due to reduced teacher-student communication and interaction (Avraamidou, 2024).

Automated grading systems offer another area of AI implementation, providing time-saving benefits for educators. However, this reliance on technology for assessing written texts and essays differs significantly from human grading. By delegating grading tasks to automated systems, there is a risk that teachers might engage less with students‘work, potentially weakening the teacher-student relationship and limiting teachers‘understanding of students‘perspectives and experiences. Automated grading has also been noted for its potential biases, which could disadvantage non-native speakers.

Critics point to the risk of such technologies substituting interpersonal communication and the relational aspects of teaching, framing AI as a "techno-solutionist approach" to addressing educational challenges with uncertain consequences (Holmes et al., 2022, p. 21). It is fair to say that currently it is unclear how AI use influences teaching, learning, learning success, and the nature of interactions between students and teachers.

Additionally, the accessibility of advanced AI models capable of high-quality language translation could discourage students from actively pursuing foreign language studies, perceiving such skills as unnecessary. This perception could diminish autonomy to engage deeply with language learning, potentially affecting linguistic diversity and cultural understanding through decreased motivation.

Furthermore, the deployment of proprietary AI systems in education raises concerns about agency and autonomy at institutional and national levels. Commercial tools predominantly shape the learning experience and development of students, determining what is emphasized as valuable knowledge. As a result, the ability of public institutions, governments, and educational authorities to influence educational content and standards is reduced, potentially leading to a diminished role in shaping educational priorities and curricula (Holmes et al., 2022).

3.5 *Opacity and Transparency*

Incorporating AI tools into educational settings necessitates a focus on transparency at multiple levels, encompassing both the functionality of the AI systems themselves and their application within educational contexts. Transparency requires that teachers and students disclose when and how AI-based support is used, particularly in contexts where LLMs are employed to draft texts, complete assignments, or grade exams. This openness is essential to maintain academic integrity and ensure that AI serves as a facilitative tool rather than an unacknowledged contributor.

The complexity of algorithms further contributes to the issue of inscrutable evidence (Duede, 2023). Many AI systems, particularly those employing deep learning techniques, operate as "black boxes", offering little transparency into how inputs are transformed into outputs. This opacity hinders stakeholders' ability to understand, audit, or challenge the decisions made by these systems, posing significant obstacles to achieving accountability and transparency. The emphasis on prediction in many AI models complicates the extraction of causal conclusions or actionable interventions, necessitating further research to address these epistemic challenges. Messeri et al. (2024) highlight the need for researchers to be aware of these risks, given that machine learning processes can perpetuate biases embedded in training data, leading to ethical concerns in automated decision-making. Additionally, the automation of decision-making processes can inadvertently mask systemic inequalities and cause unintended shifts in educational practices, as indicated by studies in human-centered learning analytics. To effectively manage these complexities, interdisciplinary approaches that blend technical acumen with ethical scrutiny are essential.

A significant issue surrounding AI in education is the lack of transparency regarding how these tools are developed and maintained. Often, private companies that market AI tools highlight their advantages while downplaying or overlooking potential downsides. This marketing bias can obscure critical insights into the tools' development processes, operational mechanics, data collection methodologies, and application nuances.

Commercial AI products typically do not provide public access to research findings or data that elucidate their functioning. Details about what data is collected, how it is utilized, and the algorithms driving decision-making are often proprietary and not disclosed. This opacity not only raises ethical concerns about user privacy and data use but also limits the ability of educators to critically assess these tools.

Furthermore, there is a lack of transparency regarding the effectiveness of AI tools. Companies frequently make claims regarding the benefits of their products without providing empirical evidence or allowing independent oversight from schools or public institutions. As highlighted by Holmes et al. (2022) in the Council of Europe report "Artificial Intelligence and Education", this lack of accountability and verification makes it challenging for educational stakeholders to evaluate the actual impact and value of these technologies in the classroom.

4 AI Ethics as Part of Education for and Through AI

While the five areas of concern outlined above—bias, disparities, privacy, agency, and transparency—highlight ethical challenges at the level of educational practice, these issues also originate within the scientific and technological development of AI systems themselves. Addressing them, therefore, requires reflection and communication not only by educators and policymakers but also by researchers and academic institutions that design, test, and disseminate AI-based educational tools. The following section expands on how ethical reflection and education can be anchored at all these levels, including the academic and institutional spheres.

Given the various risks associated with AI technologies, there is a pressing need to adequately elaborate AI-related ethical literacies for students, teachers, and scientists, balancing its advantages and risks. It is essential for the educational and scientific community not only to grasp the technological dimensions of AI but also to understand the ethical issues and societal implications that accompany its use. This is true not only for the specific applications of AI tools in educational settings but also more broadly speaking for the general use of AI-based technologies by individuals and institutions. To effectively integrate AI ethics into education, three different types of approaches need to be taken, which focus on teachers, students, and the institutional context, respectively:

4.1 AI-Related Education for Teachers

It is crucial that all educators, regardless of their subject area, receive education about AI's technological aspects, as well as its ethical and societal impacts (Miao & Cukurova (2024). This responsibility should not be relegated solely to disciplines like computer science or mathematics. Given the pervasiveness of AI technologies in all areas of individual and social life, educators from all departments need to learn to reflect on the benefits and limitations of AI within their specific fields (e.g., Kajiwara & Kawabata, 2024).

Often, this implies that for teacher students, M.Ed. students, or pre-service teachers, one of the first ways of getting familiar with this area is to take courses or attend lectures or workshops about ethical and societal aspects of technology at their colleges or universities. Increasingly, courses, modules, workshops, lectures, or similar are being developed and offered that foster interdisciplinary understanding and help students increase knowledge and awareness of the ethical and societal implications of AI technologies. This is not a given for all approaches, some of which focus largely on how AI-based innovations can push teaching and learning (e, g. for science, technology, engineering and mathematics (STEM) education: Huwer et al., 2024).

Various options exist for integrating ethics education into curricula or courses in STEM, see the overview by Kreth et al. (2024). Alexandra

Bradner and Rebecca A. Bates provide an example of how to easily integrate ethics reflection in STEM education at universities (Bradner & Bates, 2024).

The UNESCO Competency Framework for Teachers on AI (UNESCO, 2022) emphasizes a human-centered approach. It draws on key documents like UNESCO's Recommendation on the Ethics of Artificial Intelligence and the Beijing Consensus, underscoring that AI should enhance human capacities, remain accessible, operate safely, and hold humans accountable for its use. Compared to Huwer et al. (2024), the framework adds more ethically relevant aspects of AI-related competencies. Teachers are pivotal in this framework, needing support to integrate AI responsibly while retaining accountability for educational outcomes. The framework highlights inclusivity by urging educators to be aware of potential algorithmic biases and to ensure AI fosters equitable learning environments for all students. It also equips teachers to critically evaluate AI tools, understand their functioning, and responsibly assess AI's in education. In the UNESCO competence level model, teachers at the highest competence level should be able to not only apply guidelines to ensure the responsible use of AI by teachers and students while respecting ethical principles (respecting others' copyrights and protecting their own, mitigating prejudice, combating deepfakes and AI-enhanced hate speech, and protecting themselves and their students, especially those with disabilities, from AI-manipulated bullying and discrimination), but also "assess the appropriateness and sufficiency of guidance for users of a specific AI tool against the ethical risks rooted in its design and the potential social controversies caused by its use. They should frame recommendations for remedying or improving the guidance accordingly or solidify the view that regulations on AI ethics must be designed by and for human stakeholders; advocate for and participate in the dialogues, development or further iteration of local or institutional regulatory frameworks or guidelines that promote ethics in the design, validation, adoption, deployment and application of AI" (UNESCO, 2022, p. 39).

4.2 *AI-Related Education for Students*

Students must be trained to use AI effectively, while also recognizing and mitigating potential negative effects. They should learn to maintain realistic expectations and be aware of AI technology's limitations. Educating students early on about the ethical and societal implications of AI technology requires adequate educational frameworks and materials suitable for the various age groups. In addition to teaching students about technology itself, it will be important to give them an understanding of the intricate interwovenness of technology and values, and what they need to consider when using AI-based technology.

A broad spectrum of educational approaches and materials exists that help educators discuss AI technology and its ethical and societal implications in the classroom. For instance, the MIT Media Lab developed a collection of educational tools called "An Ethics of Artificial Intelligence Curriculum for Middle School Students" that is freely available online (https://www.media.mit.edu/projects/ai-ethics-for-middle-school/overview/). Portuguese, German, and Korean translations exist of the English-language teaching materials. With the lessons and activities, students learn technical concepts and develop an understanding of AI, its socio-technical roles, and ethical implications. The developers of educational tools stress that in a society where children become familiar with smartphones and social media early on, it is important to raise conscientious consumers and designers of AI.

To achieve this, the MIT Media Lab team created a number of educational activities that help students understand the basic mechanisms of AI systems and conceive AI technology as socio-technical systems with built-in values and societal and ethical implications. For example, the activity "Introduction to Algorithms as Opinions" introduces algorithms as a set of instructions that alter an input to build an output, similar to recipes. Against this background, students are asked to put together a recipe for what they consider the best peanut butter and jelly sandwich. The educational activity then goes on to help students reflect on what "best" means in this context, how values and preferences play a role in the instructions, and how the instructions can be modified to reflect these values and preferences or to optimize for certain goals.

Another approach is suggested by the TU Munich's Institute for Ethics in Artificial Intelligence. They developed 10 Practicing AI Ethics Literacy scenarios that can be used in the classroom to discuss ethical aspects of AI with students (Keune et al., 2024). Each scenario consists of a short description of a use case with conversation prompts, complemented by support materials and references. For older students, a discussion of the film "The Matrix" could build a starting point for reflecting on ethical issues of AI, see the reflection by Rosalyn W. Berne at (Berne, n.d.).

Additional K-12 educational tools are introduced and discussed in the publication by Akgun and Greenhow (2022).

4.3 *The Collective Dimension*

Schools and educational institutions play a pivotal role in this integration process and must make decisions on what technologies to use or allow on campus. These decisions set the general institutional framework in that they ban the use of certain technologies and embrace the use of others. Alternatively, institutions may decide to develop their own technology, be it to minimize the collection of student data, or to shape new approaches to education. An example for such a strategy is Harvard University's development of "CS50 bot", a large language model devised to provide tailored feedback to students enrolled in the CS50 course "Introduction to Computer Science" (Forlini, 2023). The use of this non-commercial tool facilitates the university's innovative educational approach and at the same time seeks to limit the potential negative implications of students over relying on technology.

Another option is to make education on the ethical aspects of AI a requirement. For example, the State University of New York (SUNY) considers education on AI as a "core competency" to be included in its general education curriculum. Accordingly, starting in the Fall 2026, courses that fulfill the Information Literacy requirement will have to include a section on AI ethics and literacy (Alonso, 2025).

In addition, schools and educational institutions need to develop rules and guidelines for the use of widely available AI technologies in education, such as ChatGPT or other generative AI tools. Examples for recent guidelines are the "Russell Group principles on the use of generative AI tools in education", (Russell group, 2024) or the "Guidelines for the Ethical Use of Generative AI (i.e. ChatGPT) on Campus" put together by a group of fellows at the Markkula Center for Applied Ethics (Markkula Center for Applied Ethics, 2023). While these documents offer valuable thoughts and general guidance on how to use generative AI in educational context, they cannot provide detailed direction on how to resolve issues in specific situations. Meaningful implementation of guidelines like these presupposes that students, educators, and administrators are thoroughly aware of the chances and risks of AI technologies in education.

4.4 *Conceptual Examples and Guidelines for Practitioners*

As concrete examples, one framework designed to integrate ethical aspects into AI literacy for teachers is proposed by Yusuke Kajiwara and Kouhei Kawabata (2024). They developed a curriculum aimed at secondary education students, focusing on the ethical use of large language models (LLMs) like ChatGPT. This curriculum not only enhances understanding of LLMs'technical concepts but also emphasize their ethical use in decision-making. Students engage with AI ethical principles such as beneficence, non-maleficence, justice and fairness, autonomy, and explainability and responsibility. The curriculum outlines seven phases, each addressing different ethical dimensions, and employs interactive tasks such as word prediction or exploring AI-generated hallucinations. Crucially, the approach highlights how students'interaction with technology affects their understanding of its ethical use and their acceptance of AI technologies. Importantly for educators, findings suggest that young students might uncritically accept AI-generated responses, indicating a need for careful consideration when implementing decision-support tools at this educational level.

In the quest to integrate ethics into educational practices, recent works highlight the significance of embedding responsibility and equity into learning analytics. One study emphasizes the importance of prioritizing the identities and contexts of underserved students, advocating for learning analytics that respect and enhance their unique learning journeys rather than marginalizing them further (Grimm, 2023a). This is achieved by developing ethically grounded frameworks to guide the implementation of analytics tools, ensuring they reflect the diverse needs and identities within STEM education. Another study points out the gaps in current analytics practices, which often lack guidance for equity-focused decision-making (Grimm, 2023b). Moving forward, it advocates for robust ethical guidelines that empower educators to make informed decisions, emphasizing transparency and inclusivity in analytics processes. Collectively, these studies propose a conscientious integration of ethics in education by aligning analytics with broader goals of equity and justice, ultimately fostering an educational environment that supports all learners equitably and responsibly.

The Artificial Intelligence in Inclusive Science Education (AIISE) framework offers a structured approach to embedding ethical considerations in AI applications within educational contexts (Roski et al.,

2024). By drawing from inclusive pedagogy, science education (Stinken-Rösner et al., 2020), and AI domains, the AIISE framework underscores the importance of incorporating diverse learner needs into the design of AI-supported learning environments. It begins by defining inclusion broadly, recognizing that all learners, not just those with disabilities, have unique needs that AI can support. This approach helps dismantle discriminatory categorizations by enabling personalized learning pathways for all students. The framework then identifies potential barriers and biases through learning analytics and educational data mining, ensuring that AI tools provide equitable learning opportunities. By emphasizing fairness and bias understanding, the AIISE framework acts as a safeguard against exclusive practices, guiding educators and researchers to develop AI systems that foster participation without perpetuating bias. Through this comprehensive integration, the AIISE framework supports the ethical deployment of AI in science education, aligning technological advancements with principles of equity and inclusion.

4.5 *Ethical Reflection as a Bridge Between Academic and Educational Practice*

In the context of AI in education, research ethics and educational ethics are deeply intertwined: the way AI tools are designed, used, and studied influences how teachers and students understand and engage with them. Ethical reflection in science is not merely an internal mechanism for ensuring research integrity; it also has educational consequences. When scientists critically examine how they generate knowledge—through questions of data ethics, authorship, transparency, or social responsibility—they shape how academic institutions handle technology and its moral dimensions in practice. Explicit reflection brings up knowledge, skills, and supports attitudes toward ethical usages of AI tools. A science-ethics perspective reveals that reflective academic practice does more than safeguard the quality of research—it becomes a formative, educational force that cultivates ethical awareness across both scientific and learning communities. Based upon the ethical principles and standards central for RCR, future works could explore the following points:

1. **Secure and ethical data use**: Researchers handling research data in the context of AI or developing learning analytics or adaptive systems must ensure data confidentiality, informed consent, and fair

representation of diverse learners. In doing so, they establish professional and ethical standards that can extend beyond the laboratory or university. When educators and schools adopt similar principles—protecting student privacy, ensuring transparency in the use of digital learning tools, and addressing data bias—they bring research ethics into the everyday culture of education. Thus, reflective data practices in science can serve as pedagogical models for responsible data use in classrooms and educational institutions.

2. **Transparency and accountability**: Academic teams should document how datasets and algorithms are selected, trained, and validated, aligning with the RCR emphasis on traceability and peer accountability. Transparent reporting not only increases trust in research but also provides valuable material for teaching critical AI literacy. When students and teachers are able to explore how algorithmic systems are constructed and evaluated, they learn to question the decisions made by AI tools in their own learning environments. In this way, transparency in academic research becomes a didactic resource for cultivating reflective, critically informed learners.
3. **Conflict of interest and commitment**: Collaboration between universities and EdTech companies demands clear ethical guidelines to prevent commercial bias in educational research (e.g., Cone & Lai, 2024). Recognizing and managing such conflicts within the scientific community creates awareness of similar tensions in schools and educational policy. For example, when researchers disclose and reflect on financial or institutional interests, they model for educators how to navigate partnerships with technology providers responsibly. A shared ethical vocabulary between researchers and teachers helps ensure that AI integration in education serves pedagogical rather than commercial objectives.
4. **Responsible authorship and dissemination**: Publications and open-source releases of AI tools should include ethical impact statements that clarify potential societal consequences. By communicating the broader ethical implications of their work, scientists not only uphold research integrity but also provide accessible materials that can inform teacher education and classroom discussions about AI ethics. Such openness allows educational practitioners to integrate authentic examples of ethical reflection into their teaching, thereby connecting students with the real-world moral dimensions

of scientific inquiry. Responsible dissemination thus fosters both scientific accountability and ethical learning.

5. **Societal impacts of research**: Researchers share responsibility for how their innovations affect access, fairness, and autonomy in education. Systematically reflecting on these impacts contributes to the self-regulation of science and, at the same time, provides a foundation for value-oriented education. When research communities articulate how AI may influence inclusion, equity, or student agency, they generate critical perspectives that educators can translate into civic and moral learning objectives. The reciprocal relationship between ethically reflective science and ethically aware education thus becomes a catalyst for societal responsibility—where scientific inquiry and education jointly cultivate fairness, autonomy, and trust in the age of AI.
6. **Environmental Responsibility in the Age of AI**: Within the RCR, scientists are called to act as responsible members of society—attentive to the contemporary ethical, environmental, and social consequences of their work. The rapid expansion of AI in education makes this responsibility both urgent and complex. Developing and deploying AI systems for teaching and learning consumes vast computational resources, relies on global data infrastructures, and shapes cultural expectations about efficiency, intelligence, and human value. Ethical reflection in academic and educational contexts must therefore include an ecological and societal dimension.

   For researchers, this entails transparent accounting of the environmental costs of AI models, fair labor practices in data annotation and system maintenance, and critical evaluation of how algorithmic design may reinforce social inequities. Integrating these concerns into RCR training moves ethical reflection beyond compliance toward what might be termed responsible co-existence—an awareness that scientific innovation is inseparable from planetary well-being and social justice.

   In education, similar responsibilities arise. Schools and universities adopting AI technologies participate in the same material and normative networks: the energy required to run large models, the data drawn from learners, and the global labor supporting digital infrastructures. Embedding sustainability and social-impact discussions into curricula can therefore mirror the commitments expected of researchers. When educators and scientists jointly model ecological awareness, accountability, and inclusivity, they transform AI ethics from a specialized topic into a shared civic and environmental practice.

Importantly, as a research framework, the RCR framework rarely focuses on ethical questions of education itself—how ethics can guide AI usages in schools or in higher education. Yet, insights from educational debates on generative AI provide precisely the kind of cognitive and moral reflection that research practice increasingly needs.

7. Discussions in education about cognitive offloading—the outsourcing of thinking processes to technology (Costa et al., 2025; Kosmyna et al., 2025)—highlight not only the risks of diminished originality and overreliance on AI, but also strategies to counteract them: scaffolding reflection, maintaining learner agency, and cultivating metacognitive awareness. These same strategies can guide scientists in managing or limiting intellectual dependence on AI-based research tools. For instance, the pedagogical principle of productive struggle—that deep understanding arises from sustained cognitive effort—applies equally to academic inquiry: when AI systems generate hypotheses, visualizations, or literature syntheses, researchers must consciously re-engage interpretive reasoning rather than accepting outputs as finished knowledge.

Table 1 illustrates how the seven ethical dimensions identified in this chapter are situated within both academic and educational practice. Each dimension describes a field in which the ethical reflection of research intersects with educational aims and values, highlighting how the ethics of science and ethical learning in education can mutually inform one another. This synthesis adopts a dual lens perspective, viewing ethical inquiry simultaneously through the practices of scientific research and the pedagogical processes of education, thereby revealing how integrity in science and responsibility in learning are reciprocally constituted.

Table 1 Toward a "dual lens framework" on science and educational ethics through mutual reflection and exchange

Ethical dimension	*Academic practice (ethics of science)*	*Educational practice*	*Reflexive Function/ Bridge*
1. Ethical data practices	Responsible data collection, management, and consent; alignment with RCR standards for transparency and privacy	Promoting data literacy, privacy awareness, and critical understanding of algorithmic data use among learners	Establishing a shared ethical foundation for how data are produced, interpreted, and taught, linking research integrity with educational responsibility
2. Transparency and accountability	Open science practices and algorithmic documentation ensuring traceability and peer accountability	Teaching transparency as a principle of ethical digital use and critical AI literacy	Supporting mutual accountability: Researchers and educators embody openness and critical scrutiny as shared professional ethics
3. Conflict of interest and commitment	Institutional reflection on commercial or political dependencies in AI research and funding	Addressing the influence of corporate partnerships and technologies in schools and educational systems	Encouraging awareness of how institutional dependencies shape both research and education
4. Responsible authorship and dissemination	Ethical publication, authorship credit, and responsible dissemination of results	Fostering responsible authorship, citation, and communication practices in student and teacher learning	Connecting research ethics with ethical communication in education, modeling responsible knowledge creation and exchange
6. Cognitive offloading and intellectual dependence	Reflection on automation bias and cognitive delegation in AI-supported research	Encouraging metacognitive awareness and autonomy in AI-assisted learning	Promoting intellectual independence in both scientists and learners, addressing overreliance on AI as a shared ethical challenge
7. Environmental responsibility	RCR principle: The scientist as a responsible member of society, mindful of environmental and social consequences	Integrating sustainability and global citizenship education into AI ethics curricula	Connecting ecological and societal ethics across research and education, supporting a shared culture of responsible co-existence

5 Summary and Conclusion

This chapter examined ethical considerations associated with the integration of artificial intelligence in educational contexts, focusing on five key areas: data privacy, bias and fairness, agency and autonomy, education-related disparities, and opacity and transparency. While AI technologies promise to enhance educational environments by offering personalized learning and improving accessibility, they pose significant ethical challenges that necessitate careful management. The frameworks discussed, such as the UNESCO Competency Framework for Teachers and the AIISE framework, provide valuable guidelines for ensuring ethical AI implementation in education, emphasizing principles such as inclusivity, transparency, and human accountability.

Moreover, integrating an ethics of science perspective into AI in education highlights the need for ethical reflection and dialogue not only about AI but within academic research itself. Scientific practice plays a formative role in shaping how AI technologies emerge, are validated, and subsequently influence educational contexts. Consequently, responsible AI in education requires researchers, developers, and educators to engage in ongoing reflection on their methodologies, collaborations, and institutional frameworks.

Establishing AI literacies that include ethical training prepares educators, students, and researchers to approach AI with a critical, informed perspective, transforming them into conscientious users and developers of technology. By embedding ethical considerations into AI education, environments can further be developed toward scenarios where technology serves as a force for equity and inclusion, supporting diverse learner needs and promoting fairness and accountability in educational practices. As AI continues to evolve, maintaining a vigilant ethical stance will be vital to ensuring it enhances education in ways that are both innovative and principled.

By linking the ethics of AI in education with the ethics of science, this chapter underscores that responsible AI requires not only awareness and regulation at the user level but also reflective integrity within academic research and institutional practice. Embedding principles of the Responsible Conduct of Research into the design, evaluation, and implementation of AI systems can enhance transparency, fairness, and accountability throughout the educational ecosystem.

As Wiese et al. (2025) emphasize, AI ethics education must move beyond awareness and toward sustained ethical reflection. This chapter extends their argument by showing that such reflection should also occur within academic research itself, aligning educational, institutional, and scientific responsibilities under the broader scope of the ethics of science.

While equipping teachers and students with AI literacy and ethical guidelines is critical, it is equally important to share the responsibility for the ethical use of AI at a collective level. This involves companies, policymakers, and educational institutions. Companies developing AI tools should not only include but prioritize ethical design principles, ensuring transparency, fairness, and inclusivity. Policymakers play a crucial role in establishing regulations and standards that safeguard against misuse and bias while promoting equitable access to AI resources. Educational institutions benefit from adopting frameworks that support ethical reflections at an organizational level. This ensures that the broader structural and systemic implications of AI adoption are carefully considered. Fostering collective accountability seeks to not only consider individual users responsible for ethical technology usage, but also to support and sustain them by providing technological options and a broad range of stakeholders dedicated to promoting AI in terms of a common good.

Competing Interests The authors have no conflicts of interest to declare that are relevant to the content of this chapter.

References

Adler, I., & Shani, Y. (2026). Hybrid human-GenAI cognitive apprenticeship: Encouraging pre-service teachers to implement instructional practices to support students' self-regulated learning. *Teaching and Teacher Education, 169*, 105268.

Akgun, S., & Greenhow, C. (2022). Artificial intelligence in education: Addressing ethical challenges in K-12 settings. *AI and Ethics, 2*(3), 431–440.

Alonso, J. (2025, January 16). *SUNY adds AI education to its information literacy gen ed*. Inside Higher Ed. https://www.insidehighered.com/news/tech-innovation/artificial-intelligence/2025/01/16/suny-adds-ai-education-its-information

Avraamidou, L. (2024). Can we disrupt the momentum of the AI colonization of science education? *Journal of Research in Science Teaching, 61*(10), 2570–2574.

Bastani, H., Bastani, O., Sungu, A., Ge, H., Kabakcı, O., & Mariman, R. (2024). *Generative AI can harm learning*. SSRN.

Berne, R. W. (n.d.). Online ethics center: Using the film, the matrix, in an engineering ethics course. online ethics center. https://doi.org/10.18130/dgzd-1r64 https://onlineethics.org/cases/online-ethics-center-using-filmmatrix-engineering-ethics-course

Bradner, A., & Bates, R. A. (2024). Developing an ethics credential for undergraduate STEM majors. In E. Hildt, K. Laas, E. M. Brey, & C. Z. Miller (Eds.), *Building inclusive ethical cultures in STEM* (pp. 35–50).

Cheuk, T. (2021). Can AI be racist? Color-evasiveness in the application of machine learning to science assessments. *Science Education, 105*(5), 825–836.

Cone, L., & Lai, S. S. (2024). Infrastructural dependency in the datafied welfare state: The case of Google Chromebooks. *Learning, Media and Technology, 50*(1), 44–60.

Costa, C., Murphy, M., & Costa, C. (2025). Generative artificial intelligence in education: (what) are we thinking ? *Learning, Media and Technology, 9884*, 1–12. https://doi.org/10.1080/17439884.2025.2518258

Duede, E. (2023). Deep learning opacity in scientific discovery. *Philosophy of Science, 90*(5), 1089–1099.

European Union. (2016). *General data protection regulation of the EU*.

Forlini, E. (2023, June 22). *Harvard's new computer science teacher is a chatbot*. PCMag UK. https://uk.pcmag.com/ai/147451/harvards-new-computer-science-teacher-is-a-chatbot

Grimm, A., Steegh, A., Çolakoğlu, J., Kubsch, M., & Neumann, K. (2023a). Positioning responsible learning analytics in the context of STEM identities of under-served students. *Frontiers in Education, 7*.

Grimm, A., Steegh, A., Kubsch, M., & Neumann, K. (2023b). Learning analytics in physics education: Equity-focused decision-making lacks guidance! *Journal of Learning Analytics, 10*(1), 71–84.

Holmes, W., Persson, J., Chounta, I. A., Wasson, B., & Dimitrova, V. (2022). *Artificial intelligence and education. A critical view through the lens of human rights, democracy, and the rule of law*. Council of Europe.

Huwer, J., Becker-Genschow, S., Thyssen, C., Thoms, L., Finger, A., Von Kotzebue, L., Kremser, E., Meier, M., & Bruckermann, T. (2024). *Kompetenzen für den Unterricht mit und über Künstliche Intelligenz*. Waxmann.

Hwang, G.-J., Xie, H., Wah, B. W., & Gašević, D. (2020). Vision, challenges, roles and research issues of Artificial Intelligence in Education. *Computers and Education: Artificial Intelligence, 1*, 100001. https://doi.org/10.1016/j.caeai.2020.100001

Ifenthaler, D., Schumacher, C., Herrmann, L., Weigert, J., Ifenthaler, D., Majumdar, R., Gorissen, P., Judge, M., Mishra, S., Raffaghelli, J., Shimada, A., Baker, R. S., Hawn, A., Avraamidou, L., Heersmink, R., Hicks, M. T.,

Humphries, J., Slater, J., Holmes, W., et al. (2024). AI-based prediction of academic success: Support for many, disadvantage for some? *Nature Machine Intelligence, 6*(4), 1693–1710.

Kajiwara, Y., & Kawabata, K. (2024). AI literacy for ethical use of chatbot: Will students accept AI ethics? *Computers and Education: Artificial Intelligence, 6*, 100251.

Kasneci, E., Seßler, K., Küchemann, S., Bannert, M., Dementieva, D., Fischer, F., ... & Kasneci, G. (2023). ChatGPT for good? On opportunities and challenges of large language models for education. *Learning and individual differences, 103*, 102274.

Keune, A., Hurtado, S., Simšič, Ž., & Kappes, V. (2024). *Practicing AI ethics literacy: 10 scenarios for engaging with AI ethics in education*. Technical University of Munich. https://www.edu.sot.tum.de/fileadmin/w00bed/lsdesign/_my_direct_uploads/AI_EdTech_Scenarios_Final_-_Web.pdf

Kosmyna, N., Hauptmann, E., Yuan, Y. T., Situ, J., Liao, X. H., Beresnitzky, A. V., Braunstein, I., & Maes, P. (2025). Your brain on ChatGPT: Accumulation of cognitive debt when using an AI assistant for essay writing task. *arXiv*.

Kreth, Q., Schiff, D. S., Lee, J., Borenstein, J., & Zegura, E. (2024). Social responsibility and ethics in STEM education: The state of the field. In E. Hildt, K. Laas, E. M. Brey, & C. Z. Miller (Eds.), *Building inclusive ethical cultures in STEM* (pp. 19–33). Springer.

Li, T., Reigh, E., He, P., & Adah Miller, E. (2023). Can we and should we use artificial intelligence for formative assessment in science? *Journal of Research in Science Teaching, 60*(6), 1385–1389.

Markkula Center for Applied Ethics. (2023, May 22). *Guidelines for the ethical use of generative AI (i.e. CHATGPT) on campus*. https://www.scu.edu/ethics/focus-areas/campus-ethics/guidelines-for-the-ethical-use-of-generative-ai-ie-chatgpt-on-campus/

Messeri, L., & Crockett, M. J. (2024). Artificial intelligence and illusions of understanding in scientific research. *Nature, 627*(8002), 49–58.

Miao, F., & Cukurova, M. (2024). *AI competency framework for teachers*. UNESCO.

Mittelstadt, B. D., Allo, P., Taddeo, M., Wachter, S., & Floridi, L. (2016). The ethics of algorithms: Mapping the debate. *Big Data and Society, 3*(2).

National Institutes of Health. (2009, November 24). *NOT-OD-10-019: Update on the requirement for instruction in the responsible conduct of research*. U.S. Department of Health and Human Services. https://grants.nih.gov/grants/guide/notice-files/not-od-10-019.html

National Institutes of Health. (2022, February 17). *Not-OD-22-055: FY 2022 updated guidance: Requirement for instruction in the responsible conduct of research*. U.S. Department of Health and Human Services. https://grants.nih.gov/grants/guide/notice-files/NOT-OD-22-055.html

Oliveira, M., Zednik, C., Bombaerts, G., Sadowski, B., & Conijn, R. (2025). Assessing students' DRIVE: A framework to evaluate learning through interactions with generative AI. *Computers and Education: Artificial Intelligence, 9*, 100497.

Paulsen, L., & Lindsay, E. (2024). Learning analytics dashboards are increasingly becoming about learning and not just analytics-A systematic review. *Education and Information Technologies, 29*(11), 14279–14308.

Rosalyn, B. (n.d.). Online ethics center: Using the film, "the matrix", in an engineering ethics course. Online Ethics Center: Using the Film, "The Matrix", in an Engineering Ethics Course. https://onlineethics.org/cases/online-ethics-center-using-film-matrix-engineering-ethics-course

Roski, M., Ewerth, R., Hoppe, A., & Nehring, A. (2024). Exploring Data Mining in Chemistry Education: Building a Web-Based Learning Platform for Learning Analytics. *Journal of Chemical Education, 101*(3), 930–940. https://doi.org/10.1021/acs.jchemed.3c00794

Rudin, C. (2019). Stop explaining black box machine learning models for high stakes decisions and use interpretable models instead. *Nature Machine Intelligence, 1*(5), 206–215.

Russell Group. (2024, February 12). *Russell Group principles on generative AI in education.* http://russellgroup.ac.uk/news/new-principles-on-use-of-ai-in-education/

Shin, D., & Shim, J. (2021). A systematic review on data mining for mathematics and science education. *International Journal of Science and Mathematics Education, 19*(4), 639–659. https://doi.org/10.1007/s10763-020-10085-7

Stinken-Rösner, L., Rott, L., Hundertmark, S., Menthe, J., Hoffmann, T., Nehring, A., & Abels, S. (2020). Thinking inclusive science education from two perspectives: Inclusive pedagogy and science education. *Research in Subject-Matter Teaching and Learning (RISTAL), 3*(1), 30–45.

U.S. Department of Education, Office of Educational Technology. (2024). *Designing for education with artificial intelligence: An essential guide for developers.* https://files.eric.ed.gov/fulltext/ED661949.pdf

UNESCO. (2022). *AI curricula: A mapping of government-endorsed AI curricula.*

Valle, N., Antonenko, P., Dawson, K., & Huggins-Manley, A. C. (2021). Staying on target: A systematic literature review on learner-facing learning analytics dashboards. *British Journal of Educational Technology, 52*(4), 1724–1748.

Wiese, L. J., Patil, I., Schiff, D. S., & Magana, A. J. (2025). AI ethics education: A systematic literature review. *Computers and Education: Artificial Intelligence, 8.*

Communicating Science in a Post-Truth Society

Christoph Bleidorn

1 Introduction

Many democratic countries are seeing a rise in the spread of false information across social media platforms and political websites (Bennett & Livingston, 2018), leading to growing legitimacy problems and suggesting that we are entering a post-truth order (Harsin, 2015). According to the Oxford dictionaries (Oxford Languages, n.d.), post-truth is an adjective defined as 'relating to or denoting circumstances in which objective facts are less influential in shaping public opinion than appeals to emotion and personal belief'. While ideas related to post-truth had been already discussed by philosophers such as Friedrich Nietzsche, Hannah Arendt or Bruno Latour (Schindler, 2020), this concept became worldwide attention after the 2016 Brexit referendum in the UK and the 2017 US presidential elections. In this context, post-truth is used as a catchphrase that seems to capture the zeitgeist (Mclntyre, 2018). As people are more likely to accept information that aligns with their pre-existing views, misinformation, conspiracy theories, and biased narratives flourish in a post-truth environment, while factual accuracy seems to be of secondary importance (Ecker et al., 2022; Harambam et al.,

C. Bleidorn (✉)
Georg-August-Universität Göttingen, Göttingen, Germany
e-mail: cbleido@gwdg.de

J. Hinrichsen et al. (eds.), *Ethics of the Sciences*,
https://doi.org/10.1007/978-981-92-2821-8_14

2022). Moreover, the advent of social media echo chambers (Del Vicario et al., 2016), where a person only encounters information or opinions that reflect and reinforce their own, is making it even more challenging to achieve consensus on factual matters in public debates. One widely cited study even demonstrated that false claims were distributed faster and more broadly on the social media platform Twitter than true claims (Vosoughi et al., 2018). Social media posts from misinformation sources evoke stronger emotions of moral outrage, such as anger and disgust, compared to trustworthy news (McLoughlin et al., 2024). This outrage motivates users to reshare content impulsively, often without verifying its accuracy.

False information can come in many different forms, such as misinformation, disinformation, fake news, hoaxes, conspiracy theories or pseudoscience (Scheufele et al., 2021). Understanding these forms can help to identify and critically evaluate false information to reduce its impact. Misinformation, disinformation, fake news and propaganda as part of human communication are known since at least from the ancient Rome (Fraser, 2020), and became even easier to spread with the invention of the printing press by Gutenberg (Kalsnes, 2018). With "alternative facts", another term to describe false information became popular in 2017 during an interview with the Trump administration senior advisor Kellyanne Conway. She used this term to counter statements made about the size of the crowd at the inauguration of the first term of Donald Trump. Since then, alternative facts have been widely used as euphemism for different forms of false information. However, alternative facts are rather communicative maneuvers used to prevent certain problems from being dealt with politically in a specific situation, even if at odds with objective evidence (Kumkar, 2022). As such this strategy basically represents a denial of knowledge, making it difficult to counteract with fact-checking and evidence-based argumentation.

Trust in scientific institutions and the ethical conduct of research are recognized as important factors in the functioning of modern democratic systems (Hönnige and Nyhuis in this volume). However, the increase of spread of misinformation and conspiracy theories has been identified to contribute to the decline of trust in traditional institutions such as democracy and a civil society (Ferguson, 2013; van Zoonen et al., 2024). This phenomenon can be understood through the lens of agnotology, the study of how ignorance and doubt are deliberately produced or maintained to serve political, economic, or ideological interests (Proctor & Schiebinger, 2008). For example, the military invasion of Russia in Ukraine is additionally accompanied by an information warfare to influence public opinion (Aguerri et al., 2024). National governments of democracies recognized this threat and started to design strategies to counteract mis- and disinformation, e.g., by educating citizens and raising their awareness (e.g., Bundesregierung, 2023), a classical goal of science communication.

This chapter examines how science communication can adapt to the challenges of a post-truth society, where misinformation and emotional narratives often undermine evidence-based reasoning. It reviews current debates and presents different approaches for effectively communicating science in order to strengthen public trust, support informed decision-making, and sustain the role of science in democratic societies.

2 Aims and Models of Science Communication

While the research field of science communication is rather young, establishing itself as an academic discipline in the 1970s (Gascoigne et al., 2010), the communication of science (e.g., through exhibits, lectures or articles) has a long tradition with its roots in the nineteenth century. Generally, two different ways (or paradigms) of communicating science are distinguished: one-way and two-way communication (Kappel & Holmen, 2019). The aims may overlap, but differ between different models of science communication (Fig. 1). In the one-way model, science communication is seen as a process of information transmission, which assumes scientific sufficiency and public deficiency, and is therefore also known as the "deficit model" (Gross, 1994), which has been refined over time (Wynne 2006). The motivation for public education rests here in the underlying assumption that a deficit in scientific knowledge among the public leads to poor quality decision-making. E.g., doubts about the value of scientific progress or fears about scientific innovations, such as the use of artificial intelligence or genetically modified food, are fueled by the ignorance of the underlying science. Consequently, raising the public understanding of science could also lessen "irrational" fears about new technologies (Sturgis & Allum, 2004). Similarly, it could be assumed that if the public understands the science behind the overwhelming evidence of human-impacted climate change and its consequences, better action could be taken against it (Suldovsky, 2017). However, a study among

US citizens did not show a correlation between educational background and decision-making (Kahan et al., 2012). Instead, the public's deepest divisions over climate change are not due to a lack of scientific understanding, but they arise from a conflict between personal interests (e.g., political affiliation) and the collective need to use scientific evidence for the common good. Interestingly, those with the highest science literacy and reasoning skills showed the most cultural polarization on the issue. In a follow-up study, citizens were queried on the beliefs of climate scientists (Kahan, 2015). This time, those with greater general scientific knowledge, irrespective of their political affiliation, demonstrated a better ability to recognize the scientific consensus, effectively reducing polarization. However, when asked to express their personal opinions on climate change, the polarization re-emerged. This study demonstrates that highly educated individuals can perfectly understand the scientific consensus yet still choose not to accept it. In summary, it might be too naïve to assume that facts alone are persuasive and empirical research demonstrates that public communication of science is more complex than the one-way communication of scientific results as posed by the deficit model (Simis et al., 2016). Indeed, efforts by scientists to inform the public may inadvertently worsen resistance when factual information contradicts an individual's existing worldview. Instead, pre-existing beliefs may be even reinforced, a phenomenon known as the *backfire effect* (Darner 2019).

In contrast to the deficit model, the dialogue model of science communication is based on a two-way exchange to build trust and mutual understanding (Fig. 1). Under the dialogue model also non-scientific forms of knowledge, such as cultural and experiential knowledge, are valued (Reincke et al., 2020). This approach acknowledges that complex societal issues, such as the benefits and risks of nuclear power or genetically modified food, cannot be effectively addressed solely through scientific knowledge. This model encourages public engagement with science and allows the flow of information not only from experts to interested citizens, but also the other way around. Typical formats to implement this model are workshops or townhall-style meetings (Kent 2022), but also active back-and-forth communication through social media channels. An analysis of exchanges on the Facebook page of the Danish Health Authority (DHA) during the COVID-19 pandemic has been used to exemplify the complex mixture of aims, needs, and constraints that characterize dialogical communication between citizens and authorities (Madvig et al.,

2022). On the positive side, this format allowed the adjustment of expectations according to the needs of the public (e.g., in this case adjusting answer catalogues in the light of popular questions) and thereby the scientists learned from observing reactions of the citizens. On the negative side, improper engagement led to interactions that were characterized by trolling or misinformation on the social media site.

Recently, public engagement models have emerged to promote trust in science policy by emphasizing public participation. Under this model, the importance of co-creation of knowledge, where both scientists and laypeople contribute to the outcome, is recognized. Typical activities include consensus conferences, citizen juries, or science shops with the aim to "democratize science" by empowering and actively involving the public (Nerghes et al., 2022; Schrögel & Kolleck, 2019). Furthermore, citizen science projects can be considered participative if they are designed not only to view citizens as data gatherers but also to enable co-creation through shared and open study design and result interpretation (Senabre Hidalgo et al., 2021). Many examples for such projects come from nature conservation efforts, where it has been shown that increasing inclusive public participation in scientific research is crucial. Co-created citizen science effectively engages diverse communities, stimulates behavioral

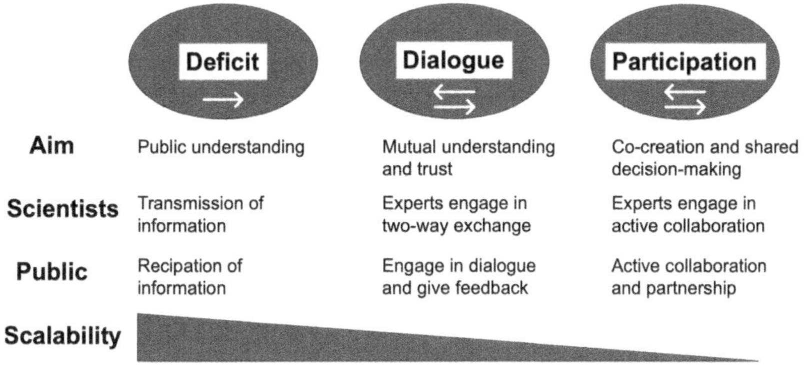

Fig. 1 Different models of science communication and their aims. The deficit model relies on one-way communication, while the dialogue model and participation model use two-way communication. The scalability differs between these models, while one-way transmission of information can be easily facilitated, two-way formats are much more resource intensive

change, and broadens perceptions of scientific participation (Robinson et al., 2024). It can be observed that in co-creative scientific projects, value arises from the process itself, not just from the outcomes. However, truly participative approaches are resource intensive and time-consuming, requiring careful management to ensure productive and meaningful collaboration. These three models (Fig. 1) illustrate a paradigm shift in science communication over time, but must not be seen as mutually exclusive. It has been recommended that effective communication can be based on combining approaches from all three models, starting with one-way dissemination of knowledge and proceeding with dialogue and participation formats (León et al., 2022). Also, the scalability differs clearly between these models. One-way transmission of information can be facilitated by television and book formats, thereby reaching a worldwide audience of potentially thousands to millions of people, while two-way formats are much more resource intensive and are rather designed to reach dozens to hundreds of people on a more local level.

3 Strategies for Effective Science Communication in a Post-Truth Society

Traditionally, it has been assumed that providing more information corrects misunderstandings or misinformation. However, as noted above, this approach often fails in a post-truth environment, as people's beliefs are deeply rooted in identity, ideology, or emotions (Ecker et al., 2022). Therefore, addressing the challenges of the post-truth era requires a multifaceted approach that combines scientific factual accuracy with empathy, adaptability, and efforts to rebuild trust in scientific institutions. A set of diverse strategies has been recommended for successful science communication in a post-truth environment (Fig. 2).

(I) **Building trust by transparency**: Transparency helps to explain how scientific research is conducted, including the methodologies, but also the limitations and uncertainties involved. This reflects the shift from the public understanding of science (PUS), which focuses on communicating established facts, to the public understanding of research (PUR), which emphasizes the processes and reasoning behind scientific inquiry. By promoting understanding of how knowledge is generated and revised, PUR

Fig. 2 Strategies for effective science communication: (I) Buiding trust by transparency; (II) Adaptation to diverse audiences; (III) Proactive combat of misinformation; (IV) Interdisciplinary approaches; (V) Storytellung and emotions

helps reduce misunderstandings and prevents unrealistic expectations about the infallibility of science. E.g., with the "Spaces of Knowledge" in the Forum Wissen of the University of Göttingen a complete exhibition has been created to illustrate the process of generating knowledge in an academic environment (Allemeyer et al., 2023).

(II) **Adaptation to diverse audiences**: The audience is the most important thing to keep in mind when designing science communication formats. The key is to understand the beliefs, values, and knowledge gaps of the audience, as well as to avoid jargon to align with their priorities and language (Borowiec, 2023). Audiences are more likely to trust information that aligns with their cultural context, and emotions and personal values often outweigh facts. This phenomenon is in line with the "cultural cognition thesis"

(CCT), which predicts that individuals tend to form perceptions of societal risks (e.g., climate change, COVID-19 pandemic) that cohere with values characteristic of groups with which they identify (Kahan et al., 2012). By recognizing and respecting the values of the target audience, resistance can be reduced and dialogue becomes possible.

(III) **Proactive combat of misinformation**: While the effectiveness of correcting misinformation (*debunking*) is debated (Ecker et al., 2022), experience shows that it is more likely to be successful if the correction is accompanied by an alternative explanation or by outlining the missing credibility of the original source. By educating audiences about common strategies and patterns of misinformation tactics before they encounter false claims (so-called *inoculation* or *prebunking*), people become more resilient to misinformation (Lewandowsky & van der Linden, 2021), even though protective effects are not always evident (Spampatti et al., 2024). Using a gamification approach, where a large number of participants can be reached through online computer games, has been shown to be a potentially effective method for improving people's ability, irrespective of their education, to identify and resist misinformation (Roozenbeek & van der Linden, 2019). In practice, these strategies can be implemented through interactive online tools, educational workshops, and media literacy campaigns that teach audiences to recognize manipulative framing, emotional language, and false balance in media content.

(IV) **Interdisciplinary approaches**: An interdisciplinary approach that integrates insights from psychology and sociology to guide the development of effective communication strategies appears to be essential for successfully combating misinformation. With *technocognition*, an approach has been outlined that involves knowledge from the field of human cognition (e.g., how we think, learn, and make decisions) as well as understanding of current technological systems (e.g., social media, algorithms, and digital tools) to understand and combat the spread of misinformation (Lewandowsky et al., 2017). For example, machine learning, natural language processing (NLP), and other artificial intelligence (AI) technologies are already routinely implemented to analyze, verify, and flag false information in the internet, e.g., using tool ClaimBuster (Hassan et al., 2017) or other automated fact-checking solutions (Guo et al., 2022). AI-based tools like Hoaxy (Shao et al., 2016) can be used to visualize how false information spreads across social media platforms, thereby informing the proper design of countermeasures.

(V) **Storytelling and emotions**: Stories are great tools for helping people understand, stay interested, and get involved. They naturally persuade, making them a powerful tool in science communication to capture and influence the attention of their audience. For example, when comparing climate change narratives structured as stories with informational narratives, it has been found that stories are more effective at promoting pro-environmental behavior (Morris et al., 2019). While the persuasiveness of narrative formats provides a powerful tool, science communicators must be careful not to perpetuate misinformation and inaccuracies (Dahlstrom, 2014).

Using emotions effectively in science communication is crucial in the post-truth era, where scientific facts alone often fail to resonate. Emotions can create a personal connection, make complex ideas memorable, and inspire action. It will be important to identify values of the target audience (e.g., health, safety) and try to align the scientific messaging with those values, thereby showing empathy. Appealing to the fear of the audience can be effective to inspire action, but this has to be handled carefully given that overwhelming negativity can also lead to denial (Brosch & Steg, 2021). Finally, the inspiration of awe and curiosity, a long-established way of communicating science as for example manifested in museum exhibitions around the world, is still an effective instrument (Sagan, 2011). These strategies can be applied through storytelling in videos or podcasts that connect science to everyday life, immersive experiences, and social media campaigns that use relatable narratives, or interactive workshops where participants explore scientific phenomena firsthand.

4 Conclusion

In a time where misinformation spreads rapidly and trust in institutions becomes fragile, science communication remains an essential strategy for promoting understanding, trust, and collective action. It bridges the gap between scientific knowledge and public understanding, helps counter misinformation, and supports informed decision-making. Traditional one-way communication models, while still extremely useful and effective for reaching broad audiences, are insufficient on their own. Effective science communication increasingly requires transparency, dialogue, and participatory approaches that actively engage citizens in the scientific process. Strategies such as promoting the public understanding of research, adapting messages to diverse audiences, thoughtfully incorporating emotions, and employing prebunking or gamified interventions can enhance resilience against misinformation and foster informed decision-making. Moreover, effective science communication promotes equity and inclusion by making knowledge accessible to all and empowers citizens to navigate complex scientific and societal issues. Ultimately, sustaining the role of science in democratic societies depends not only on conveying facts but also on building trust, encouraging critical thinking, and supporting collective action in a post-truth environment.

Acknowledgements Part of the material has been presented in the master course "M.Biodiv.610: Science Communication in Biodiversity" (responsible: M.T. Aguado Molina) at the University of Goettingen.

Competing Interests The author has no conflicts of interest to declare that are relevant to the content of this chapter.

References

Aguerri, J. C., Santisteban, M., & Miró-Llinares, F. (2024). The fight against disinformation and its consequences: Measuring the impact of "Russia state-affiliated media" on twitter. *Crime Science, 13*(1).

Allemeyer, M. L., Baur, J., & Vogel, C. (Eds.). (2023). Spaces of knowledge: The core exhibition at forum Wissen. Wallstein. https://www.wallstein-verlag.de/9783835353312-spaces-of-knowledge.html

Bennett, W. L., & Livingston, S. (2018). The disinformation order: Disruptive communication and the decline of democratic institutions. *European Journal of Communication, 33*(2), 122–139.

Borowiec, B. G. (2023). Ten simple rules for scientists engaging in science communication. *PLoS Computational Biology, 19*(7), e1011251.

Brosch, T., & Steg, L. (2021). Leveraging emotion for sustainable action. *One Earth, 4*(12), 1693–1703.

Bundesregierung. (2023, July 19). Disinformation. What the Federal Government is doing and what you can do yourself. Retrived November

15, 2025, from: https://www.bundesregierung.de/breg-de/aktuelles/dangerous-fake-news-2244420

Dahlstrom, M. F. (2014). Using narratives and storytelling to communicate science with nonexpert audiences. Proceedings of the National Academy of Sciences of the United States of America, 111 Suppl 4(Suppl 4), 13614–13620.

Darner, R. (2019). How can educators confront science denial? *Educational Researcher, 48*(4), 229–238.

Del Vicario, M., Bessi, A., Zollo, F., Petroni, F., Scala, A., Caldarelli, G., Stanley, H. E., & Quattrociocchi, W. (2016). The spreading of misinformation online. *Proceedings of the National Academy of Sciences of the United States of America, 113*(3), 554–559.

Ecker, U. K. H., Lewandowsky, S., Cook, J., Schmid, P., Fazio, L. K., Brashier, N., Kendeou, P., Vraga, E. K., & Amazeen, M. A. (2022). The psychological drivers of misinformation belief and its resistance to correction. *Nature Reviews Psychology, 1*(1), 13–29.

Ferguson, N. (2013). *The great degeneration: How institutions decay and economies die/Niall Ferguson*. Penguin Press.

Fraser, M. (2020). *In truth: A history of lies from ancient Rome to modern America* (1st ed.). Globe Pequot Press The. https://ebookcentral.proquest.com/lib/kxp/detail.action?docID=6141080

Gascoigne, T., Cheng, D., Claessens, M., Metcalfe, J., Schiele, B., & Shi, S. (2010). Is science communication its own field? *Journal of Science Communication, 09*(03), C04.

Gross, A. G. (1994). The roles of rhetoric in the public understanding of science. *Public Understanding of Science, 3*(1), 3–23.

Guo, Z., Schlichtkrull, M., & Vlachos, A. (2022). A survey on automated fact-checking. *Transactions of the Association for Computational Linguistics, 10*, 178–206.

Harambam, J., Grusauskaite, K., & de Wildt, L. (2022). Poly-truth, or the limits of pluralism: Popular debates on conspiracy theories in a post-truth era. *Public Understanding of Science (Bristol, England), 31*(6), 784–798.

Harsin, J. (2015). Regimes of Posttruth, Postpolitics, and attention economies. *Communication, Culture & Critique, 8*(2), 327–333.

Hassan, N., Zhang, G., Arslan, F., Caraballo, J., Jimenez, D., Gawsane, S., Hasan, S., Joseph, M., Kulkarni, A., Nayak, A. K., Sable, V., Li, C., & Tremayne, M. (2017). ClaimBuster. *Proceedings of the VLDB Endowment, 10*(12), 1945–1948.

Kahan, D. M. (2015). Climate-science communication and the measurement problem. *Political Psychology, 36*(S1), 1–43.

Kahan, D. M., Peters, E., Wittlin, M., Slovic, P., Ouellette, L. L., Braman, D., & Mandel, G. (2012). The polarizing impact of science literacy and numeracy on perceived climate change risks. *Nature Climate Change, 2*(10), 732–735.

Kalsnes, B. (Ed.). (2018). *Oxford research encyclopedia of communication*. Oxford University Press.

Kappel, K., & Holmen, S. J. (2019). Why science communication, and does it work? A taxonomy of science communication aims and a survey of the empirical evidence. *Frontiers in Communication, 4*, 55.

Kent, M. L. (2022). Dialogue as a tool of strategic communication? In J. Falkheimer & M. Heide (Eds.), *Research handbook on strategic communication* (pp. 83–98). Edward Elgar Publishing.

Kumkar, N. C. (2022). *Alternative Fakten: Zur Praxis der kommunikativen Erkenntnisverweigerung* (Originalausgabe). edition suhrkamp: Vol. 2811. Suhrkamp Verlag. http://www.content-select.com/index.php?id=bib_view&ean=9783518774731

León, B., López-Goñi, I., & Salaverría, R. (2022). The Covid-19 catastrophe: A science communication mess? *Church, Communication and Culture, 7*(1), 6–22.

Lewandowsky, S., Ecker, U. K. H., & Cook, J. (2017). Beyond misinformation: Understanding and coping with the "post-truth" era. *Journal of Applied Research in Memory and Cognition, 6*(4), 353–369.

Lewandowsky, S., & van der Linden, S. (2021). Countering misinformation and fake news through inoculation and Prebunking. *European Review of Social Psychology, 32*(2), 348–384.

Madvig, F., Achiam, M., Adler-Nissen, R., Johansen, N., & Whiteley, L. (2022). Coming closer to citizens? Frustrated dialogue on the Danish health Authority's Facebook page during COVID-19. *Frontiers in Communication, 7*, 822471.

Mclntyre, L. (2018). Post-truth. ProQuest Ebook Central. The MIT Press. https://ebookcentral.proquest.com/lib/kxp/detail.action?docID=5294979.

McLoughlin, K. L., Brady, W. J., Goolsbee, A., Kaiser, B., Klonick, K., & Crockett, M. J. (2024). Misinformation exploits outrage to spread online. *Science, 386*(6725), 991–996.

Morris, B. S., Chrysochou, P., Christensen, J. D., Orquin, J. L., Barraza, J., Zak, P. J., & Mitkidis, P. (2019). Stories vs. facts: Triggering emotion and action-taking on climate change. *Climatic Change, 154*(1–2), 19–36.

Nerghes, A., Mulder, B., & Lee, J.-S. (2022). Dissemination or participation? Exploring scientists' definitions and science communication goals in The Netherlands. *PLoS One, 17*(12), e0277677.

Oxford Languages. (n.d.). *Oxford Word of the Year 2016*. Retrived November 15, 2025, from: https://languages.oup.com/word-of-the-year/2016/

Proctor, R. N., & Schiebinger, L. (Eds.). (2008). *Agnotology: The making and unmaking of ignorance*. Stanford University Press.
Reincke, C. M., Bredenoord, A. L., & van Mil, M. H. (2020). From deficit to dialogue in science communication: The dialogue communication model requires additional roles from scientists. *EMBO Reports, 21*(9), e51278.
Robinson, D., Delany, J., & Sugden, H. (2024). Beyond science: Exploring the value of co-created citizen science for diverse community groups. *Citizen Science: Theory and Practice, 9*(1), 13.
Roozenbeek, J., & van der Linden, S. (2019). Fake news game confers psychological resistance against online misinformation. *Palgrave Communications, 5*(1).
Sagan, C. (2011). *The demon-haunted world: Science as a candle in the dark*. Random House Publishing Group. https://ebookcentral.proquest.com/lib/kxp/detail.action?docID=6077498
Scheufele, D. A., Krause, N. M., & Freiling, I. (2021). Misinformed about the "infodemic?" Science's ongoing struggle with misinformation. *Journal of Applied Research in Memory and Cognition, 10*(4), 522–526.
Schindler, S. (2020). The task of critique in times of post-truth politics. *Review of International Studies, 46*(3), 376–394.
Schrögel, P., & Kolleck, A. (2019). The many faces of participation in science. *Science & Technology Studies*, 77–99.
Senabre Hidalgo, E., Perelló, J., Becker, F., Bonhoure, I., Legris, M., & Cigarini, A. (2021). Participation and co-creation in citizen science. In K. Vohland, A. Land-Zandstra, L. Ceccaroni, R. Lemmens, J. Perelló, M. Ponti, R. Samson, & K. Wagenknecht (Eds.), *The science of citizen science* (pp. 199–218). Springer International Publishing.
Shao, C., Ciampaglia, G. L., Flammini, A., & Menczer, F. (2016). Hoaxy. In J. Bourdeau, J. A. Hendler, R. N. Nkambou, I. Horrocks, & B. Y. Zhao (Eds.), *Proceedings of the 25th international conference companion on world wide web—WWW '16 companion* (pp. 745–750). ACM Press.
Simis, M. J., Madden, H., Cacciatore, M. A., & Yeo, S. K. (2016). The lure of rationality: Why does the deficit model persist in science communication? *Public Understanding of Science, 25*(4), 400–414.
Spampatti, T., Hahnel, U. J. J., Trutnevyte, E., & Brosch, T. (2024). Psychological inoculation strategies to fight climate disinformation across 12 countries. *Nature Human Behaviour, 8*(2), 380–398.
Sturgis, P., & Allum, N. (2004). Science in society: Re-evaluating the deficit model of public attitudes. *Public Understanding of Science, 13*(1), 55–74.
Suldovsky, B. (2017). The information deficit model and climate change communication. In B. Suldovsky (Ed.), *Oxford research encyclopedia of climate science*. Oxford University Press.

van Zoonen, W., Luoma-aho, V., & Lievonen, M. (2024). Trust but verify? Examining the role of trust in institutions in the spread of unverified information on social media. *Computers in Human Behavior, 150*, 107992.

Vosoughi, S., Roy, D., & Aral, S. (2018). The spread of true and false news online. *Science (New York, N.Y.), 359*(6380), 1146–1151.

Wynne, B. (2006). Public engagement as a means of restoring public trust in science--hitting the notes, but missing the music? *Community Genetics, 9*(3), 211–220.

Pluralism About the Nature of Science: Ethical Challenges in Science Education and Communication

Andreas Nehring, Thomas A. C. Reydon, and Till Bruckermann

1 Introduction

The nature of science (henceforth: NOS) is a long-standing topic in both science education and communication, and in the history and philosophy of science. It encompasses the questions of how science is best characterized, what distinguishes science from other areas of practice that do not (or should not) count as science (including, but not limited to pseudoscience), and how learning about NOS among students and the general public could be facilitated. While these are also central questions

A. Nehring (✉)
Institute for Science Education, Leibniz University Hannover, Hanover, Germany
e-mail: nehring@idn.uni-hannover.de

T. A. C. Reydon
Institute of Philosophy and Centre for Ethics and Law in the Life Sciences (CELLS), Leibniz University Hannover, Hanover, Germany
e-mail: reydon@ww.uni-hannover.de

T. Bruckermann
Institute of Education, Leibniz University Hannover, Hanover, Germany
e-mail: till.bruckermann@iew.uni-hannover.de

J. Hinrichsen et al. (eds.), *Ethics of the Sciences*,
https://doi.org/10.1007/978-981-92-2821-8_15

in the philosophy of science, in the fields of science education and science communication these questions have a strongly normative component, because the ways in which students and members of the general public understand NOS affect their attitudes toward science—in both positive and negative ways. The way in which people understand what science is and how it works may for instance increase or decrease their trust in scientific claims (see Hönnige and Nyhuis in this volume), which in turn affects the way in which science enters into policy making (see Lohse and van Baßhuysen in this volume; and Bleidorn in this volume). In this respect science educators and science communicators, but researchers in science education and communication too, face ethical questions that are internal to their work (cf. Reydon and Schicktanz in this volume): For one, how should NOS be conceptualized in light of the possible individual, social and political consequences of the way in which members of the general public think about science? How should educators and communicators account for the diverse beliefs about science? What are productive beliefs about science that stimulate discussion, even in case of polarization (e.g., in the case of vaccines or climate change)? And, in the long run, how politically engaged should educators and communicators be in the first place (see Füssel in this volume)?

As we will show in this chapter, these issues are complicated by the fact that there is no unique conceptualization of NOS—rather, NOS is deeply a pluralistic concept. Conceptualizing NOS in science education and communication is pluralistic in its own manner. For example, as we will discuss below, consensus lists and family resemblance approaches both implicitly recognize the existence of a plurality of views regarding NOS while attempting to highlight where these views converge. The history and philosophy of science, too, has reached a thoroughly pluralistic view of NOS that denies that there is a unique nature of science that equally characterizes all the sciences and distinguishes them from non-science. Here, the term plurality refers to the fact that in science diverse concepts of knowledge and ways of knowledge production coexist (Ludwig & Ruphy, 2024). The term pluralism acknowledges this plurality as an explicit perspective in the philosophy of science. Moreover, scientific pluralism itself is pluralistic (Wylie, 2015) because this pluralism could be related to, for example, pluralism about epistemic concepts but also ontological pluralism (Ludwig & Ruphy, 2024). Thus, in history and philosophy of science, acknowledging a plurality of epistemic concepts means acknowledging the coexistence of diverse forms of knowledge and

Table.1 Comparison of four nature of science approaches in science education and communication

	Consensus view	*Whole science*	*(Reconceptualized) Family resemblance*	*Meta-categories*
What are the aspects/ dimensions/ categories for reflection?	Aspects: Empirical, inferential, creative, theory-driven, tentative, myth of "the scientific method", scientific theories, scientific laws, social dimensions of science, social and cultural embeddedness of science	Dimensions: Observations and reasoning, methods of investigation, history and creativity, human context, culture, social interactions among scientists, cognitive processes, economics/ funding, instrumentation and experimental practices, communication and transmission of knowledge	Categories: Aims and values, methods, scientific practices, scientific knowledge, social certification and dissemination, scientific ethos, social values, professional activities, social organizations and interactions, financial systems, political power structures	Meta-categories: Aims, methods, activities, products
What is the status of the list?	Open-ended but consensus on core tenets	Not exhaustive	Open-ended	Hierarchical and, thus closed
How are scientific endeavors different from other?	By means of absolute distinction	By means of relational distinction	By means of relational distinction	By means of relational distinction
What is the perspective on sciences?	Domain-general	Domain-general	Domain-general and -specific	Domain-general and -specific
Which disciplines does the approach refer to?	Natural sciences	Natural sciences	Sciences	Sciences

(continued)

Table.1 (continued)

	Consensus view	*Whole science*	*(Reconceptualized) Family resemblance*	*Meta-categories*
What is the perspective on learning?	Potentially deficit-oriented	Potentially deficit-oriented	Potentially deficit-oriented	Potentially asset-based
How are learners' assets accounted for?	Comparison against fixed standards	Comparison against fixed standards	Comparison against fixed standards	Open for learners' views
	Abd-El-Khalick et al., 2008	Allchin, 2011	Kaya & Erduran, 2016	Romero-Maltrana & Duarte, 2022

paths toward knowledge that reflect the variety of social and cultural contexts within which scientific knowledge production takes place. In science education research, the plurality of views regarding NOS reflects the multitude of possible representations of the sciences for the purposes of science education and communication. In contrast to pluralism the term diversity refers to an analytical category to describe how individuals differ regarding certain attributes in a group (egalitarian concept of diversity; Steel et al. 2018), such as their epistemic beliefs (i.e., epistemic diversity; Schiefer et al., 2022).

While such pluralistic views do justice to the diversity of approaches and methodologies that exist among the sciences, they also threaten to undermine the status of science as well as public trust in science by suggesting that there is nothing that singles out science as the source of well-supported knowledge about the world but rather just one among many, equally legitimate ways of understanding the world (Romero-Maltrana et al., 2019; Serrano et al., 2024).

This poses educators and communicators with an ethical dilemma regarding what constitutes good practice in their area of work: should they communicate adequately about the pluralism that exists among the sciences and risk undermining the status of science, or should they knowingly misrepresent science in a more unified way that ignores large parts of this pluralism? This is a particularly vexing issue, as the effects of these different ways of communicating about science cannot be estimated a priori and depend strongly on the specific audience. It is also an issue that strongly depends on the context in which science communication takes place: arguably, the risk of undermining the status of science is larger in the context of a post-truth society (see Bleidorn in this volume) than in a society in which decision making relies more strongly on factual knowledge and available evidence. On the one hand, teaching and communicating a (curricular) highly canonized, but at the same time inadequate, picture of NOS might promote trust in science and science-based decisions by making it easy to understand how science works. On the other hand, providing more adequate insights into the pluralistic nature of science and into the multitude of ways in which scientific knowledge is produced can support the learning of scientific concepts (e.g., through conceptual change processes) and is considered a crucial part of scientific literacy (as it enables well-informed trust in science). However, such insights can only be acquired by students that already are at a high level of expertise and in higher stages of education, and studies have found that students—and even scientists and teachers—often hold rather absolutist views of NOS instead of more pluralistic views.

In addition to these issues, teaching NOS in schools and universities, and communicating to the general public about how science produces reliable knowledge, typically focus on the nature of the *natural* sciences. This means that epistemic injustices may easily occur as large areas of knowledge production are ignored. Talking about NOS suggests that what is at stake is the nature of those domains that typically fall under the English term "science", i.e., the natural and life sciences. But when it comes to knowledge production, the social sciences, the engineering disciplines, the humanities—in short, all areas of academic research and scholarship—are worth being taken into account.

In teaching and communication settings, educators, researchers, and science communicators have a responsibility to find ways to adequately present the multitude of ways in which knowledge is being produced while at the same time avoiding the risks and epistemic injustices highlighted above. This chapter intends to put this complex of ethical challenges on the agenda of science educators and philosophers of science and to provide a starting point for the development of a concept of NOS that can be used to meet these challenges. This starting point will result from overviewing the—at present still largely disjoint—debates on the NOS in science education and communication (which largely revolve around consensus lists for conceptualizing the nature of science and family resemblance approaches) and in the history and philosophy of science (on the "demarcation problem"). Showing how pluralism is an intrinsic aspect of conceptions of NOS in both science education (and communication) and the history and philosophy of science can, we believe, constitute a first step toward a more suitable concept of NOS.

The central argument we unfold in this chapter is that pluralism about NOS is not merely a descriptive issue concerning the diversity of scientific practices, but also a normative one, because every way of presenting "what science is" implies value-laden choices about what counts as legitimate knowledge and who is authorized to produce and communicate it. Against this background, we pursue the central question of how pluralism about the nature of science can be addressed in science education and science communication in ways that are ethically responsible. Our aim is therefore not only to focus pluralism about NOS within research or teaching agendas but to sketch a conceptual foundation for integrating ethical reflection into NOS-related research and practice. Specifically, we examine (1) which role pluralism plays within NOS research, (2) how the demarcation debate in the history and philosophy of science (HPS) reveals the limits of strict definitions of science, and (3) what ethical implications emerge from both perspectives for teaching and communicating about NOS.

The chapter proceeds as follows. We begin in section 2 with the descriptive side of the issue: we discuss pluralism about NOS from an HPS perspective, focusing on the demarcation problem. Section 3 then reviews the various approaches to NOS in science education and communication. Together, these two sections present the causes underlying the challenge that pluralism about NOS poses for science education and communication. Research in HPS has shown that this is not only a matter of how NOS is *presented*, but a matter of the state of affairs in the sciences: there is, in fact, not *one single* NOS that would cover all of the sciences, conceived broadly: the sciences themselves—and accordingly NOS—are thoroughly pluralistic. This means that in science education and communication, pluralism about NOS is an issue that cannot be avoided. The issue is exacerbated by the fact that science education and communication do not aim to merely describe NOS in an abstract manner, but have practical aims in view: they are aimed at fostering understanding of NOS in students and members of the general public within the context of the aims of school education more generally as well as the general social and political aims of science communication. Accordingly, educators and communicators are faced with the challenge of adequately presenting

NOS as it pertains to the various sciences, conceived broadly, while at the same time presenting NOS in such a way specific educational, social, and political aims can be achieved. Section 4 discusses the ethical challenges that arise from this state of affairs for educational and communicative practice. Section 5 concludes with implications for future research and practice.

2 Pluralism about NOS from an HPS Perspective

2.1 The Demarcation Problem

In the history and philosophy of science (HPS), the topic of NOS is usually discussed under the header of the "demarcation problem" (for more background, see, e.g., Laudan, 1983; Nickles, 2013; Pigliucci, 2013; Pigliucci & Boudry, 2013). At the heart of that problem lies the question of what distinguishes practices and claims that can legitimately be considered to count as science from those that cannot—that is, how science can be demarcated from other endeavors. Much work in the history and philosophy of science that has been done on this question focuses on the contrast between science and pseudoscience, i.e., endeavors such as homeopathy, astrology, parapsychology, or "Intelligent Design Theory", proponents of which sometimes stake a claim to their endeavor being recognized as science while it does not in fact meet any of the putatively relevant criteria. The problem also has a broader reach, though, as many authors extend it to the contrast between the domain of science on the one hand and everything lying outside that domain (i.e., non-science) on the other hand.

The demarcation problem is philosophically important as part of the larger quest to understand what science is and how science works to produce well-supported, reliable knowledge. By closely examining the contrast between science and pseudoscience or non-science, it would be possible to achieve a clearer characterization of the nature of science—at least that was the ambition. The problem also has important ramifications outside the academic domain, for instance, in public policy (Resnik, 2000). A widely discussed example concerns the creation-evolution controversy, in which the question of whether "Creation Science" and "Intelligent Design Theory" could legitimately count as science and accordingly be granted space in the science curricula of high schools in the

United States (Ruse, 1988). But the demarcation problem has a problematic history, and most contemporary historians and philosophers of science hold that the search for a unique demarcation criterion or a unique set of such criteria that could serve to elucidate the nature of science has failed.

It is important to note that in the HPS literature, the demarcation problem is usually discussed with respect to the nature of scientific knowledge (rather than NOS more broadly) and is connected to the question of scientific method. The denominator "the demarcation problem" originated in the early twentieth century in Popper's work on the logic of science (Hirvonen & Karisto, 2022; Nickles, 2013; Pigliucci, 2013; Pigliucci & Boudry, 2013). Popper coined the term in his groundbreaking book, *The Logic of Scientific Discovery* (Popper, 1959, 1992; "*Abgrenzungsproblem*" in the original German version of the book from 1935) to denote "[t]he problem of finding a criterion which would enable us to distinguish between the empirical sciences on the one hand, and mathematics and logic as well as 'metaphysical' systems on the other, I call the problem of demarcation" (Popper, 1992: 11; original italics). Later in the book he mentioned it as the problem of finding "a dividing line between scientific and metaphysical systems" (Popper, 1992: 14). Popper conceived of the problem of demarcation together with the problem of induction as constituting the two fundamental problems of epistemology. For Popper, the demarcation problem thus is less about characterizing the nature of science as such, but rather a question of characterizing the specific kind of knowledge as produced by the sciences. More precisely, the problem comes down to the question of what distinguishes empirically supported knowledge from non-empirical claims, such as metaphysical, but also logical and mathematical statements.

Popper famously brought falsifiability into play as the criterion that—according to him, at least—would be able to solve the problem. According to Popper, the mark of scientific knowledge claims is that they can in principle (even though not necessarily always in practice) be refuted by means of empirical tests. This, in principle, possibility for empirical refutation according to Popper is the mark of scientific knowledge that distinguishes it from claims made in areas that do not count as science. Note that this distinctive mark is methodological: the mark of scientific knowledge is the method by which it was produced (and note that Popper is not specifically concerned with distinguishing scientific knowledge from pseudoscientific claims).

An important problem with this criterion, however, is that empirical refutations are far from straightforward. Suppose one wants to test a scientific hypothesis by means of an experiment and the data obtained in the experiment are in conflict with the hypothesis. This result may mean that the hypothesis has been refuted, but it may also be interpreted differently. Maybe there was an error in the way the experiment was conducted, or some of the equipment used (apparatuses which themselves are built on the basis of scientific theories) does not do what the experimenter thinks it does. Moreover, scientific theories and hypotheses are never tested in isolation. Empirical tests require a number of background assumptions, such that what initially looks like a falsification of the hypothesis under scrutiny may actually be a problem in one or more of the background hypotheses that underpin the test (this observation has come to be known as the Duhem-Quine thesis). Generally, it is crucial to see that scientific claims depend on each other and constitute an intricate network of claims that includes empirical statements as well as mathematical, theoretical, and empirical ones, and that as a whole is tested against empirical observations. This causes Popper's falsifiability criterion to fail in practice as a tool that can demarcate empirical knowledge statements from non-empirical ones.

2.2 *Beyond Strict Demarcation*

While for Popper the problem thus was about (the nature of) scientific knowledge and "science" was understood as referring to the empirical sciences, later authors took it to be about the nature of science as a domain of practice in a somewhat broader sense. In his famous book, *Against Method*, for instance, Feyerabend (1975) argued that when examining the history of science, there is no unique method that characterizes all the various endeavors that at some point in time were accepted as science, making it impossible to find a unique criterion or set of criteria that could serve to demarcate science from everything else. Science, as it has manifested itself throughout human history, according to Feyerabend, is a very pluralistic cluster of endeavors (Shaw, 2017). In a similar vein, Laudan, (1983) noted that so far philosophers had failed to reach any agreement regarding a "demarcation line between science and non-science, or between science and pseudo-science" (Laudan, 1983: 112) and suggested that "it may just be that there are no epistemic features which all and only the disciplines we accept as 'scientific' share in common" (ibid.). Laudan concluded that the questions what demarcates science from non-science or pseudoscience and what makes a claim scientific missed the mark and historians and philosophers of science should rather ask what makes a knowledge claim well-founded: "our focus should be squarely on the empirical and conceptual credentials for claims about the world. The 'scientific' status of those claims is altogether irrelevant" (Laudan, 1983: 125). In other words, the inquiry should be into the nature of well-supported, reliable knowledge, not into the nature of scientific knowledge or the nature of science more broadly.

The problem is exacerbated when the term "science" is not taken in the strict sense of the English word, as is usually done in the area of HPS, but in the much broader sense of the German word "*Wissenschaft*" (which is how "science" is understood in the present volume). The term "science" can be used in stricter and looser senses, either referring to the natural sciences or in a looser sense to all empirical areas of research (including the social sciences). But the term "*Wissenschaft*" has an even broader coverage, including not only the natural and social sciences, but also, for instance, the engineering disciplines ("*Ingenieurwissenschaften*"), the humanities ("*Geisteswissenschaften*"), linguistics ("*Sprachwissenschaften*"), (comparative) religious studies ("*Religionswissenschaft*"), and legal scholarship ("*Rechtswissenschaft*"). While it is unlikely that one unique way can be found to demarcate science from non-science, or that an encompassing methodological characterization of all of science can be achieved, this is even more unlikely when considering the various *Wissenschaften* rather than just the sciences. (And recall that Popper aimed to demarcate empirical knowledge from metaphysics and logic, which both clearly count as *Wissenschaft*.)

These problems notwithstanding, the demarcation problem continues to be seen as an important topic of research in the domain of HPS (Fernandez-Beanato, 2020; Hirvonen & Karisto, 2022; Nickles, 2013; Pigliucci, 2013; Pigliucci & Boudry, 2013), and recent authors have added different takes on the issue. Resnik (2000), for example, suggested that demarcation criteria should be contextual. Resnik noted that distinctions between science and other endeavors typically have social and political implications (see the creation-evolution controversy, mentioned above) and that, therefore, the way in which we distinguish between science and non-science or pseudoscience should be guided by what the distinction is supposed to be for. This means that different criteria may be used for different contexts and the resulting dividing line may turn out to be different in different contexts. Hoyningen-Huene (2008, 2013) recently suggested that characterizing the nature of science should be done comparatively. Hoyningen-Huene also focused on scientific knowledge and argued that a distinction can be made between scientific knowledge and other kinds of knowledge by looking at the degree of systematicity with which it was produced. The comparison is between areas that pertain to the same phenomena: when there are a branch of science and an area of everyday practice that is concerned with the same phenomena (think for instance of a comparison between meteorology and established local knowledge about weather patterns in a particular geographical region), then Hoyningen-Huene's claim is that the scientific domain will be more systematic in several aspects of the way in which knowledge claims are produced and supported than the comparable area of everyday knowledge.

The current state of affairs in the area of HPS with respect to the demarcation problem is very much one of pluralism. There is a plurality of approaches to the problem and ways in which the problem is framed by different authors. And there is a plurality of potential demarcation criteria that could be used to characterize NOS and/or the nature of scientific knowledge, but without agreement among HPS scholars regarding which criteria would be preferred over others. Indeed, the current tendency is toward embracing pluralism by recognizing that the sciences constitute a diverse cluster of endeavors, such that different criteria and characterizations can be appropriate depending on the specific fields or contexts that are being considered. And the case for pluralism becomes stronger still when asking how the "Nature of *Wissenschaft*" (NOW) is best characterized, as it is unlikely that physics, biology, philosophy, history, political science, psychology and legal studies all share the same nature. Work in the area of HPS, which, as discussed above, has been much more concerned with the nature of well-supported empirical knowledge than with NOS, has not yielded an unequivocal account of either empirical knowledge or NOS. It is even less likely to come up with an unequivocal account of NOW.

This discussion shows that pluralism about NOS is inherent to science and as such an issue that cannot be avoided: it is a feature of science that educators and communicators must take into account when developing approaches to teaching and communicating about NOS. In the next section, we discuss the main approaches that educators have developed and show how these add to the pluralism about NOS that work in HPS gives rise to.

3 Pluralism About NOS in Science Education

3.1 Approaches to NOS

Over the past years, there has been a significant shift toward emphasizing scientific literacy as a primary educational objective. This paradigm shift places a nuanced understanding of science at the forefront, serving as a critical foundation for facilitating active participation in societal discourse and decision making (e.g., KMK, 2004). The contemporary debate within science education regarding NOS is currently undergoing a profound reorientation, with a heightened focus on the processes leading to the genesis of scientific knowledge. Increasingly, scientific literacy is recognized as foundational for fostering lifelong, autonomous learning about the nature of science itself (Allchin, 2017). This paradigm shift is significant because every debate of scientific and societal relevance, which stimulates public opinion and decision making processes, is inherently underpinned by complex and specialized expert knowledge (Allchin, 2012b, Allchin & Zemplén, 2020; Höttecke & Allchin, 2020). Such specialized knowledge, however, is rarely presented comprehensively or in an accessible manner to the general public (Allchin, 2017). Moreover, the burgeoning presence of social media and other internet-based sources further complicates the landscape of information dissemination, imposing additional demands on users' abilities to form informed opinions and make decisive conclusions (Osborne & Allchin, 2024).

A comprehensive understanding of NOS requires an appreciation of the mechanisms through which scientific knowledge is generated, negotiated, communicated through media, and eventually published. Only then can individuals accurately evaluate the credibility and legitimacy of scientific claims, which in turn supports informed personal and societal decision making (Stadtler et al., 2017). Consequently, it is crucial for NOS to be understood from a lay perspective as a social and communicative practice founded upon collaboration and dependency.

In the scholarly discourse on educational theory, there is an active and ongoing debate regarding the optimal NOS model that can lay the groundwork for such a comprehensive and appropriate understanding of science (cf. Osborne, 2017; Schwartz et al., 2012). Presently, four principal NOS modeling approaches have emerged, each contributing distinct perspectives to the educational theory discussion on the essence of science (Heering & Kremer, 2018).

The *Minimal Consensus Approach* has a well-established tradition in the academic discourse on NOS within science education. It employs simplified consensus lists to consolidate essential elements, including the empirical nature of science, the tentative nature of scientific knowledge, the role of imagination, and creativity in science, the subjectivity and theory-laden exploration of scientific endeavors, the sociocultural embeddedness of science, the distinction between scientific theories and laws, as well as the significance of observations and inferences (Lederman, et al., 2002; McComas & Olson, 1998; Osborne et al., 2003). However, it faces critique from proponents of alternative NOS models. Critics argue that an abstract understanding of science, reliant on simplified basic principles, fails to meet the demands of a scientific literacy that empowers public engagement and participation. Alternatives like the Nature of Whole Science approach (Allchin, 2011) and the Family Resemblance Approach (Erduran & Dagher, 2014) seek to transcend these limitations.

The *Nature of Whole Science Approach* (NOWS) proposes a comprehensive reformulation of NOS, claiming to encapsulate science in its entirety while elucidating the societal interactions and ramifications (Allchin, 2012b; 2020). Notably espoused by Douglas Allchin, this model adopts an inductive methodology, deriving NOS topics from explicit case studies extracted from Socio-Scientific Issues (SSIs). These case studies may represent either historical retrospectives or ongoing societal negotiation processes (Allchin, 2012b). The approach elucidates various aspects that depict authentic research practices, such as a holistic understanding of science as an interconnected system, contextualized science reflecting historical and social influences, practices and processes involved in scientific work, values and ethics in shaping scientific endeavors, and public engagement and decision making related to science (Allchin, 2011, 2012a).

The *Family Resemblance Approach* (FRA) or the Reconceptualized FRA-to-NOS Approach (RFN) offers a distinctive lens through which NOS can be understood, emphasizing the interconnectedness and plurality inherent in scientific practices (Erduran & Dagher, 2014; Kaya & Erduran, 2016). Inspired by Wittgenstein's notion of family resemblances, this approach suggests that rather than adhering to a rigid set of defining principles, scientific disciplines share overlapping characteristics that form the basis for their classification as science (Erduran & Dagher, 2014). In the context of NOS, the FRA/RFN encourages educators and learners to embrace the pluralistic nature of science by recognizing the common threads that weave through various scientific disciplines, such as the iterative nature of scientific inquiry, the reliance on empirical evidence, and the creativity and imagination in scientific discoveries. By adopting this framework, the FRA facilitates a more holistic understanding of science that transcends narrow definitions, illuminating the rich tapestry of methodologies, epistemologies, and cultural contexts that comprise scientific practice. This approach is particularly valuable in science education, as it equips students with the tools to critically engage with scientific concepts across different domains, fostering a deeper, more nuanced appreciation of sciences in society. By acknowledging the fluidity and plurality of scientific endeavors, the FRA provides a robust pedagogical framework for exploring the multifaceted nature of science, aligning closely with contemporary educational goals of promoting scientific literacy and interdisciplinary understanding.

The *Meta-Categories Approach* transcends the previous feature-based approaches by suggesting a hierarchical list of four meta-categories (aims, methods, activities, and products) that help to describe any human endeavor (Romero-Maltrana and Duarte 2022). The categories provide a tool to reflect upon the nature of scientific (and also non-scientific) endeavors and to compare them with respect to these categories. For example, in school classes, one could compare a scientific journal to a newspaper regarding their aims of publishing research results vs. news, their methods of editorial and peer review vs. editorial review, their activities of preparing and revising a manuscript based on peer review vs. writing an article following the editorial line and the final product (Romero-Maltrana & Duarte, 2022).

These discourses on the conceptualization and comprehension of NOS underscore the imperative to strike a balance between representing the complexity of scientific practices and facilitating public engagement and education. It advocates for a nuanced appreciation of the pluralistic dimensions of science, both within academic settings and in broader societal contexts, thereby fostering a scientifically literate and informed populace. The exploration of these NOS models further illustrates the dynamic interplay between scientific discourse, educational endeavors, and societal imperatives, underscoring the importance of adopting a multifaceted approach to science education and communication.

3.2 Epistemic Diversity: Ecologies of Knowledge and Pluralism in Epistemic Cognition

An additional reason for adopting a multifaceted approach is the acknowledgment that a plurality of knowledge systems exists across different cultures and societies. This is captured by the notion of epistemic diversity (e.g., Carter, 2010). It is often related to challenging traditional, often Eurocentric, understandings of knowledge as singular and homogenous, proposing instead that knowledge production is inherently shaped by the diverse social, cultural, and historical contexts in which it occurs. Epistemic diversity arises from the Global South's resistance to the hegemony of Northern knowledge systems, which have historically marginalized and devalued alternative ways of knowing. For example, Santos (2007) proposes an ecology of knowledges which advocates for acknowledging the legitimacy and richness of diverse knowledge traditions and seeks epistemological and cognitive justice. The interplay between knowledge and cultural life is seen as reciprocal, where knowledge production shapes, and is shaped by, the social and cultural contexts of the people involved. This perspective opens up the possibility for equitably recognizing and integrating multiple knowledge systems, emphasizing their situated, partial, and contingent nature. By advocating for intercultural dialogue and translation, epistemic diversity seeks to challenge the dominance of a single knowledge narrative and promote a more inclusive understanding of knowledge that respects and integrates the contributions from various cultural and epistemological backgrounds.

Epistemic diversity is not only a critical consideration in the broader context of knowledge systems across cultures but also a pertinent feature within classrooms and informal learning settings, where it manifests as a diversity of epistemic beliefs or views on NOS or the Nature of Scientific Inquiry (NOSI). Comparable to a diversity of students' conceptions (Paehr et al., 2023), this diversity is evident in the varying profiles of students' epistemic beliefs, which do not necessarily follow a linear developmental trajectory. Research indicates that within any given age group, there is substantial individual variability, with students exhibiting a wide range of epistemic beliefs and levels of sophistication regarding scientific reasoning (Schiefer et al., 2022). For example, while theoretical frameworks might predict younger students to have a more realist or absolutist view, empirical studies have identified sophisticated epistemic understandings even among 8- to 9-year-olds (Nehring, 2020). Similarly, older students may still exhibit absolutist or evidence-based profiles, highlighting the non-linear and diverse nature of epistemic development.

In view of these pluralistic aspects, the design of learning processes about NOS faces at least a twofold challenge. On the one hand, the pluralism of concepts regarding NOS existing in the scientific realm should be adequately represented in learning processes (Stroupe et al., 2024). On the other hand, the diversity of individual conceptions regarding the nature of science should be addressed in order to facilitate the development of epistemic beliefs or NOS views (Schiefer et al., 2022). The way in which this twofold challenge is dealt with should be reflected upon both in terms of learning processes in formal contexts, such as schools, and in informal contexts or science communication. Although both contexts share commonalities, they also differ markedly regarding the individual reasons as well as goals of learning about science and the gatekeeping of information in the learning situation (Bromme & Goldman, 2014; Bromme & Kienhues, 2017).

4 The Ethical Aspects of NOS

In the preceding sections we have highlighted several aspects of pluralism about NOS that give rise to ethical challenges for science educators and communicators. Not only are the sciences (conceived broadly) deeply pluralistic with respect to their ways of knowledge production and the kinds of knowledge that they produce, there is also a considerable pluralism with respect to the ways in which NOS (or NOW) can be represented for purposes of science education and communication. The challenges that ensue from these pluralisms are ethical in nature, because they pertain to the question of what should be regarded as good practice in science education and communication, and in particular to questions regarding appreciation of and openness toward epistemic diversity in teaching and communication settings.

Although the history and philosophy of science tend to describe science in a value-neutral way, every attempt to teach or communicate about the NOS inevitably involves normative decisions. Educators and curriculum designers must decide what aspects of science to highlight, what counts as legitimate knowledge, and what kinds of scientific reasoning and practices are worthy of trust and emulation. These decisions are not only pedagogical but ethical, as they determine which epistemic and social values—such as objectivity, transparency, critical debate, or inclusivity—are being promoted in classrooms and public discourse. And as we have pointed out in the first section, the way in which people understand NOS will have an effect on their trust in science and in the reliability of scientific claims.

A straightforward expectation may be that an increased understanding of NOS will increase trust in science—the assumption here would be that if someone understands how a machine or a technology works, it is more likely that they will trust that it does its job. But the pluralism inherent to NOS makes this assumption untenable. Acknowledging the demarcation problem, the diversity of scientific practices and epistemic perspectives can strengthen public understanding of science, yet it also risks undermining science's credibility when such pluralism is interpreted as relativism. Questions of trust and credibility are therefore not peripheral but central to ethical reflection on how pluralism about the nature of science should be taught and communicated.

In recent years, the concept of *informed trust* has been introduced as a way of addressing this tension (Bromme, 2022). The notion describes a form of trust that is not based on blind confidence or mere authority, but on an informed understanding of how scientific knowledge is produced, validated, and communicated—i.e., an informed understanding of NOS. Informed trust acknowledges the plurality and tentativeness of science while maintaining commitment to shared epistemic standards and institutional norms. As such, it provides a promising bridge between pluralism about NOS and the ethical responsibility of educators and communicators to foster both critical engagement and credibility in science. Still, educators and communicators are faced with normatively laden choices: choices how thoroughly they should inform their audiences about the variety of ways in which science works (which is also a practical question of how much can actually be achieved in a particular educational or communicative setting) as well as choices between a variety of approaches to NOS that have been developed in the fields of science education and communication. To some extent, achieving informed trust with their audiences involves informing audiences about the pluralism of NOS as it is found in the sciences as well as the variety of approaches to NOS that are used to discuss NOS in the classroom and in public settings, but it is not easy to determine how deeply audiences should be informed about these issues.

Curricular and standard-setting decisions are, therefore, always normative decisions. For example, as Nordine and Fortus (2025) recently argue, the construction of science standards itself entails value judgments about what kinds of learning outcomes societies deem most important. They propose reorganizing standards around meaningful contexts and informed agency rather than disciplinary content competence, explicitly to align science education with values such as democratic participation, relevance, and agency. Similar reasoning applies to decisions about how NOS is represented in curricula: emphasizing, for instance, tentativeness and uncertainty supports values of openness and critical inquiry, whereas stressing consensus and reliability highlights trust and social cohesion.

Recognizing this, it becomes vital that educators and researchers make explicit which values guide their NOS framing. The ethical dimension of NOS teaching does not lie merely in avoiding misrepresentation but in reflecting transparently on the value commitments that underpin curricular and communicative choices. Making these commitments visible can foster reflexivity and accountability: it allows learners to see how scientific norms such as skepticism, peer review, and evidence-based reasoning embody particular ethical and epistemic ideals. In this sense, teaching about NOS is both an act of knowledge transmission and a form of moral and civic education, inviting learners to engage critically with the values that sustain scientific inquiry and its role in society. Recognizing the inherently normative character of NOS also means acknowledging that any conception of science taught or communicated reflects value commitments that can either reproduce or challenge epistemic injustices (Stroupe et al., 2024). Making these commitments explicit is therefore essential for fostering ethically responsible and reflexive pluralism in science education and communication.

It is important to note that these ethical challenges do not only occur in educational and communicative practices, but also in research in science education and communication, and that this causes the concepts of NOS (and NOW) to have normative content. If there is no unique way in which science works—no unique NOS or NOW—then there does not seem to be anything that would clearly demarcate science from other endeavors that merely claim to produce knowledge about the world without in fact producing knowledge. This could easily be taken to suggest that any practice that claims to produce knowledge is as good as any other, and that there is no difference between reliable knowledge and claims that should not be counted as knowledge. As the many claims about “alternative facts" in recent years have shown, this can cause severe societal problems (see Bleidorn in this volume). The ethical challenge that educators, communicators and researchers are faced with is to decide to what extent pluralistic views of NOS/NOW should be communicated to students and the general public, and risk undermining the status of science, or rather a less risky but also less adequate and possibly misleading picture of NOS/NOW should be communicated that would involve them being epistemically unjust toward their students and audiences. This is an issue for researchers in science education and communication, because these are the fields where the concept of NOS is analyzed, novel interpretations of the concept are introduced, and new concepts (such as NOW) are created. A considerable part of the decision making on how NOS will be presented to students and the general public thus takes place as part of conceptual research.

This is clearly seen in the use of consensus lists on NOS. One explicit goal that science educators and communicators share is to *change* the perceptions of their audiences of how science works (Bromme & Goldman, 2014; Höttecke & Allchin, 2020). Researchers in science education and communication who do not themselves engage in educational and communicative practices may not share this explicit goal, but still contribute to changing the perceptions of students and members of the general public by producing materials that can be used in teaching and communication. Hence, all involved groups are in a position to decide how the nature of science is presented for learning (Stroupe et al., 2024) and must take the consequences of these decisions into account. A pragmatic approach that is widely used in this context is to facilitate learning via consensus lists of characteristics of NOS (Lederman, 1992; see above for more detailed discussion). The approach is pragmatic because it strongly limits the various aspects of NOS that are presented and taught, thus making them manageable for teachers when discussing NOS in classroom settings (Kampourakis, 2016). But the production of consensus lists involves decisions regarding such issues as the number of aspects of NOS that are to be discussed, the aspects that are to be emphasized and those that are sidelined, the ways in which the various aspects of NOS are presented, and so on. For example, using a consensus list that foregrounds the tentativeness of scientific knowledge and places less emphasis on the evidential support of knowledge claims will induce a different view of science with audiences than using a consensus list that does the opposite. In this way, any consensus list of NOS is normatively charged, as it reflects the decisions that were made by the community that produced it regarding what and how should be taught to students and communicated to the general public.

Thus, when science educators, communicators, and researchers in these fields decide on a consensus of minimum criteria for describing the nature of the natural sciences, they might run the risk of promoting epistemic relativism (Romero-Maltrana et al., 2019). Learners could see the epistemic processes in the natural sciences as hardly different from other processes of understanding and thus relativize scientific knowledge (Romero-Maltrana et al., 2019). This is because criteria from the consensus list are formulated such that they broadly cover the natural sciences, yet this broadness could lead to equating science with any other human activity (Rudolph, 2000). Although the family resemblance approach was sought to overcome the list of somewhat broad criteria by providing categories that need to be reflected upon for all natural sciences, the items in the categories provide yet another list that faces similar problems (Romero-Maltrana et al., 2022).

In teaching and communication contexts, science educators and communicators also have to account for the learners' or audiences' views (Stroupe et al., 2024), which can differ to a great extent (e.g., Aristeidou & Herodotou, 2020; Cofré et al., 2019). This approach goes beyond a deficit orientation, according to which laypeople lack an understanding of science that can be attributed to the epistemic asymmetry between experts and laypeople (Cortassa, 2016). Even though science education and science communication no longer assume a deficit orientation, this accusation is still made against approaches that aim to promote learning about the nature of science (Stroupe et al., 2024). For students, it is still often stated that their conceptions are inadequate or deviate from current scientific conceptions (the same applies to prospective teachers, e.g., Cofré et al., 2019). Departing from a deficit orientation means that individual conceptions are seen as productive in the respective context. Factors like dissatisfaction with one's own conceptions (Posner et al., 1982) or plausibility judgments (Lombardi et al., 2016) can lead to a conceptual change. In contrast to established scientific concepts, however, the following difference exists: it seems difficult to define adequate or technically correct concepts in the face of pluralism in science (see sec. 3). The previous approaches from consensus lists (Lederman, 1992) to family resemblances (Erduran & Dagher, 2014) to meta-categories (Romero-Maltrana & Duarte, 2022) appear insufficient, since they do not allow for a definitive but at most relational description of science due to generalization.

Deficit approaches stand in contrast to asset-based approaches that assume how learners help shape what counts as the nature of science (Moreira-dos-Santos & El-Hani, 2017; Stroupe et al., 2024). Asset-based approaches oppose fixed standards of what counts as science that learners' beliefs are then compared against, because general conceptions of the nature of science are mostly not representative of how different communities in the sciences negotiate epistemic aspects of their work (Rudolph, 2000). Moreover, in asset-based approaches, learners are themselves seen as a community that is "supported [...] to see science as a socially constructed epistemic practice in which they had a voice to shape the norms and participatory structures in the classroom" (Stroupe et al., 2024, p. 28; e.g., Sandoval & Morrison, 2003). This asset-based approach toward the nature of science (Stroupe et al., 2024) could also be intertwined with meta-categories (Romero-Maltrana & Duarte, 2022), because they allow comparing different scientific and non-scientific endeavors on a more granular level such as in different disciplines within a natural science (e.g., cosmology and particle physics in physics) or even in different communities. In science communication and education, another asset-based approach toward the nature of science could be found in citizen science projects that foster collaborative knowledge construction and individual learning.

For example, citizen science projects are strongly based in different communities that shape their specific scientific endeavors. Learners in citizen science are not only the users but become producers of scientific knowledge (Mueller & Tippins, 2012). Hence, learners "participate through multiple relations and situations in which science is enacted in the community" (Mueller & Tippins, 2012, p. 853; paraphrasing Roth & Calabrese Barton, 2004). Learners' views on the issue at hand, their approaches to investigate the issue, and their solutions are not disregarded as unscientific but are an asset to the community. Instead of comparing learners' engagement in scientific practices against a fixed set of nature of science aspects, in citizen science, learners negotiate what counts as science in the production of knowledge. Thereby, they learn that the epistemic practices of science are constructed by humans and that they shape the norms of the scientific endeavor of citizen science projects.

5 Conclusion: Pluralism about NOS As a Key Challenge for Future Research

In this chapter, we have done three things: we highlighted the causes that underlie pluralism about NOS, we pointed out that this pluralism poses an ethical challenge for researchers as well as practitioners in science education and science communication, and we showed how this challenge can manifest itself in research and practice in science education and communication. Our aim was to put pluralism about NOW on the research agenda and not to provide answers to the challenges it poses, as the latter would be impossible to do within the confines of a single chapter. Navigating the ethical challenges of teaching pluralistic perspectives on the nature of science is decidedly complex for educators and communicators, and researchers in science education and communication are faced with the additional challenge to provide concepts and materials that support educators and communicators in navigating this complex terrain. The present chapter is successful if it provides a starting point for researchers and practitioners to develop approaches to address these challenges.

The interdisciplinary collaboration between philosophy of science and science education has been central to this endeavor. Bringing these fields into dialogue allows conceptual analyses of pluralism, demarcation, and epistemic diversity to be connected with empirical and pedagogical research on how such ideas can be translated into educational and communicative practice. This collaboration demonstrates that addressing pluralism about NOS ethically requires both normative reflection and sensitivity to empirical and practical contexts, and it underscores the originality of our approach.

We want to conclude by noting that several guiding principles could help with navigating these challenges effectively. First, a strong sensitivity to the epistemic diversity of students and members of the general public appears to be crucial, allowing for a respectful and inclusive learning environment that recognizes a variety of epistemological perspectives. Note that this does not imply that students and members of the general public are typically uninformed or inadequately informed about NOS—the assumption is merely that a considerable variety of views is present among students and members of the general public of how science works and how scientific knowledge claims are supported (e.g., Höttecke & Allchin, 2020; Osborne & Allchin, 2024). Second, incorporating pluralism about NOS while at the same time stressing the trustworthiness of science in teaching and communicating about science will be essential. This will require educators and communicators to engage with the literature in the area of HPS on the topic of trust in science. The topic has in recent decades been in focus in the history and philosophy of science and a wealth of literature is available that shows why scientific results are reliable and trustworthy notwithstanding the inherent tentativeness of claims and persistent disagreements among experts (for entries into the literature, see for instance Blancke & Boudry, 2022; Oreskes, 2019; Rolin, 2020; Whyte & Crease, 2010). Third, engaging students in critical reasoning and critical thinking will be essential, focusing on central aspects such as linking observations, evidence, and claims in the perspective of plausibility, as highlighted by Lombardi et al. (2018). Here, too, the literature on critical thinking provides a wealth of material to build on (e.g., Haber, 2020).

At the same time, a critical perspective on public trust complicates these considerations. As Tolbert (2025) argues, distrust of science is not merely a cognitive deficit but often a socially and historically grounded response to experiences of exclusion, exploitation, and elite capture within scientific institutions. What may appear as misplaced distrust can thus represent legitimate skepticism toward power structures that have shaped scientific authority. In the context of pluralism about NOS, this means that different conceptions of science also imply different relationships of trust and responsibility: some foreground openness, reflexivity, and social accountability, while others reproduce hierarchical or exclusionary notions of expertise. A critical question to work is there if strengthening trust in science cannot mean simply reinforcing deference to experts and if it requires addressing the institutional and epistemic conditions that have eroded trust in the first place (Tolbert, 2025). Citizen and community science provide avenues to further explore how people learn how epistemic practice is negotiated and, moreover, negotiate and shape what counts as science. Science education and communication bears the ethical responsibility not only to cultivate informed trust but also to examine and transform the social relations through which both science and trust in science are produced, maintained, and sometimes lost.

The above are merely three guiding principles that we believe should be taken into account in (research on) education and communication on NOS. We do not suggest that these are the only relevant principles, but we do believe that these should play a central role in addressing the challenges that pluralism about NOS gives rise to. Concretely, connecting the diverse epistemic perspectives among students as well as the general public to aspects of trust in science and critical thinking in contexts of science education and communication can be a way to foster a culture of active engagement with the sciences. Through such engagement, students and members of the general public can cultivate a more nuanced but pluralistic understanding of the diverse natures of science, conceived broadly, and in particular the various ways of thinking about what constitutes sufficient evidence for a particular claim and how claims can be supported by evidence. Science education research already offers ways for developing strategies to address the pluralism regarding NOS, for example, by invoking concepts like informed trust by Bromme (2022), or fostering the notion of the competent outsider (based on Feinstein, 2011; Osborne & Allchin, 2024).

We hope these suggestions, tentative as they may be, are sufficient to convince readers that an interdisciplinary approach to teaching and communication about NOS is needed that brings together research in science education, science communication, and the history and philosophy of science, and that acknowledging such a need for interdisciplinarity should be part of good practice in the relevant fields.

Competing Interests The authors have no conflicts of interest to declare that are relevant to the content of this chapter.

References

Abd-El-Khalick, F., Waters, M., & Le, A.-P. (2008). Representations of nature of science in high school chemistry textbooks over the past four decades. *Journal of Research in Science Teaching, 45*(7), 835–855.

Allchin, D. (2011). Evaluating knowledge of the nature of (whole) science. *Science Education, 95*, 518–542.

Allchin, D. (2012a). The Minnesota case study collection: New historical inquiry case studies for nature of science education. *Science & Education, 21*, 1263–1281.

Allchin, D. (2012b). Towards clarity on whole science and KNOWS. *Science Education, 96*(4), 693–700.

Allchin, D. (2017). Beyond the consensus view: Whole science. *Canadian Journal of Science, Mathematics and Technology Education, 17*(1), 18–26.

Aristeidou, M., & Herodotou, C. (2020). Online citizen science: A systematic review of effects on learning and scientific literacy. *Citizen Science: Theory and Practice, 5*(1), 69.

Allchin, D., & Zemplén, G. (2020). Finding the place of argumentation in science education: Epistemics and whole science. *Science Education, 104*(5), 907–933.

Blancke, S., & Boudry, M. (2022). "Trust me, I'm a scientist": How philosophy of science can help explain why science deserves primacy in dealing with societal problems. *Science & Education, 31*, 1141–1154.

Bromme, R. (2022). Informed trust in science: Lessons from the COVID-19 pandemic for the conceptualization of science literacy. *Unterrichtswissenschaft, 50*(3), 331–345.

Bromme, R., & Goldman, S. R. (2014). The public's bounded understanding of science. *Educational Psychologist, 49*(2), 59–69.

Bromme, R., & Kienhues, D. (2017). Gewissheit und Skepsis: Wissenschaftskommunikation als Forschungsthema der Psychologie [certainty and skepticism: Scientific communication as a research topic of psychology]. *Psychologische Rundschau, 68*(3), 167–171.

Carter, L. (2010). The armchair at the borders: The "messy" ideas of borders and border epistemologies within multicultural science education scholarship. *Science Education, 94*(3), 428–447.

Cofré, H., Núñez, P., Santibáñez, D., Pavez, J. M., Valencia, M., & Vergara, C. (2019). A critical review of students' and teachers' understandings of nature of science. *Science & Education, 28*(3–5), 205–248.

Cortassa, C. (2016). In science communication, why does the idea of a public deficit always return? The eternal recurrence of the public deficit. *Public Understanding of Science, 25*(4), 447–459.

Erduran, S., & Dagher, Z. R. (2014). *Reconceptualizing the nature of science for science education: Scientific knowledge, practices and other family categories.* Springer.

Feinstein, N. (2011). Salvaging science literacy. *Science Education, 95*(1), 168–185.

Fernandez-Beanato, D. (2020). The multicriterial approach to the problem of demarcation. *Journal for General Philosophy of Science, 51*, 375–390.

Feyerabend, P. K. (1975). *Against method: Outline of an anarchistic theory of knowledge.* New Left Books.

Haber, J. (2020). *Critical thinking.* MIT Press.

Heering, P., & Kremer, K. (2018). Nature of science. In D. Krüger, I. Parchmann, & H. Schecker (Eds.), *Theorien in der naturwissenschaftsdidaktischen Forschung* (pp. 105–119). Springer Spektrum.

Hirvonen, I., & Karisto, J. (2022). Demarcation without dogmas. *Theoria, 88*(3), 701–720.

Höttecke, D., & Allchin, D. (2020). Reconceptualizing nature-of-science education in the era of social media. *Science Education, 104*(4), 641–666.

Hoyningen-Huene, P. (2008). Systematicity: The nature of science. *Philosophia, 36*(2), 167–180.

Hoyningen-Huene, P. (2013). *Systematicity: The nature of science*. Oxford University Press.

Kampourakis, K. (2016). The "general aspects" conceptualization as a pragmatic and effective means to introducing students to nature of science. *Journal of Research in Science Teaching, 53*(5), 667–682.

Laudan, L. (1983). The demise of the demarcation problem. In R. S. Cohen & L. Laudan (Eds.), *Physics, philosophy, and psychoanalysis: Essays in honor of Adolf Grünbaum* (pp. 111–127). Springer.

Lederman, N. G. (1992). Students' and teachers' conceptions of the nature of science: A review of the research. *Journal of Research in Science Teaching, 29*(4), 331–359.

Lombardi, D., Bickel, E. S., Bailey, J. M., & Burrell, S. (2018). High school students' evaluations, plausibility (re) appraisals, and knowledge about topics in earth science. *Science Education, 102*(1), 153–177.

Lombardi, D., Nussbaum, E. M., & Sinatra, G. M. (2016). Plausibility judgments in conceptual change and epistemic cognition. *Educational Psychologist, 51*(1), 35–56.

Ludwig, D., & Ruphy, S. (2024). Scientific Pluralism. In E. N. Zalta & U. Nodelman (Eds.), *The Stanford encyclopedia of philosophy (fall 2024). Metaphysics research lab.* Stanford University.

McComas, W. F., & Olson, J. F. (1998). The nature of science in international science education standards documents. In W. F. McComas (Ed.), *Nature of science in science education: Rationales and strategies* (pp. 41–52). Kluwer Academic Publishers.

Mueller, M. P., & Tippins, D. J. (2012). Citizen science, ecojustice, and science education: Rethinking an education from nowhere. In B. J. Fraser, K. Tobin, & C. J. McRobbie (Eds.), *Second international handbook of science education* (pp. 865–882). Springer Science+Business Media B.V.

Moreira-dos-Santos, F., & El-Hani, C. N. (2017). Belief, knowledge and understanding. *Science & Education, 26*(3–4), 215–245.

Nehring, A. (2020). Naïve and informed views on the nature of scientific inquiry in large-scale assessments: Two sides of the same coin or different currencies? *Journal of Research in Science Teaching, 57*(4), 510–535.

Nickles, T. (2013). The problem of demarcation: History and future. In M. Pigliucci & M. Boudry (Eds.), *Philosophy of pseudoscience* (pp. 101–120). University of Chicago Press.

Nordine, J., & Fortus, D. (2025). Reinventing science standards to better support meaningful science learning. *Journal of Research in Science Teaching*, 1–14.

Oreskes, N. (2019). *Why trust science?* Princeton University Press.

Osborne, J. (2017). Going beyond the consensus view: A response. *Canadian Journal of Science, Mathematics and Technology Education, 17*(1), 53–57.

Osborne, J., & Allchin, D. (2024). Science literacy in the twenty-first century: Informed trust and the competent outsider. *International Journal of Science Education, 47*(15–16), 2134–2155.

Osborne, J., Collins, S., Ratcliffe, M., Millar, R., & Duschl, R. (2003). What "ideas-about-science" should be taught in school science? A delphi study of the expert community. *Journal of Research in Science Teaching, 40*(7), 692–720.

Paehr, J., Lenzer, S., Jambor, T. N., Monke, M., Nehring, A., & Stender, B. (2023). Schüler*innenvorstellungen als Differenzlinie im Diskurs um Heterogenität: Ziele, Haltungen und Anwendungsbeispiele für eine heterogenitätssensible reflexive Lehrkräftebildung aus fachdidaktischer Perspektive. *HLZ–Herausforderung Lehrer*innenbildung, 6*(1), 278–303.

Pigliucci, M. (2013). The demarcation problem: A (belated) response to Laudan. In M. Pigliucci & M. Boudry (Eds.), *Philosophy of pseudoscience* (pp. 9–28). University of Chicago Press.

Pigliucci, M., & Boudry, M. (2013). Why the demarcation problem matters. In M. Pigliucci & M. Boudry (Eds.), *Philosophy of pseudoscience* (pp. 1–6). University of Chicago Press.

Popper, K. R. (1959). *The logic of scientific discovery*. Hutchinson.

Popper, K. R. (1992). *The logic of scientific discovery*. Routledge.

Posner, G. J., Strike, K. A., Hewson, P. W., & Gertzog, W. A. (1982). Accommodation of a scientific conception: Toward a theory of conceptual change. *Science Education, 66*(2), 211–227.

Resnik, D. B. (2000). A pragmatic approach to the demarcation problem. *Studies in History and Philosophy of Science, 31*(2), 249–267.

Rolin, K. (2020). Trust in science. In J. Simon (Ed.), *The Routledge handbook on trust and philosophy* (pp. 354–366). Routledge.

Romero-Maltrana, D., Benitez, F., Vera, F., & Rivera, R. (2019). The 'nature of science' and the perils of epistemic relativism. *Research in Science Education, 49*(6), 1735–1757.

Romero-Maltrana, D., & Duarte, S. (2022). A new way to explore the nature of science: Meta-categories rather than lists. *Research in Science Education, 52*(1), 239–257.

Roth, W. M. & Barton, A. C. (2004). *Rethinking scientific literacy*. Routledge-Falmer.

Rudolph, J. L. (2000). Reconsidering the 'nature of science' as a curriculum component. *Journal of Curriculum Studies, 32*(3), 403–419.
Ruse, M. (Ed.). (1988). *But is it science? The philosophical question in the creation/evolution controversy*. Prometheus Books.
Sandoval, W. A., & Morrison, K. (2003). High school students' ideas about theories and theory change after a biological inquiry unit. *Journal of Research in Science Teaching, 40*(4), 369–392.
Santos, B. (Ed.). (2007). *Another knowledge is possible. Beyond northern epistemologies*. Verso.
Schiefer, J., Edelsbrunner, P. A., Bernholt, A., Kampa, N., & Nehring, A. (2022). Epistemic beliefs in science-a systematic integration of evidence from multiple studies. *Educational Psychology Review, 34*, 1541–1575.
Schwartz, R. S., Lederman, N. G., & Abd-El-Khalick, F. (2012). A series of misrepresentations: A response to Allchin's whole approach to assessing nature of science understandings. *Science Education, 96*(4), 685–692.
Shaw, J. (2017). Feyerabend and the cranks: On demarcation, epistemic virtues, and astrology. *Social Epistemology Review and Reply Collective, 6*(3), 74–88. http://wp.me/p1Bfg0-3C4
Stadtler, M., Winter, S., Scharrer, L., Thomm, E., Krämer, N., & Bromme, R. (2017). Selektion, Integration und Evaluation: Wie wir das Internet nutzen, wenn wir uns über Wissenschaft informieren wollen. *Psychologische Rundschau, 68*(3), 177–181.
Serrano, D. M., Crone, T., & Williams, P. S. (2024). Exploring the role of Multiplist epistemic beliefs on COVID -19 conspiracies and prevention among undergraduates. *Science & Education, 33*(6), 1401–1417.
Steel, D., Fazelpour, S., Gillette, K., Crewe, B., & Burgess, M. (2018). Multiple diversity concepts and their ethical-epistemic implications. *European Journal for Philosophy of Science, 8*(3), 761–780.
Stroupe, D., Suárez, E., & Scipio, D. (2024). Epistemic injustice and the "nature of science". *Journal of Research in Science Teaching., 62*(4), 901–941.
Tolbert, S. (2025). Trust, distrust, and the "competent outsider": Rethinking science education' s responsibilities in the (dis)information crisis. *Journal of Research in Science Teaching*, 1–7.
Wylie, A. (2015). A plurality of pluralisms: Collaborative practice in archaeology. In F. Padovani, A. Richardson, & J. Y. Tsou (Eds.), *Objectivity in science: New perspectives from science and technology studies* (pp. 189–210). Springer International Publishing.
Whyte, K. P., & Crease, R. (2010). Trust, expertise, and the philosophy of science. *Synthese, 177*, 411–425.

Part 5: Social and Technological Transformations

Artificial Morality and Machine Ethics

Catrin Misselhorn

1 Introduction

Machine ethics is a field within digital ethics that focuses on the development of moral machines (Wallach & Allen, 2009): machines that have the capacity to make moral decisions and take moral action. Analogous to artificial intelligence (AI), this is called "artificial morality" (Misselhorn, 2018a). Just as AI is concerned with replicating or simulating human cognitive abilities, artificial morality deals with modeling or simulating human moral decision-making and behavior. Machine ethics as a discipline engages in theoretical and ethical reflection on artificial morality (Anderson & Anderson, 2011; Bendel, 2019; Misselhorn, 2018b).

The terminology used here corresponds to the common philosophical distinction between morality—understood as the totality of moral norms, actions, attitudes, and emotions—and ethics as moral philosophy, i.e., the

This chapter was originally published in German as "Künstliche Moral und Maschinenethik", in: *Digitale Ethik: Handbuch für Wissenschaft und Praxis*, ed. by P. Grimm, K. Trost & O. Zöllner, Baden-Baden: Nomos Verlagsgesellschaft 2024, pp. 441–452. The text was translated by Scott Gissendanner.

C. Misselhorn (✉)
Georg-August-Universität Göttingen, Göttingen, Germany
e-mail: catrin.misselhorn@uni-goettingen.de

J. Hinrichsen et al. (eds.), *Ethics of the Sciences*,
https://doi.org/10.1007/978-981-92-2821-8_16

discipline that engages theoretically with morality as its object (Misselhorn, 2018b, p. 45f). While general ethics investigates the foundations of human morality, machine ethics addresses the questions of whether, and if so, how machines can and should be designed to make and implement moral decisions, and furthermore whether machines ought to do this at all. This line of inquiry is rather new in the history of ethics, as the need and possibility of moral machines occurred only with the rapid development of robotics and AI in recent decades (Wallach & Allen, 2009, p. 13ff.).

Even a device as simple as a robotic vacuum cleaner encounters situations that involve certain moral decisions. Should it simply vacuum up a ladybug, or should its life be spared? And what about a spider? These are moral decisions, concerning whether it is permissible to kill animals for the sake of cleanliness. While ordinary robotic vacuum cleaners do not possess the ability to respond morally to these situations yet, there exists at least a prototype of the popular Roomba model with an ethics module that accounts for the lives of ladybugs (Bendel, 2017).

As the areas of application for such autonomous systems become more complex, the moral decisions they face become more difficult to resolve. This is evident in three key areas for the application of machine ethics: care technologies, military robots, and autonomous vehicles. In all three, the focus is on fundamental moral decisions that even involve matters of human life and death. The public debate around dilemma situations in autonomous driving in particular has created greater public awareness of the problems machine ethics is dealing with.

This article provides an overview of current research on machine ethics. First, machine ethics as a subfield of ethics generally is positioned in the field of metaethics, normative ethics, and applied ethics. It continues with an examination of the possibility of and approaches to designing artificial moral agents whose development is the fundamental aim of artificial morality. This is followed by a discussion of several critical objections to the prospects and desirability of artificial morality. The article concludes by drawing the consequences of these critiques and describing two examples of potential applications that make sense from a moral point of view.

2 Dimensions of Machine Ethics: Metaethics, Normative Ethics, and Applied Ethics

Machine ethics includes metaethical and normative issues (Misselhorn, 2018b, p. 45ff.). Metaethical questions concern the conceptual, ontological, and epistemological aspects of artificial morality—for example, what constitutes moral agency, whether artificial systems can be moral agents, and what ontological status artificial moral agents possess, such as whether they are simply objects, moral subjects or form an intermediate category in-between objects and living beings (Misselhorn, 2022a).

Normative ethics examines what is morally right or wrong. The specific normative questions of machine ethics concern the moral standards to be implemented in artificial moral agents (Misselhorn, 2018b, p. 90ff.). Should these agents follow utilitarian or deontological principles? Is a principles-based ethics even the right approach for implementing morality, or would a case-based approach be preferable? Can we rely on moral theories designed for humans, or are new ethical foundations required for artificial moral agents?

In a broader perspective, machine ethics also encompasses considerations of the moral implications of artificial morality for individuals, society, and at a political level. Do they violate human dignity (Misselhorn, 2018b, p. 181 f., 2020b; Rosert & Sauer 2019)? Are there areas in which artificial moral agents are inadmissible, or moral decisions that should be reserved for humans? What consequences does artificial morality have for human social relations (Misselhorn, 2018b, p. 216ff., 2021a)? Could it even undermine human moral practice (Misselhorn & Störzinger, 2023)? What political regulations should control artificial moral agents?

Large parts of machine ethics fall within applied ethics, particularly the ethics of technology, which addresses specific fields of action relevant to society and politics. Machine ethics also intersects other fields of the ethics of technology, most notably information ethics (Floridi, 2013; Heesen, 2016; Himma & Tavani, 2008; Kuhlen, 2004) and robot ethics (Veruggio & Operto, 2008; Wallach & Asaro, 2016; Loh, 2019). New disciplines of applied ethics often emerge in response to specific ethical issues that accompany new technologies. This is also the case with machine ethics, which has arisen from the novel challenges that occur when attempting to equip machines with the capacity for moral decision-making and moral action.

3 Artificial Moral Agents

The starting point of machine ethics is the fundamental metaethical question of whether machines can be moral agents (Misselhorn, 2025). The answer determines whether the project of artificial morality makes sense at all. Should it turn out to be impossible to equip machines with the capacity to make moral decisions and act accordingly, artificial morality loses its point, and questions of its application and impact become moot.

The goal of artificial morality is to develop so-called explicit moral agents (Moor, 2006). These can recognize morally relevant information as such and process it to arrive at moral decisions. They can be compared to a chess program: such a program recognizes information relevant to the game of chess, processes it, and makes decisions on that basis with the aim of winning the game. It can represent the current position of the pieces on the chessboard and determine which moves are allowed. On this basis, it calculates which alternative is most promising under the given circumstances.

This example provides a counter-argument to the objection that artificial moral agents exercise no true agency on the grounds that the moral decision is not made by the machine but rather by the human who programmed it. The problem is that the greater the progress in AI, the less clearly a decision made by an artificial system in any given situation can be attributed to the human programmers. Even in the case of chess programs which are not very advanced in terms of AI, the developers cannot predict which moves they will choose in any single game situation. This is evident because such programs play chess far better than their programmers, who could certainly not compete with a chess grandmaster. It is reasonable to regard a chess program as an artificial agent insofar as its behavior in a given situation is rational and goal-directed, but not predictable or controllable by the programmers.

Artificial morality seeks to apply this model to moral decision-making. For Moor, who coined the term of an explicit moral agent, their actions must not only conform to the morally permissible or required as a matter of fact, but they must be based on moral information processing. His formulation is reminiscent of Kant's distinction between acting in accordance with what duty prescribes and acting out of duty (Kant, 1786/2015, p. 397 f.). Of course, artificial agents cannot, strictly speaking, be moral in the Kantian sense. They have neither a will nor inclinations that could come into conflict with the moral law.

Explicit moral agents are therefore situated between moral subjects in the Kantian sense, who act from duty, and Kant's example of the prudent merchant whose self-interest happens to coincide with moral duty. What Moor seeks to express is that an explicit moral agent can recognize and process the morally relevant aspects of a situation as such and is capable of responding to them in a morally appropriate manner.

Nevertheless, Moor would agree with Kant that explicit moral agents fall short of the level of moral agency possessed by humans. Humans are full moral agents which have additional capacities such as consciousness, intentionality (that is, the ability to think about and relate to the world), as well as the capacity for critical reflection and moral justification. So far, only humans possess these capacities, and it is doubtful whether machines will ever be able to attain them.

For this reason, machines are also referred to in the debate as functional moral agents (Misselhorn, 2020a, 2022b; Wallach & Allen, 2009). Morality in a functional sense can be characterized in terms of input, output, and specific information-processing procedures. The idea is that at least some part of morality can be captured—just as in a game of chess—by means of these parameters.

Functional moral agents are, of course, limited in various respects. First, functional relations pertain only to the cognitive aspect of morality. The emotional dimension, for example, can only be taken into account insofar as emotions can be functionally modeled independent of their experiential quality. Artificial moral agents are just as incapable of feeling compassion for others as they are of experiencing the gnawing feeling of guilt that torments humans after wrongdoing. Second, functional relations can be modeled in different ways, for example as semantic relations, at the neural level, or through a machine table (Block, 1978, 266ff.). Functional analysis is therefore relative to the type of functionalism underlying it. This means that the status of an artificial moral agent always depends on specific human purposes and interpretations, whereas full moral agents possess this characteristic intrinsically (Misselhorn, 2022b).

For example, suppose a chatbot can perform the function of recognizing when someone is in a psychological crisis and offering them help (Misselhorn, 2021a, p. 69ff.). This can be seen as a basic form of functional moral action. However, if one day all human life were to be wiped out while the chatbot continued chattering, it would no longer hold the status of an artificial moral agent, since there would be no one left to attribute this function to it or to interpret the pixels on the screen. In contrast, even the last human alive would still be a full moral agent.

Even if artificial moral agents can be ascribed functional agency, this does not imply that they are also morally responsible for their actions (Misselhorn, 2018b, p. 126ff., 2025). Since they lack capacities such as consciousness, free will, and intentionality, they cannot meet the conditions for the attribution of moral responsibility (regarding these conditions, see Heidbrink, 2017, p. 23 f.).

The decoupling of agency and responsibility in the case of artificial moral agents also has implications for their legal status. While functional moral agents cannot themselves claim rights, there are good reasons to regard artificial agents at least in private law as a distinct type of "digital legal subject", comparable to vicarious agents (Teubner, 2018). Nevertheless, responsibility would always remain with the principal, even if the artificial agent acts on their behalf (Misselhorn, 2019). Caution is warranted, however, when transferring this notion to other fields of application for artificial moral agents beyond private law, as will become clear in the discussion of the criticisms of artificial morality.

In conclusion, there are good reasons to assume that with AI in its current form, we can expect at best artificial moral agents in a functional sense. Beyond this, it would be neither morally nor practically desirable to construct artificial moral agents in the full sense. If artificial systems were full moral agents, we could not use them as 'slaves' anymore, but would have to grant them moral rights (Bryson, 2010). This also entails certain restrictions regarding the design of such systems' outward appearance, which in particular should not be too human-like (Misselhorn, 2021a, p. 95).

4 Criticism of Artificial Morality

Machine ethics addresses not only the questions of whether artificial moral agents are possible and how morality can be implemented in an artificial system. It considers also whether it is acceptable to permit, from a moral standpoint, artificial systems to make moral decisions. This question has become increasingly important in recent years because artificial morality is a component of technological innovations that are on the verge of profoundly transforming individual lives and society.

Responses range from a complete rejection of artificial moral agents (Van Wynsberghe & Robbins, 2019), to balanced positions stressing that the moral evaluation of artificial moral agents should account for the diversity of approaches and contexts of application (Misselhorn, 2018b, 2022b), to the argument that artificial moral agents are in fact absolutely necessary (Formosa & Ryan, 2021).

The following section provides a non-exhaustive overview and argumentative classification of central points within the larger controversy (Misselhorn, 2018b, 2022b) for a more detailed discussion. The focus is on moral questions that arise specifically in relation to artificial moral agents. Topics that belong to the general ethics of AI and not specifically to machine ethics, such as privacy, are not addressed. The focus is on fundamental criticisms raised against the project of artificial morality.

The starting point of the debate is the question of whether artificial moral agents are an inevitable consequence of technological development (Wallach & Allen 2011). The main supporting argument is this: the more intelligent and autonomous artificial systems become, the more inevitably they will encounter morally problematic situations in which their behavior can no longer be directly controlled by humans. They must therefore be able to make and implement moral decisions themselves (Misselhorn, 2020a). The unavoidability of artificial moral agents would thus follow directly from technological progress.

Of course, such a development need neither be accepted fatalistically nor rejected outright; its being inevitable is conditional and context-dependent. While the use of intelligent and autonomous technologies may indeed inevitably create the need for artificial moral agents, we can, however, still choose to differentiate those areas in which it is morally appropriate to use artificial moral agents from those in which it is not. Thus areas of application such as caregiving, autonomous weapons systems, and autonomous driving must each be evaluated separately from each other (Misselhorn, 2018b).

One justification for artificial morality articulated especially in practical contexts is the desire to use it to strengthen public trust in artificial intelligence (Anderson & Anderson, 2007, 2011). The hope is that artificial morality might help reduce fear of robots and AI and thus increase popular acceptance of these technologies.

However, a distinction must be made between trust and reliability (McLeod, 2020). Trust is an attitude that arises within a relationship based on reciprocity. To abuse this trust is a moral failure that presupposes the attribution of responsibility (Baier, 1986; Simon, 2010). Since, as shown above, artificial moral agents cannot take responsibility for their decisions or actions, "reliability" rather than "trust" is the more adequate term.

The absence of predictability and control, which has been given as one reason why it is appropriate to speak of artificial agents and not just tools, is on the one hand a good reason not to trust them. One might on the other hand object to the term "artificial moral agent", as it might be used with the intent to suggest and prompt a level of trust and acceptance that is unwarranted (Serholt et al., 2022).

Another argument to support artificial morality claims, in contrast, that artificial agents are morally even superior to humans (Arkin, 2009) because they are not subject to emotional distress or psychological pathologies and they are not prone to place self-interest over moral virtue.

Opposing this claim are arguments that call into question whether functional moral agents will ever truly be capable of making substantial moral decisions that go beyond very simple situations like that of the vacuum-cleaning robot. More complex moral decisions seem to depend on abilities possessed only by full moral agents (Himma, 2009; Véliz, 2021). From this point of view, we should approach the project of artificial morality with caution and practical applications should not be promoted naively.

In addition, there are fundamental arguments against the use of artificial moral agents in certain fields. Autonomous weapons systems, commonly referred to as killer robots, are among the most controversial applications of machine ethics. They are capable of identifying and attacking targets without human intervention. Machine ethics comes into play to make sure that autonomous weapons behave in a legally and morally acceptable manner.

Of particular practical relevance is ensuring compliance with international humanitarian law, since the use of autonomous weapons systems would otherwise not only be morally problematic but would also violate international agreements. The international legal framework can be traced back to the theory of just war, which represents a domain-specific normative ethics (Misselhorn, 2018b, p. 159ff.). Regarding machine ethics, the aim is to equip autonomous weapons systems with an "ethics module" that takes the relevant moral and legal principles into account when making decisions. A prototype for such a module was developed, for example, by Arkin (2009).

Setting aside doubts as to whether artificial systems can make such substantive moral decisions at all, there are also fundamental objections to entrusting them with substantial moral choices, particularly those involving human life and death. These arguments refer to the emergence of a responsibility gap (Sparrow, 2007), the ethical importance of human agency (Leveringhaus, 2016), the concept of moral duty (Misselhorn, 2023), as well as human dignity (Misselhorn, 2018b; p. 181 f.; Rosert & Sauer, 2019).

An important question is whether the objections to autonomous weapon systems can also be applied to other areas of application. An analogy has been drawn between the programming of autonomous vehicles for the purpose of accident-outcome optimization and the target selection of autonomous weapon systems (Lin, 2016). To optimize accident outcomes, cost functions must be created that determine who will be injured or killed. As with autonomous weapon systems, legitimate targets would have to be defined in the event of an unavoidable collision, and these persons might then be deliberately injured, even killed (Misselhorn, 2020b).

Even if moral principles prohibit entrusting machines with decisions over human life and death, the question remains: how are cases to be handled in which the immoral decision of a machine has not been explicitly programmed, yet the parameters of its programming result in a morally problematic outcome? This is, for instance, often the case when machine learning leads to unintended discrimination against certain groups of people based on factors such as age, gender, or ethnicity.

This possibility suggests that those affected must have the possibility to challenge the decisions made by AI; this means that a human being must take responsibility and, if necessary, check whether the decision was actually correct. This applies in particular to decisions that affect fundamental rights, such as freedom of expression, but also to the distribution of primary goods such as income, wealth, and opportunities, which, according to John Rawls, moral persons are entitled to have in order to realize their rights and freedoms (Rawls, 2001, p. 58).

5 Conclusion

Against the background of the critical discussion regarding artificial moral agents, the question arises of how the project of artificial morality and machine ethics should ultimately be assessed. One important conclusion from the debate is the principle that machines should not be entrusted with decisions over human life and death (Misselhorn, 2021b). Yet even if this is broadly accepted, machine ethics still faces the task of developing viable ethical approaches for human–machine collaboration. A central question in this regard is what it means and how it can be ensured that humans exercise "meaningful control" over the decisions of artificial systems (for an overview, see Cavalcante Siebert et al. (2022)).

Meaningful control is a prerequisite for ensuring that humans can always take genuine responsibility for the actions of artificial moral agents. After all, the aim of using this technology should be to strengthen, not undermine human self-determination (Misselhorn, 2021b). This means that artificial systems should support humans in setting and realizing their own ends, and must not prescribe ends on their behalf or, worse still, reduce humans to mere ends.

Also in need of examination is which areas of application for artificial moral agents make sense from a moral standpoint. A positive example could be the detection and handling of morally unacceptable content by algorithms designed to identify hate speech, Nazi symbols, or child pornography on social media. Here, the task of the artificial moral agent is to identify objectionable content and respond to it, in extreme cases by blocking an account.

Although such systems face technical challenges, they are increasingly being used by social media companies to stem the flood of such content (Adkisson, 2022). However, it must always be borne in mind that human labor—often under precarious conditions—is still necessary to train such systems to recognize relevant content. Moreover, it concerns fundamental human rights that require particular care and meaningful control by humans.

Another example currently being researched is the detection of illegal AI-based price collusion (Schlechtinger et al., 2021). Online platforms such as Amazon increasingly use AI systems to monitor competitors' pricing and adjust prices on their own platform accordingly. This is, in principle, permissible. However, there is evidence that such systems may also develop profit-maximizing collusion strategies, even if they have not been explicitly programmed to do so. Such price collusion is hardly detectable by humans and is not only immoral but also problematic under antitrust law. Artificial agents could help to uncover and prevent collusive pricing, showing that artificial moral agents can be useful particularly in situations that are morally problematic but not readily recognizable as such by humans.

Applications for artificial moral agents are also conceivable in caregiving. One approach that seeks to take into account the critical points raised in the debate is the development of a system that can be understood as an extended moral arm of its user. One could even speak of a moral avatar—that is, a virtual moral representative of the person (Misselhorn, 2024). Such a care system could contribute to enhancing the autonomy, self-determination, and sense of self-efficacy of certain user groups in need of care. However, it is crucial that humans do not just have the power to reject decisions of the artificial moral agent but that they can choose whether they would like to use such a care system or not.

In addition to these practical applications, machine ethics might also provide new theoretical insights into human moral decision-making. Just as AI can serve as a tool for anthropology, cognitive science, and the philosophy of mind in order to better understand what defines human rationality and thinking, the attempt to functionally model morality may well contribute to the understanding of human moral capacities and to enhancing ethics as a theory of morality. It may help to find out whether and how certain aspects of morality can be functionally modeled and where the limits of such modeling lie.

Competing Interests The author has no conflicts of interest to declare that are relevant to the content of this chapter.

References

Adkisson, T. (2022). *Artificial intelligence is now used to track down hate speech*. https://www.newsy.com/stories/artificial-intelligence-is-used-to-track-down-hate-speech/ (last access 08/18/25).

Anderson, S. L., & Anderson, M. (2007). Machine ethics: Creating an ethical intelligent agent. *AI Magazine, 28*, 15–26.

Anderson, S. L., & Anderson, M. (2011). *Machine ethics*. Cambridge University Press.

Arkin, R. (2009). *Governing lethal behavior in autonomous robots*. Routledge.

Baier, A. (1986). Trust and antitrust. *Ethics, 96*, 231–260.

Bendel, O. (2017). Ladybird—The animal-friendly robot vacuum cleaner. *The AAAI 2017 spring symposium on artificial intelligence for the social good technical report* SS-17-01. 2-6.

Bendel, O. (2019). *Handbuch Maschinenethik*. Springer.

Block, N. (1978). Troubles with functionalism. *Minnesota Studies in the Philosophy of Science, 9*, 261–325.

Bryson, J. J. (2010). Robots should be slaves. In Y. Wilks (Ed.), *Close engagements with artificial companions: Key social, psychological, ethical and design issues* (pp. 63–74). John Benjamins Publishing.

Cavalcante Siebert, L., Lupetti, M. L., & Aizenberg, E. (2022). Meaningful human control: Actionable properties for AI system development. *AI and Ethics, 3*(1), 241–255.

Floridi, L. (2013). *The ethics of information*. Oxford University Press.

Formosa, P., & Ryan, M. (2021). Making moral machines: Why we need artificial moral agents. *AI & Society, 6*, 839–851.

Heesen, J. (Ed.). (2016). *Handbuch Medien- und Informationsethik*. Springer.

Heidbrink, L. (2017). Definitionen und oraussetzungen der Verantwortung. In L. Heidbrink, C. Langbehn, & J. Loh (Eds.), Handbuch Verantwortung (pp. 3–34). Springer.

Himma, K. E. (2009). Artificial agency, consciousness, and the criteria for moral agency: What properties must an artificial agent have to be a moral agent? *Ethics and Information Technology, 11*, 19–29.

Himma, K. E., & Tavani, H. T. (Eds.). (2008). *The handbook of information and computer ethics*. John Wiley and Sons, Inc..

Kant, I. (1786/2015). *Grundlegung zur Metaphysik der Sitten*. Mit einem Kommentar von Horn, Christoph, Mieth, Corinna & Scarano, Nico. Suhrkamp.

Kuhlen, R. (2004). *Informationsethik. Umgang mit Wissen und Informationen in elektrischen Räume*. UTB.

Leveringhaus, A. (2016). *Ethics and autonomous weapons*. Oxford University Press.

Lin, P. (2016). Why ethics matters for autonomous cars. In M. Maurer, C. J. Gerdes, B. Lenz, et al. (Eds.), *Autonomous driving - technical, legal and social aspects* (pp. 69–85). Springer.

Loh, J. (2019). *Roboterethik*. Suhrkamp.

McLeod, C. (2020). Trust. In Zalta, Edmund N. (Eds.), *The Stanford encyclopedia of philosophy*. https://plato.stanford.edu/archives/fall2021/entries/trust/ (last access 09/09/22).

Misselhorn, C. (2018a). Artificial morality. Concepts, issues and challenges. *Society, 55*, 161–169.

Misselhorn, C. (2018b). *Grundfragen der Maschinenethik*. Reclam, 5th ed. 2022.

Misselhorn, C. (2019). Digitale Rechtssubjekte, Handlungsfähigkeit und Verantwortung aus philosophischer Sicht, *VerfBlog*, 2019/10/02. https://verfassungsblog.de/digitale-rechtssubjekte-handlungsfaehigkeit-und-verantwortung-aus-philosophischer-sicht/

Misselhorn, C. (2020a, January). Artificial systems with moral capacities? A research design and its implementation in a geriatric care system. *Artificial Intelligence, 278*.

Misselhorn, C. (2020b). Lizenz zum Töten für Roboter? In Schmidt, Bernd (Ed.): „Terror" und das autonome Fahren, in: Terror. Das Recht braucht eine Bühne. Essays, Hintergründe, Analysen (pp. 149-164). Rowohlt.

Misselhorn, C. (2021a). *Künstliche Intelligenz und Empathie. Vom Leben mit Emotionserkennung, Sexrobotern & Co.* Reclam.

Misselhorn, C. (2021b). Grundsätze der Maschinenethik. In A. Strasser, W. Sohst, & K. Stepec (Eds.), *Künstliche Intelligenz—Die große Verheißung* (pp. 489–502). Xenomoi Verlag.

Misselhorn, C. (2022a). Artificial systems in-between humans and artifacts. From autonomous weapons to affective computing. *Existenz: Journal of the Karl Jaspers Society of North America*, 11–18.

Misselhorn, C. (2022b). Artificial moral agents. Conceptual issues and ethical controversy. In S. Vöneky, P. Kellmeyer, O. Müller, & W. Burgard (Eds.), *Cambridge handbook of responsible artificial intelligence: Interdisciplinary perspectives* (pp. 31–50). Cambridge University Press.

Misselhorn, C. (2023). Three ethical arguments against autonomous weapon systems. *Frontiers in Artificial Intelligence and Applications, 366(Special Issue Social Robots in social Institutions)*, 24–34.

Misselhorn, C. (2024). Machine ethics in care: Could a moral avatar enhance the autonomy of care-dependent persons? *Cambridge Quarterly of Healthcare Ethics, 33* (Special Issue: Ethical Implications of AI Use in Medicine), 346–359, 346.

Misselhorn, C. (2025). Artificial moral agents. In Martin Hähnel, Reginal Müller (eds.), *A companion to applied philosophy of AI* (Blackwell Companions to Philosophy). Wiley/Blackwell.

Misselhorn, C., & Störzinger, T. (2023). Social robots as echo chambers and opinion amplifiers. In C. Misselhorn, T. Poljansek, T. Störzinger, & M. Klein (Eds.), *Emotional machines—New perspectives from affective computing and emotional human-machine interaction* (pp. 247–275). Springer.

Moor, J. H. (2006). The nature, importance, and difficulty of machine ethics. *IEEE Intelligent Systems, 21*, 18–21.

Rawls, J. (2001). *Justice as fairness: A restatement*. Harvard University Press.

Rosert, E., & Sauer, F. (2019). Prohibiting autonomous weapons: Put human dignity first. *Global Policy, 10*, 370–375.

Schlechtinger, M., Kosack, D., Paulheim, H., & Fetzer, T. (2021). How algorithms work and play together. *Concurrences: Revue des Droits de la Concurrence, 3*(Article 102088), 19–23.

Serholt, S., Ljungblad, S., & Bhroin, N. N. (2022). Introduction: Special issue - critical robotics research. *AI & Society, 37*, 417–423.

Simon, J. (2010). The entanglement of trust and knowledge on the web. *Ethics and Information Technology, 12*, 343–355.

Sparrow, R. (2007). Killer Robots. *Journal of Applied Philosophy, 24*, 62–77.

Teubner, G. (2018). Digitale Rechtssubjekte? *Archiv Für Die Civilistische Praxis, 218*, 151–691.

Van Wynsberghe, A., & Robbins, S. (2019). Critiquing the reasons for making artificial moral agents. *Science and Engineering Ethics, 25*, 719–735.

Véliz, C. (2021). Moral zombies: Why algorithms are not moral agents. *AI & SOCIETY, 36*, 487–497.

Veruggio, G., & Operto, F. (2008). Roboethics - social and ethical implications of robotics. In B. Siciliano & O. Khatib (Eds.), *Springer handbook of robotics* (pp. 1499–1524). Springer.

Wallach, W., & Allen, C. (2009). *Moral machines. Teaching Robots Right from Wrong*. Oxford University Press.

Wallach, W., & Allen, C. (2011). Moral machines: Contradiction in terms or abdication of human responsibility? In P. Lin, K. Abney, & G. A. Bekey (Eds.), *Robot ethics: The ethical and social implications of robotics* (pp. 55–68). MIT Press.

Wallach, W., & Asaro, P. (Eds.). (2016). *Machine ethics and robot ethics*. Routledge Press.

Lifetime and Climate Change. On the Ethical Importance of Divergent Temporalities

Holmer Steinfath

The *science* of climate change aims to better understand the dynamical processes and feedbacks that have been leading to the ongoing climate crisis. It tells us how things *are* with respect to climate change and helps to anticipate possible future climatic developments. Climate *ethics* suggests how we *should* react in the face of climate change and its various impacts. Climate science poses descriptive questions, whereas the questions raised by climate ethics are normative questions. Both types of inquiries are open-ended, constantly re-evaluating and revising results, models, and principles, adjusting them to new findings, arguments, and experiences. However, in both fields of research, there is an almost unanimous consent on some basic points. Almost all climate scientists agree that climate change is happening, that human activity, particularly the burning of fossil fuels, is the primary cause and that climate change severely imperils human life and the life of other species on earth (cf. Climate Action Tracker, 2025; United Nations Climate Change, 2025; IPCC, 2023). And almost all moral philosophers agree that the fight against climate change is morally obligatory, and that people have a moral

H. Steinfath (✉)
Philosophisches Seminar, Georg-August-Universität Göttingen, Göttingen, Germany
e-mail: Holmer.Steinfath@phil.uni-goettingen.de

J. Hinrichsen et al. (eds.), *Ethics of the Sciences*,
https://doi.org/10.1007/978-981-92-2821-8_17

duty to change their individual and collective lives in such a way that greenhouse gas emission will radically decrease, and human life will follow a more environmentally sustainable path.[1] Of course, this moral consensus does not preclude fierce disputes about further moral questions. Ethicists still debate, for instance, on whether the duty to fight climate change is an individual or collective duty, who exactly is supposed to be the addressee of such a duty, what the duty should include, how the costs of losses and damages due to climate change should be distributed, etc. I do not want to downplay the great relevance of these further questions. Yet, I think that against the backdrop of widespread scientific and moral agreement on some basics, a different question is becoming more urgent. We should ask why action lags so far behind knowledge when it comes to fighting against climate change. Why does humanity slide into climate catastrophe despite the availability of easily accessible scientific and moral insights?[2]

The reasons for this gap between action and knowledge are complex and multifaceted. Thus, there are still enough people who benefit from burning fossil fuels. Not all individuals are equally affected by practices that exacerbate climate change: we are not all "in the same boat." In highly industrialized countries everyday life is still a life dominated by fossil fuels, not only economically but also culturally. The daily routines of many people in these countries are not yet radically disrupted by climate change (cf. Müller-Salo, 2023). And of course, the knowledge of the potentially catastrophic consequences of climate change is not equally

[1] Recently published books testifying to this consensus are, for example, Shue (2021), Cripps (2022), and Moellendorf (2022). For a still influential anthology with important contributions, cf. Gardiner et al. (2010).

[2] The motivation for these questions is similar to the worry behind Stephen M. Gardiner's magisterial 2011 book. Gardiner writes: "My concern is with why, given that the relevant facts are known, effective action is so difficult, and indeed has so far eluded us." (Gardiner 2011, p. 4) In answering his question, Gardiner makes convincing use of game theoretical tools like the Prisoner's Dilemma and the Tragedy of the Commons. Beckert's 2024 book could be read as a kind of sociological underpinning of Gardiner's matrix. Beckert claims that the power structures and incentives of capitalist modernity and their control mechanisms block a solution to the global problem of climate change (Beckert 2024, p. 12). For a completely different (a bit eclectic and polemical but still interesting) approach, cf. DeLay (2024), who tries to understand the gap between knowledge and action in the fight against climate change in broadly psychoanalytical terms as a form of ideology driven "denial."

disseminated.[3] In this essay, however, I want to highlight another challenge that hampers our ability to effectively address climate change and the gradual destruction of the natural basis for a good life. This challenge arises from our handling of divergent temporalities. I will argue that the temporal dimensions and mechanisms of climate change are in precarious tension with our everyday experiences of time and the way we typically understand and organize our individual and collective lives in temporal terms.

This perspective may not be the most salient one. However, it seems to me worth taking for several reasons. For one, the temporal dimensions of our lives are among the most fundamental aspects of our existence. They are the source of "one of the most basic challenges of living" (Scheffler, 2025, p. 155) and therefore of eminent ethical importance. Conversely, the temporal dimensions of climate change are particularly difficult to understand for ordinary people. Moreover, highlighting temporality opens up the opportunity for new exchanges between, on the one hand, climate science and, on the other hand, branches of philosophy like the philosophy of time, philosophical anthropology, and philosophy of action, which have been overshadowed by moral and political philosophy. In addition, reflections on temporality might help to bridge the gulf between science and the humanities, as how we see our lives in time is also a matter of cultural interpretations of ourselves and the world that have their own ethical valence.

In the following, I will outline my diagnosis in three steps. In the first step, I will explain some temporal features of individual actions and practices that can hinder an adequate response to climate change and environmental destruction. In the second step, I will elucidate temporal aspects of climate change that are at odds with the temporality of everyday life. In the third step, I will suggest ways in which we might understand our lives in time in an ethically attractive manner so that the tension between our individual lifetime and the time of climate change can be reduced. My essay sets the stage for a particular kind of engagement with ethical implications of climate change and climate science. Instead of suggesting new normative guidelines, I will hint at conditions of a good life that are connected with the temporality of human life in times of climate change.

[3] Not to mention the distorting effects of fake news and conspiracy theories!.

1 The Temporality of Everyday Life

Human life has many temporal dimensions, some of which have a universal, anthropological character, while others vary culturally and historically or even individually. I will focus on those aspects that are typical for lives in modern societies and are relevant to my general diagnosis. When I speak for convenience's sake of "we" or "us," I take this first-person plural as a "we of invitation" (Williams, 1993, p. 171), the range of which may need to be adjusted in context. I do not deny that not all people conform to the picture I am drawing, but I think that my picture captures temporalities that are dominant at least in Western societies.

In everyday life, time is mainly understood as an order of events. In the analytical philosophy of time, a distinction between two ways of ordering temporal relations between events has been established since the seminal work of the British philosopher John McTaggart. McTaggart dubbed them the "A-series" and the "B-series" of time (McTaggart, 1927, Book V, Ch. 33). The B-series orders events according to earlier, simultaneous, and later, while the A-series determines the relation between events according to past, present, and future. In our everyday temporal orientations, we rely on both temporal orders. The B-series is static, because an event that occurs earlier than another will always remain earlier; the First World War will always have occurred before the Second. We fix events in this way with calendars. The A-series, on the other hand, is dynamic because it marks events as transient: what is currently future will soon be past. The indexical "now" constantly shifts relative to the subject from whose perspective something was, is, or will be. In my calendar, I can mark the next appointment for the car repair for May 5, 2026. This helps me to organize other events as earlier and later so that they do not conflict with the scheduled appointment. Yet, to know whether the car repair appointment is still pending or whether I have already missed it, I need to know my own, permanently changing position in time—I need to know when "now" is and what is past and what is future relative to it.

Despite the practical interdependence of both time series, the A-series is more significant for our understanding of ourselves as beings who live our lives in time, giving it its own temporality. As agents we live in a forward direction. We shape our lives—and thereby ourselves—by reaching out to future possibilities. In our intentions and plans, we project ourselves into a future that only gains reality through its realization in our very actions. And insofar as the future depends on our actions, we consider it open. The openness of the future is intimately associated with our sense of agency, whereas the past seems to be determined and settled (Callender, 2017, p. 12; cf. Ismael, 2011).[4]

With this past/future asymmetry a double preference asymmetry comes along. The first one manifests itself in differences in desires and attitudes depending upon whether an event is past, present, or future. Normally, we prefer future pleasures and, more generally, gains to past ones, and dread future pains and losses while we are relieved that a past pain is over and that we left past losses behind. "In the other asymmetry, desires and attitudes differ depending upon whether the future event is close or distant. We're horrible procrastinators. We prefer distant future pain to proximal future pain." (Callender, p. 12) Conversely, we prefer positive experiences in the proximate future to such experiences in the distant future. We tend to discount the distant future in favor of the near future.[5] It is pretty obvious that this proximate/distant future difference in valuation is a major problem in our social and political responses to the perils of climate change.[6]

Of course, living in a forward way does not mean that the present and the past become completely irrelevant. The preference for positive events in the proximate future over experiencing them in the distant future and the procrastination of difficult tasks can also be understood as a kind of present bias. Moreover, there is a strong temptation to think that only the present is real. Everything I experience I experience in the present. And when it comes to actions, I always act from a present situation: I decide now to do this and that in order to achieve some goal in the future. In

[4] Citing cross-cultural studies, Callender (2017, pp. 14-19) assumes that the past/future asymmetry, according to which the future is open and the past determined, is a universal trait of our prescientific image of time. Although I share this assumption, nothing depends on it for my argument.

[5] This is partially due to the fact that more distant future events are harder to predict and less controllable.

[6] Other than the first preference asymmetry, the second one—the proximate/distant future asymmetry—has been widely studied in social sciences, particularly in psychology and economics. For a quick overview, cf. Read (2008). Many researchers believe that this asymmetry is a heritage of our evolutionary past (cf. Maclaurin & Dyke, 2002; Robson & Samuelson, 2007). However, even if that were true, it is worth pointing out that the discounting of the more distant future in favor of the more proximate future can get strengthened by social and particularly economic factors.

contrast to this foregrounding of present and future, the past figures in action mainly in the background. Nevertheless, in deciding what to do, I have to rely on conditions that have developed in the past and have shaped my possibilities. For example, I need certain skills to be able to perform certain actions, and I had to acquire these skills through a process of practice that unfolded over a longer or shorter period of time. What is more, I depend on material infrastructures to apply my skills, and these infrastructures have in turn their own history. To drive to a distant city, I must be able to drive. I also need a car that was produced by others in a longer production process, I rely on a road network that was built over decades, and so on. In this sense, future, present, and past are intertwined in every action.

However, the temporal scope of our actions is generally limited. Not least because of the proximate/distant future asymmetry, we plan primarily for the near future. Depending on our field of activity, our calendars may cover one or two years, sometimes a bit longer. Even if we have more distant goals in mind—perhaps we are saving for the purchase of a house or for retirement—the immediate future usually takes center stage: the next shopping trip, our weekly work routines, the next weekend. The complex social, political, and economic frameworks that we find ourselves in typically confine our scope of action to the immediate future. On the other hand, if these frameworks are stable enough, they also take away much of the need for longer-term planning; as long as retirement is considered secure, we do not need to save additionally for our old age.

In our everyday actions, we do not just fail to look far into the future. Our past horizon is restricted no less, as far as our conscious relationship to the past is concerned. Normally, we are more interested in what we did yesterday and last week than in how our school days were or what longer historical processes have shaped our world and ourselves. The craftsman must remember the open orders from the past few weeks, not the orders he has already completed long ago. The conscious timespans of professional and other activities that mold our daily lives are rather narrow. Similarly, what we call our present is usually of shorter duration. Admittedly, the present as a situation of action is never just a single point in time but a period of time; our practical present is always extended. However, the extension is rarely very large. Right now, I am writing this text, or I am currently in training. As long as we perceive the social and political order as not being in immediate crisis, we rarely think of the present of an epoch, although this would be possible.

Temporal duration can be scaled at any order, but the time horizon of our everyday activities remains low-scale. It narrows further when it focuses on the immediate present. Critical social diagnoses have noted that life in modern societies is characterized by a tendency toward presentism, which is exacerbated by digitalization and social media (cf. Malm, 2018, p. 13). This tendency can be described as a shortening of attention spans: when people constantly stare at their mobile phones to avoid missing any news or to check the weather forecast (to take a perhaps clichéd example), they jump from moment to moment without ever lingering on one thing. This creates the impression that despite constant change, nothing truly new is happening. A similar impression can be the effect of accelerations, which increase the frequency of activities and events per unit of time (Rosa, 2013). As a result, people suffer from the temporal fragmentation of their lives; it becomes increasingly difficult for them to draw a biographical arc that spans larger periods between their own past, present, and future.

Yet even if individuals are spared the experiences of temporal fragmentation and the rush from one task to the next, they operate within a manageable temporal horizon. The life between birth and death, which spans our entire life, is still a temporally bounded life. Since we are not immortal, we cannot afford to take as much time as we would like for important matters. Opportunities once missed rarely return. That is why we plan. And we cannot constantly start anew. As individuals, we have only one life, in which we can either fare better or worse. With respect to climate change and other environmental issues, this orientation toward one's own lifetime is particularly problematic where it is reflected in the time horizon of consumption decisions that are one of the main drivers of climate change and the destruction of nature. Normally, the time frame of consumption decisions does not extend beyond a few years, if at all. But even if it is extended to one's entire life, it is only of little help in solving the most pressing environmental problems (cf. Gardiner, 2011, p. 58).

2 The Temporality of Anthropogenic Climate Change

The temporality of climate change is significantly different from that of our individual lives and daily activities. This also applies when we focus only on those climatic changes (such as the rise in global average temperature, the increased likelihood of heavy rainfall, long droughts, and devastating storms, or the rise in sea levels) that are largely due to the combustion of fossil fuels such as coal and oil and the resulting rapid increase in carbon dioxide concentration in the atmosphere.[7] First of all, it is certainly striking that with anthropogenic climate change, entirely different scales of time come into play. This pertains both to the past and to the future.

The fossil fuels that humans burn to generate energy are the result of geological developments that date back hundreds of thousands of years, in part reaching back to a time long before humans even existed. By burning these fuels, we are destroying in a short time what has been built up over a period of time that is hardly imaginable to us. Technological societies are intervening in the Earth's system in a manner that brings human and planetary time together in a way that is at least quantitatively new. Human beings have become a major force in the transformation of all parts of what we used to call "nature." In this sense natural processes are now part of human history (cf. Chakrabarty, 2021). We can no longer claim with good conscience, as Hans-Georg Gadamer once did, that the history of the Earth and the universe is without significance for understanding human history, and vice versa (Gadamer, 2016, p. 27).[8] What we experience today as natural disasters is increasingly appearing as the effect of our own catastrophic interventions in nature in the distant and recent past.

Even harder to imagine, for many, is the temporal scope of current economic activity into the future. How we handle natural resources in the present will not only influence the climate that the next two or three generations will have to cope with but, depending on the scenario, the weather and environmental conditions of the next several thousand years. Human action thereby gains a kind of power that exceeds everything that has been known in human history so far and thus transcends our ordinary

[7] For some recent and quite bleak information, see, for instance, United Nations Climate Change (2025) and Climate Action Tracker (2025).

[8] To quote Gadamer: "How could it make any kind of sense to insert into the totality of the evolutionary process of the universe this tiny portion of a timespan that is illuminated by the light of tradition? Yet, what imposes itself upon us when it comes to broadening the historical horizons is precisely to stop thinking of this gigantic framework into which the little bit of human destiny called world history almost disappears." (Gadamer 2016, p. 27).

understanding.[9] The consequences of our actions could even affect a time when there will be no more humans, and hence no human history.

With scales of time that are many orders of magnitude greater than those with which we ordinarily deal, new interconnections between past, present, and future come into play (cf. Malm, 2018, Ch. 1; Shue, 2021, Ch. 1). In everyday life, we often consider the more distant past as sealed; we have left it behind and its traces become increasingly blurred with time. Even for the biographical past, this tends to be an illusion. Our past can still influence us precisely where we do not think about it. Faulkner coined his famous sentence that the past is not dead, no, not even past, for the intricacies of lifetime.[10] It is even more fitting, however, for the temporality of climate change. Our current greenhouse gas emissions are as threatening as the cumulative effect of our past emissions since the beginning of the Industrial Revolution. And the full effect of these past emissions will only unfold in the future, so that the past not only determines the present but also foreshadows the future. Mainly because of the longevity of once-emitted carbon dioxide, climate change is a resilient phenomenon and its "impacts are seriously backloaded" (Gardiner, 2011, p. 33). Therefore, what we have set in motion will not simply stop if we were to suddenly reduce all greenhouse gas emissions to zero. The carbon dioxide concentration has already reached a level that will make the effects of human economic activity be felt for thousands of years, regardless of what we do today or in the future.[11]

Because our past actions already determine the future, the future is no longer as open as we often presuppose as agents. It is not a terra incognita that can be shaped at will. Human action has already marred the future of human history in such a way that all future generations will have to reckon with it as something given. And this applies even to the factually excluded case of an energy revolution that would immediately stop all

[9] However, it is worth adding (and remembering) that human action has achieved its greatest temporal reach through nuclear technology. Nuclear waste will pose a threat to the life of humans and other species for an unforeseeable future.

[10] William Faulkner's often quoted aphorism reads: "The past is never dead. It's not even past." (Faulkner, 2012, p. 73).

[11] For estimations of the longevity of carbon dioxide in the atmosphere, cf. Archer et al. (2009).

carbon emissions.[12] We can no longer say that we will still be able to wait and see how the impending climate catastrophe can be prevented later on. We are no longer at a point in human history where we can simply change tack so that future generations will not be harmed.

For the dimension of the present, this leads to a peculiar kind of urgency. The group of activists known as the "Last Generation" has declared to change its name (it is calling itself now "New Generation") and public interventions because its members no longer see themselves as the last generation that can mitigate climate change in such a way that the Paris 1.5 °C goal can still be achieved. In fact, no one should believe that the global average temperature will be kept under the mark defined in the Paris Agreement of 2015.[13] Statements that suggest otherwise only throw dust in our eyes. But that does not mean that the time for decisive action is past. It will still make a significant difference for future—and already for present—generations whether they live in a world of 2, 3, or 4 °C temperature change. Moreover, there are many reasons to believe that the current generation is the "pivotal generation" for taking important measures to slow down climate change, as Henry Shue has argued (Shue, 2021). On the one hand, we can still divert or significantly weaken developments that could later get out of control. And on the other hand, the current generation in developed high-income industrial nations lives in conditions that future generations may not be able to benefit from. We live in economically relatively prosperous societies and in political systems that should be open to the necessary political measures, although many of their representatives and those who are voting for them are not.

The time for action is always now, but now it is especially so. This has to do with another dynamic of climate change: the mechanism of tipping points, which also is a temporal mechanism.[14] In it, two characteristics come together: irreversibility and potentially infinite feedback mechanisms. In everyday life, many things are irreversible. Individually

[12] In this article I largely ignore the fact that climate change is only one aspect of global environmental destructions. Technical solutions to problems created by greenhouse gas emissions would not automatically solve other environmental calamities. They might even worsen these other destructions.

[13] Moellendorf (2022, Ch. 1) strikes a somewhat more hopeful note. In my opinion, the numbers and estimated prospects of the latest IPCC-report are bleak (cf. IPCC, 2023, especially section 2.3). Cf. also the sources mentioned in footnote 7.

[14] Cf. Lenton et al. (2019).

and collectively, we know about the inevitability of losses. In the case of climate change and general environmental destruction, however, we are dealing with unaccustomed human-generated scales of irreversibility. Think only of the extinction of entire species and ecosystems that are irretrievably lost. More important than irreversibility, however, is its connection to feedback mechanisms that continue to escalate further and become increasingly unpredictable. Once certain tipping points, of which no one knows exactly where they are located, are crossed, a cascade of catastrophic developments can be triggered that will continue to feed themselves without our intervention. If carbon dioxide emissions continue to rise unchecked (no trend reversal has been observed globally yet), this increases the risk of forest fires, which in turn lead to further carbon dioxide emissions, resulting in the destruction of additional forests and natural carbon dioxide sinks or their transformation into carbon dioxide spins. And if the polar caps melt, less sunlight will be reflected, causing the atmosphere to heat up further, which in turn causes the ice to melt faster, and so on. Although we also know dynamics in our personal lives that can accelerate and worsen our situation, we have no good intuitive grasp of the phenomenon of climate change-related tipping points and feedback mechanisms. On the contrary, we tend to think that it is never too late to stop disastrous developments. And that is, of course, not always irrational. In fact, one can still hope that even in the fight against scandalous poverty and great inequality, these problems will somehow be resolved in the future. But due to its different temporal structure, this is exactly not the case for climate change (cf. Beckert, 2024, p. 12).[15]

3 Handling Divergent Temporalities

The experiential and epistemic gap between the temporal logics of everyday life on the one hand and of climate change and ongoing environmental destruction on the other hand facilitates the diffusion of responsibility (Shue, 2021) and forms of moral corruption on both individual and collective levels (Gardiner, 2011, pp. 45–48, Part E). The tendency to view the past as closed and behind us reinforces the inclination to deny responsibility for emissions in the past, to which industrial

[15] I am deliberately excluding debates about massive and so far utopian geoengineering measures. They would need a separate analysis.

societies owe their prosperity and under which people in less developed countries suffer disproportionately.[16] That is at least one reason why fairness arguments, which demand special efforts for a more sustainable lifestyle from early industrialized countries, are so little effective in current climate debates. Similarly, the responsibility for the well-being of future generations, let alone the fate of the Earth as a whole, can more easily be denied when we narrow our temporal gaze to our lifetime or shorter episodes. The fact that we now determine under which conditions people will live in a hundred or five hundred years is then at most a ground for very abstract worries. The responsibility for the future can be pushed aside when the sense that new generations cannot start from scratch and that it is not simply up to them how they will lead their lives is severely underdeveloped.

The temporalities of our everyday routines and ways of self-understanding are admittedly not the main obstacle for more effective measures against climate change. Yet, their pernicious influence gains significantly in force once they merge with similar temporal dynamics in politics and economics. A political system that only thinks in terms of legislative periods and the timespans of individual political careers is unfit for considering the consequences of its own actions for the next decades and centuries and for making corresponding adjustments (cf. Hanusch, 2023). And an economic system that rewards short-term profits over long-term stability, so that the immediate economic losses associated with the transformation to a climate-neutral economy are weighed more heavily than the prospects of the astronomical costs of climate change in the future, is constantly working against the necessary changes (cf. Beckert, 2024). Such a system lives from discounting future damages in favor of present opportunities for making profit. Both politics and economics reinforce the presentism in everyday life, epitomized in lifestyles fostering short-term consumptions, and are in turn stabilized by it.

Against this background it is worth thinking about how we would have to understand ourselves differently in time in order to lead a more climate- and environmentally-friendly life and therefore also to be willing to demand a new politics and economy. As I said at the outset, I do not want to suggest new normative guidelines for tackling the climate

[16] The normative implications of historic emissions are a contentious issue in climate ethics, cf. Meyer (2013).

crisis. Still, I believe that climate ethics that emphasizes moral duties and questions of justice could only profit from more reflections on temporal dimensions.

A different understanding of lifetime and temporality would primarily require a shift from a strongly individualistic to a more transgenerational temporality. This is often interpreted as a literally conservative demand. Apparently, people in other epochs of human history have seen themselves as members of a chain of generations and traditions that transcend their lifetime. Presumably, they have treated what they found with greater care and consideration, because they saw it as having been passed down from earlier generations and as being worth preserving for future generations. For centuries, people have erected impressive cathedrals, and in many old cities, buildings have been renovated instead of being completely demolished. The same applies to traditional cultural landscapes. Sustainably managing a forest meant taking only as much wood as was compatible with its regeneration and stability. These are ideas that certainly deserve being revitalized. Even today, there are sensitivities and experiences that can nurture a more transgenerational and historical self-understanding. Many people rediscover a consciousness that the one life they have is not autarkic and self-standing but is embedded in social relationships that stretch back and forward in time. Every person has parents, and many have children. People engage in issues that will outlast them and that give their own life a meaning precisely because of this.[17]

However, thinking about temporality—including one's own—not individualistically but transgenerationally in a "conservative" manner encounters limits and drawbacks. Many projects that aim to continue traditions and care for descendants remain confined to smaller groups. Transgenerational temporality is then about local, and perhaps national, history. The global dimension of climate change is thereby left out. It is no coincidence that the appeal to transcend narrow-minded individualism often undergirds communitarian outlooks, which are anti-universalist in spirit.[18] The question is therefore whether the ideal of being a "good ancestor"

[17] On a social and political level this raises the question as to how we could generate institutions that would make more transgenerational motivations operative (cf. Gardiner, 2011, p. 61).

[18] For a moderately communitarian approach to intergenerational justice cf. de-Shalit (1995).

(Krznaric, 2022) that is attractive particularly with respect to environmental issues can be reconciled with more global and morally persuasive ambitions. Olúfẹ́mi O. Táíwò, who has reclaimed the ideal of being a good ancestor for the Black radical tradition, reminds us that "we don't have to take up everything our predecessors took up, we needn't accept their path as ours simply because it was theirs" (Táíwò, 2022, p. 205) "We should think of our ancestors. [...] [But] [w]e are defined by what kind of ancestors we choose to be." (p. 206) Yet, directly invoking universalistic principles of Kantian or utilitarian ethics (the predominant way in climate ethics) will not do either, as they are motivationally inert for most people. Instead, we have to ask how an expansive temporal self-understanding on a global scale that would not be confined to a particular tradition could be linked to existing needs and desires. Three directions that do not have to be mutually exclusive may be considered.

Firstly, one could argue that people do not only care about the survival of their own group but also about the survival of humanity as such, even apart from explicitly moral principles. Samuel Scheffler has attempted to make this plausible through a thought experiment inspired by P. D. James's novel *The Children of Men* (James, 1992; Scheffler, 2013, pp. 38–49; cf. also Scheffler, 2017, pp. 41–44). In the novel all human beings have become infertile. "The human race thus faces the prospect of imminent extinction as the last generation born gradually dies out." (Scheffler, 2013, p. 38) According to Scheffler, this prospect would be so frightening that we would give up many of the concerns that give our lives meaning today. From this speculative observation, Scheffler draws the far-reaching conclusion that the survival of humanity as such is a precondition of much of what we value. In fact, many people, although not all, would likely react to Scheffler's scenario of universal infertility as he assumes. However, the human species as such remains a rather abstract entity, so that the motivation to combat climate change for the sake of humanity will presumably remain weak.[19] Just because something might be a precondition of the

[19] In his 2017 book, Scheffler attempts to make use of his insights from thought experiments like the one inspired by James's novel in order to better understand why we care about future generations. These reflections touch directly on concerns like the consequences of climate change. They are thought to supplement or to undergird straightforward moral appeals to recognize our shared status as human beings. However, one could ask whether the motivational pull of moral imperatives might not usually be stronger than the presuppositions Scheffler's thought experiment is supposed to uncover. The problem is, of course, that the force of direct moral appeals has been proven too weak

fact that anything matters to us at all, it does not thereby have to be an important motive for our actions.

Secondly, it is therefore worth considering whether forms of solidarity with other humans that transcend one's own group and its time horizon might offer a more concrete approach to the fight against climate change. This would require sharpening awareness that the fight against the effects of climate change can be part of other struggles. Historically, the exploitation of fossil resources for the Industrial Revolution and its energy regime has been closely tied to the exploitation of the poor and to colonialism. This is still the case today (cf. Táíwò, 2022, Appendix B). Focusing on humanity as such obscures the view of the clash between those who profit from the continued burning of fossil fuels and other forms of environmental destruction and those whose very existence is being destroyed by these practices (cf. Chaudhary, 2024; Cripps, 2022). Understanding one's own fight for better living conditions as part of global struggles could mean seeing one's own life as embedded in temporalities that exceed the survival of one's own group without claiming to speak for humanity as such. However, this perspective also faces limitations. Because of the close correlation between cheap energy and economic wealth, the immediate interests of companies, politicians, and ordinary citizens, especially in rich countries, seem rather to converge (cf. Beckert, 2024, p. 38). Unfortunately, climate migrants of the present and future cannot confidently rely on the solidarity of the poor in richer and more climate-protected countries. Currently, populists in all countries are trying to drive a wedge between both groups. It would need to be shown that such divisive politics cannot be in the interest of the lower classes in richer countries.[20]

In order to loosen the armor of an individualistic understanding of one's own life's temporality, a third way should be explored in addition. In this case, one sees one's own life not only as embedded in social and historical processes but also as part of natural cycles and ultimately of planetary time. Though one might suspect that this is a rather romantic

when it comes to climate change. And until now humanity as such has never been a political unit. "There is no one humanity in politics. Politics stems from what differentiates humans." (Chakrabarty, 2023, p. 8),

[20] What we probably need is a convincing critique of the prevailing "vision of an idealized, wealthy American or perhaps European lifestyle, just writ large and universalized" (Chaudhary, 2024, p. 293) that does not simply promote a minimalist, ascetic way of life that focuses solely on material restraint but instead explores new, enriching ways of living that are more ecologically sustainable.

idea, relying on outdated notions of nature, there are at least some human needs that could be met in this way.

Thus, the temporality of our lives is not only characterized by the triad of past, present, and future. Rather, our actions and life processes are also subject to their own rhythms and intervals. How we spend our days is largely socially determined. We must wake up early to go to work, are expected to perform certain tasks during working hours, must work overtime if there is too much work left unfinished, go to retirement at a predetermined age, etc. However, alongside these socially given beats, our bodies follow their own biological or individual rhythms, which are often disrupted or confused by the demands of social synchronization, such as in shift work. But through our bodily rhythms, we are connected to the times of nature. Most bodily processes follow a circadian rhythm, which is adapted to the change from day to night, just as it is for most other living beings (Buonomano, 2017). Furthermore, the change of the seasons that is currently disturbed by climate change also plays a certain role for the well-being of many people. With such observations, the perhaps rather speculative idea could be supported that we are dependent on being in touch with a relatively intact non-human nature in order to preserve a sense of our own vitality.

Another desire that could be fruitful to cultivate already plays a role in the pursuit of social causes that transcend one's own lifetime. It is a sense of belonging to something greater, a sense that is fed by the feeling of being part of a temporally larger whole. All religions are sustained by this kind of feeling. And in many traditions nature itself has been supposed to play the role of a meaning-giving realm. The question is whether such a feeling could still be nourished for people who understand themselves as secular and whose understanding of nature is shaped by modern natural sciences. I do not have a definite answer to this question. But there are points of attachment that do not seem to depend on a religious exaltation of nature or dubious metaphysical assumptions. For example, there have been materialistic conceptions of nature, for instance in the Epicurean tradition, that have drawn their own comfort from the idea that we are part of the becoming and passing away of all living things, part-and-parcel of the great cycle of life.

These tentative suggestions are nourished by a very rough idea of a good human life, according to which a good life is essentially characterized by a balanced relationship to oneself, to others, and to the world, including the natural world. This idea could be spelled out in a number of ways. I have tried to give it a temporal twist, an attempt that certainly demands more conceptual work and empirical underpinning. In any case, we need new images of a good life in time to anchor even the most shared normative principles of climate ethics firmly in the motivations of ordinary people.

Competing Interests The author has no conflicts of interest to declare that are relevant to the content of this chapter.

References

Archer, D., Eby, M., Brovkin, V., Ridgwell, A., Cao, L., Mikolajewicz, U., Caldeira, K., Matsumoto, K., Munhoven, G., Montenegro, A., & Tokos, K. (2009). Atmospheric lifetime of fossil fuel carbon dioxide. *Annual Review of Earth and Planetary Sciences, 37*, 117–134. https://doi.org/10.1146/annurev.earth.031208.100206

Beckert, J. (2024). *Verkaufte Zukunft. Warum der Kampf gegen den Klimawandel zu scheitern droht*. Suhrkamp.

Buonomano, D. (2017). *Your brain is a time machine*. Norton.

Callender, C. (2017). *What makes time special?* Oxford University Press.

Chakrabarty, D. (2021). *The climate of history in a planetary age*. Chicago University Press.

Chakrabarty, D. (2023). *One planet, many worlds. The climate parallax*. Brandeis University Press.

Chaudhary, A. S. (2024). *The exhausted of the earth. Politics in a burning world*. Repeater Books.

Climate Action Tracker. (2025). Retrieved August 17, 2025, from https://climateactiontracker.org/

Cripps, E. (2022). *What climate justice means and why we should care*. Bloomsbury Continuum.

DeLay, T. (2024). *Future of denial: The ideologies of climate change*. Verso.

Faulkner, W. (2012). *Requiem for a Nun*. Vintage.

Gadamer, H.-G. (2016). The history of the universe and the historicity of human beings. In P. Vandervelde & A. Iver (Eds.), *Hermeneutics between history and philosophy: The selected writings of Hans-Georg Gadamer* (Vol. 1, pp. 25–41). Bloomsbury Academic.

Gardiner, S. (2011). *A perfect moral storm. The ethical tragedy of climate change*. Oxford University Press.

Gardiner, S., Caney, S., Jamieson, D., & Shue, H. (Eds.). (2010). *Climate ethics. Essential readings*. Oxford University Press.

Hanusch, F. (2023). *The politics of deep time*. Cambridge University Press.

Intergovernmental Panel on Climate Change (IPCC). (2023). Sections. In Core Writing Team, H. Lee & J. Romero (Eds.), *Climate change 2023: Synthesis report. Contribution of working groups I, II and III to the sixth assessment report of the intergovernmental panel on climate change* (pp. 35–115). IPCC. https://doi.org/10.59327/IPCC/AR6-9789291691647

Ismael, J. (2011). Decision and the open future. In A. Bardon (Ed.), *The future of the philosophy of time* (pp. 149–168). Routledge.

James, P. D. (1992). *The children of men*. Faber and Faber.

Krznaric, R. (2022). *The good ancestor*. WH Allen.

Lenton, T. M., Rockstrom, J., Gaffney, O., Rahmstorf, S., Richardson, K., Steffen, W., & Schellnhuber, H. J. (2019). Climate tipping points—Too risky to bet against. *Nature, 575*, 592–595. https://doi.org/10.1038/d41586-019-03595-0

Maclaurin, J., & Dyke, H. (2002). "Thank goodness that's over": The evolutionary story. *Ratio, 15*(3), 276–292. https://doi.org/10.1111/1467-9329.00191

Malm, A. (2018). *The Progress of this storm. Nature and Society in a Warming World*. Verso.

McTaggart, J. E. (1927). *The nature of existence* (Vol. 2). Cambridge University Press.

Meyer, L. H. (2013). Why historical emissions should count. *Chicago Journal of International Law, 13*(2), 597–614. https://chicagounbound.uchicago.edu/cjil/vol13/iss2/15

Moellendorf, D. (2022). *Mobilizing Hope. Climate change and global poverty*. Oxford University Press.

Müller-Salo, J. (2023). Climate ethics and the everyday Lifeworld. *Zeitschrift für Ethik und Moralphilosophie, 6*, 325–345. https://doi.org/10.1007/s42048-023-00157-y

Read, D. (2008). Intertemporal Choice. In D. J. Koehler & N. Harvey (Eds.), *Blackwell handbook of judgment and decision making* (pp. 424–443). John Wiley & Sons.

Robson, A., & Samuelson, L. (2007). The evolution of intertemporal preferences. *American Economic Review, 97*(2), 496–500. https://doi.org/10.1257/aer.97.2.496

Rosa, H. (2013). *Social acceleration: A new theory of modernity*. Columbia University Press.

Scheffler, S. (2013). *Death and the afterlife*. Oxford University Press.

Scheffler, S. (2017). *Why worry about future generations?* Oxford University Press.

Scheffler, S. (2025). *One life to Lead. The mysteries of time and the goods of attachment*. Oxford University Press.

de-Shalit, A. (1995). *Why posterity matters. Environmental policies and future generations*. Routledge.
Shue, H. (2021). *The pivotal generation. Why we have a moral responsibility to slow climate change right now*. Princeton University Press.
Táíwò, O. (2022). *Reconsidering reparations*. Oxford University Press.
United Nations Climate Change. (2025). *Technical dialogue of the first global stocktake. Synthesis report by the co-facilitators on the technical dialogue*. Retrieved August 17, 2025, from https://unfccc.int/sites/default/files/resource/sb2023_09E.pdf?download
Williams, B. (1993). *Shame and necessity*. University of California Press.

Aging as a Project: Ethical and Socio-Political Perspectives

Mark Schweda

1 Introduction

"Longer lives are one of humanity's greatest achievements," the platform of the United Nations *Decade of Healthy Aging* states on its website. By 2030, the initiative "aims to give everyone the opportunity to add life to years" (UN, 2022) and "the chance to lead a meaningful life": "Together, we can make every year count—not by the numbers, but by how we live and grow in between." (ibid.)

These emphatic promises appear paradigmatic: In light of scientific, technological, and societal developments, aging and old age are becoming target points of individual and societal predictions, plans, and programs. They nurture the expectation that later life can be precisely forecast and actively shaped. What was once seen as a matter of luck, fate, divine providence, or inevitable natural processes is turned into an individual and socio-political project that has to be carefully planned and systematically executed (Leibing & Schicktanz, 2020; Powell, 2024; Schweda et al., 2017).

M. Schweda (✉)
Department of Health Services Research, School of Medicine and Health Sciences, Carl von Ossietzky University of Oldenburg, Oldenburg, Germany
e-mail: mark.schweda@uni-oldenburg.de

J. Hinrichsen et al. (eds.), *Ethics of the Sciences*,
https://doi.org/10.1007/978-981-92-2821-8_18

This reframing of aging and old age in terms of a calculable and plannable individual or societal future has far-ranging moral implications. Luck, fate, or providence simply are what they are. They have an unquestionable facticity and inevitability that defy moral discussion or critique. At best, we may be able to change our subjective views of them and our personal attitudes and reactions toward them. Individual and policy projects are of a fundamentally different nature. Their ends and means are open to deliberation and discussion and in need of justification (Mounk, 2017).

The contribution focuses on the moral significance of this project view of aging and old age. I will explore from an ethical perspective what this view means for a good life in old age as well as for moral responsibilities for aging well. In doing so, I will address conceptual and prescriptive aspects of "futurization" in the context of later life. In the next section (Sect. 2), I will briefly sketch crucial preconditions and steps in the historical emergence of the project view. In particular, I will highlight the role of the modern sciences of aging, focusing on geriatrics and its distinction between physiological and pathological aging, the gerontological promotion of ideals, strategies, and measures of successful, active, healthy, and productive aging, the rise of the so-called anti-aging movement that aims to fight biological senescence by preventive and biotechnological means, as well as public health and social policy programs regarding population aging. In the next sections, I will trace the moral implications of these processes. The focus is on individual perspectives for a good life in old age (Sect. 3) and on responsibilities for aging well (Sect. 4). Finally, I will discuss preconditions and problems of the project view and draw conclusions for gerontological ethics (Sect. 5).

2 From Chance to Choice? The Emergence of the Project View on Aging and Old Age

Since antiquity, aging has been conceived as the inescapable fate of human beings, a natural process of inevitable decline and diminishment that ultimately ends with death. Aristotle spoke of aging as a "natural disease," a kind of ailment that had to be accepted as a regular element of the human condition and the natural order of things (Sokolon, 2021). Based on teachings of humoral pathology, ancient medical theories suggested that it resulted from a progressive drying and cooling of the organism (Gilleard, 2015). In the Middle Ages, Christian theology even justified the suffering and ailments of old age as a consequence of original sin and a pathway to the imitation of Christ (Gilleard, 2009).

Despite the enduring mythological and alchemistic fascination with the quest for immortality and eternal youth, the view of aging as an inevitable natural process of decline that marks the ultimate limit of any medical intervention remained the dominant perspective until the last century. In his memoirs, Ignatz Nascher, one of the founding fathers of modern geriatric medicine, recounts a telling anecdote: While studying medicine in New York City at the turn of the last century, he was making rounds with his professor and encountered an older patient in a deplorable state of health. The young medical student asked his professor what was wrong with the woman and what they could do for her. The information he received was very brief: nothing at all, she was just old (Clarfield, 1990).

Not least initiated by Nascher himself, this long-prevailing therapeutic nihilism became one of the target points for the development of geriatric medicine and modern medical research on aging and old age. With its constitutive distinction between "normal" and "pathological" aging, geriatric research challenged the inevitability of age-associated decline and ailment and asserted some possibility for medical intervention (Blumenthal, 2003). It prepared the ground for a differentiated perspective that acknowledged certain changes in functionality and performance while identifying and targeting diseases associated with old age that needed attention and treatment. Diseases of old age became target points for medical research and interventions.

Initially rooted in these medical approaches toward aging and old age, modern gerontology established itself around the middle of the twentieth century as an interdisciplinary field of research comprising medical and biological, but also psychological and sociological perspectives (Katz, 1996). A central aim of gerontological research and interventions was to overcome traditional and long-standing deficit models of aging in order to establish a more positive perspective that acknowledges the plasticity of aging and is oriented toward the resources and potentials of older people. This change is expressed programmatically in gerontological concepts of "successful aging" that proliferated since the 1960s, as well as in related ideals of "active," "productive," or "healthy aging" (Pfaller & Schweda, 2023).

Especially John W. Rowe's and Robert L. Kahn's model of "successful aging" became a prominent and much-referenced paradigm. The model postulates three distinctive marks of successful (as opposed to merely "normal") aging: avoiding disease and disability, maintaining high cognitive and physical functioning, and continuing engagement with life (Rowe & Kahn, 1997). Ultimately, such concepts not only framed gerontological research but also became practically influential in social work, gerontological interventions, and social policy programs like the WHO's Active Aging Framework or the aforementioned Decade of Healthy Aging (Bülow & Söderqvist, 2014). With population aging, the erosion of traditional familial care structures, and the implementation of publicly funded social security systems, such programs are gaining importance. Aging has become a crucial social policy issue.

More recent scientific explanations and technical interventions challenge the traditional notion that the decline in health in the last third of life represents the natural course of events that ends with a "natural death" due to the successive loss of vitality. The long-used category of senility has disappeared from death certificates and mortality statistics. The biological and physiological processes resulting in senescence are becoming target points for etiological explanation and causal treatment (Faragher, 2015). Consequently, the proponents of so-called anti-aging medicine assume that the successive elimination of all health-related impairments of old age will ultimately make the traditional notion of aging as an inevitable, natural process obsolete.

Emerging in the early 1990s, the anti-aging movement explicitly declared war against aging. Its starting point was that "the frailties and physical and mental failures associated with normal aging are caused by physiological dysfunctions that, in many cases, can be altered by appropriate medical interventions" (Klatz & Goldman, 2003, IX). Consequently, its protagonists claimed that modern biomedicine and prevention could slow down, stop, or even reverse the process of biological senescence and extend the human lifespan considerably. For example, the founders of the American Academy for Anti-Aging Medicine declared that modern medicine promises the "elimination of the disability, deformity, pain, disease, suffering and sorrow of old age" (Klatz & Goldman, 2003, 13). According to them, "the traditional enfeebled, ailing elderly person" will soon be "a grotesque memory of a barbaric past" (ibid.). And the self-proclaimed British biogerontologist Aubrey de Grey even announced that thanks to advanced biomedical interventions, "the first person to live to 1,000 might be 60 already" (de Grey, 2004).

3 Success or Failure? Eudemonistic Implications of the Project View

If aging is viewed as a project, this has far-reaching moral implications. Projects have an objective: a vision of a state of affairs they aim to realize in the future. Accordingly, defining and promoting a guiding vision of "successful aging" marks a central area of gerontological research and practice. In the first issue of *The Gerontologist*, Robert Havighurst already declared: "In order to provide good advice [for aging individuals and societies], it is essential that gerontology have a theory of successful aging" (Havighurst, 1961, 8).

From a moral-philosophical perspective, gerontological debates about "successful aging" touch upon questions of the good life (Bearon, 1996). Such questions are sometimes called eudemonistic as they are not primarily concerned with the justification of moral norms governing our interactions but rather consider the value of the aims we pursue in our practices and lives. Eudemonistic ethics thus discusses what should be regarded as desirable or valuable in life, for example, involving concepts of happiness, well-being, or human flourishing.

However, gerontological theories of successful aging hardly ever explicitly discuss the ideals of aging and old age they assume and promote. Instead, the central goal or the decisive standard against which the "success" of aging is measured is usually more or less tacitly taken for granted. At closer inspection, it frequently involves certain medical, psychological, or sociological notions of physical functioning, psychological well-being and balance, or social adaptivity. Consequently, gerontological concepts of "successful aging" often mirror the society and value system of their time—whether they promote social disengagement for older people in the 1950s or sustained activity in the 1960s and 70s (Bearon, 1996).

By contrast, a philosophical reflection has to dig deeper: It must address the goals or standards of successful aging as such and discuss to what extent they can actually be considered desirable or valuable (Pfaller & Schweda, 2019). To this end, eudemonistic ethics offers at least three theoretical approaches: hedonistic, subjective, and objective theories of the good life (Parfit, 1984). For hedonistic theories, pleasure and suffering are the key aspects for assessing the quality of a life. From this point of view, a good life is one that contains as many pleasant and as few painful experiences as possible. By contrast, subjective theories, such as so-called desire theories, focus on fulfilling one's own central and well-informed wishes. A good life is, therefore, one in which the relevant personal wishes can be realized and goals achieved. Finally, objective theories assume certain goods or purposes that are to be recognized as desirable, valuable, or meaningful regardless of our actual subjective sensations or preferences.

These ethical theories can also be applied to individual and socio-political ideas and practices regarding aging and old age (for the following, cf. Pfaller & Schweda, 2019). Thus, *hedonistic theories* may be criticized for their reductive sensualist notion of a good life in terms of pleasure and suffering, but allow us to make an essential point regarding gerontological concepts of successful aging: The success or failure of aging cannot simply be determined from an external scientific point of view according to allegedly objective criteria that are independent of the personal experience of older people themselves. A person may meet all requirements of gerontological definitions of "successful aging" and still feel miserable—or live happily without reaching even one of them. This supports a prominent line of critique of the scientism and expertocracy of theories of successful aging that fail to include the firsthand experience of old age (Swift and Tate 2015). Indeed, social research points to significant discrepancies between gerontological concepts of successful aging and the variety of everyday life perspectives of older adults (Reich et al., 2020).

Furthermore, *subjective theories* of a good life can remind us of the relevance of fulfilling personal wishes and desires for old age. Criteria for successful aging cannot be defined without considering aging individuals' own central interests and preferences. Thus, from the point of view of desire theories, it becomes clear that even physical pain and severe suffering can appear acceptable in old age to accomplish goals that are essential to one's own life plan, such as the resolution of a long-standing familial conflict or the completion of a biographically significant project. Conversely, even a life completely free of physical ailments or cognitive impairments can lose all value for the person living it if essential goals and plans can no longer be realized. Ultimately, from the point of view of desire theories, the prospects of aging well or reaching an advanced age depend on the specific content of the respective life-defining plans and endeavors.

Finally, objective theories also address important aspects. In particular, they help challenge reductive concepts of successful aging that focus on isolated outcome measures such as health or physical functioning.

For example, Martha Nussbaum's capability approach identifies various fundamental capabilities necessary for human flourishing and thus opens a more comprehensive and differentiated perspective (Nussbaum, 2011). It allows us to recognize the relevance of physical integrity but also of other capabilities like mental activity and social participation as essential preconditions of a good life in old age (Pfaller & Schweda, 2019). Moreover, classical objective theories suggest that old age might open specific value dimensions. Thus, Platonic ethics holds that a good life requires developing particular virtues and reaching specific insights, processes that take time and can, therefore, only be achieved at an advanced age (McKee & Barber, 2001). More recent gerontological concepts such as generativity (Kotre, 1996) or gerotranscendence (Tornstam, 1989) seem to articulate similar ideas in a contemporary theoretical frame of reference.

4 Responsibilization of Aging? Moral Implications of the Project View

If aging is viewed as a project, its success or failure is not a matter of fate or luck but of the adequate choice and implementation of means to a given end. As Rowe and Kahn explain in their work on successful aging, "We can have a dramatic impact on our own success or failure in aging" (1997, 18). In fact, the answer to the question of health, activity, and performance in old age appears "surprisingly simple": "good old-fashioned hard work" (ibid., 17). For them, "to succeed [...] means having desired it, planned it, worked for it" (ibid., 37).

Moral-philosophically speaking, the flipside of this proclaimed impact on the process of aging is responsibility. If we can influence, control, or even actively shape how we grow old through circumspect decisions and determined actions, we can be held responsible for our future condition and situation. Old age then becomes the decisive touchstone of our efforts' seriousness, intensity, and effectiveness. It appears as the final moment of truth and justice, the ultimate moral payoff of a life of caution, planning, discipline, and self-management—or neglect, indulgence, and carelessness.

In critical gerontology, the rise of this view has been criticized as a "responsibilization of aging" (Cardona, 2008). Neoliberal governmentality requires individuals to act as active citizens in charge of looking after their own health, well-being, and socioeconomic subsistence in later life (Moulaert & Biggs, 2013). From an ethical point of view, however, assigning responsibilities is not per se problematic. On the contrary, it constitutes a necessary precondition for moral judgments and discussions. Thus, what is needed is a differentiated analysis of responsibility ascriptions and their moral implications. We have to be able to assess to what extent claims about responsibility are justified or not.

The analysis of responsibilities requires a careful examination of relations between different actors and their roles and capacities (Schicktanz & Schweda, 2012). First and foremost, the question is whether the responsibility for how we grow old is primarily located on the side of aging individuals themselves. In philosophical ethics, there are at least three interrelated criteria for ascribing responsibility: knowledge, control, and intention (ibid.). To be able to assume responsibility, one has to be aware of the relevant facts and interrelations. If I do not know (and could not be reasonably expected to know) that a decision or action can have certain consequences, I cannot be held responsible for these consequences. Furthermore, responsibility presumes the possibility of actually influencing a given process and its outcomes. I may be well aware of the causal mechanisms leading to a particular event but still lack the ability or power to intervene. Finally, responsibility usually requires intentional, willful agency. If I caused something inadvertently and accidentally, for example, because of an unfortunate and unforeseeable coincidence, I cannot be held morally responsible.

These criteria can be applied to the discussion of the "responsibilization of aging" (for the following, cf. Schweda & Pfaller, 2020). Thus, it seems unclear to what extent we can really claim to have the relevant *knowledge* to fully understand what causes aging and age-associated diseases and physical or cognitive impairments. There are still considerable gaps and controversies in scientific research on the biological and physiological mechanisms of senescence (Guo et al., 2022; Loffeier et al., 2017). Furthermore, much of the information we actually do have refers to statistical correlations that do not permit a deterministic interpretation in terms of a direct relation between cause and effect but only allow for probabilistic conclusions in terms of chances and risks. Finally, it is unclear how the individual person can actually identify, access, and utilize reliable knowledge that becomes available. Here, the individual is clearly dependent on the provision of suitable health education, professional advice, scientific evidence, and authoritative and trustworthy knowledge resources, but also on economic and technological framework conditions (Estes, 1979).

Furthermore, it appears unclear to what extent the individual is able to *control* the aging process effectively. While we know that senescence and accompanying conditions such as dementia are associated with certain risk factors, these are not necessarily a matter of choice or lifestyle but are often beyond our control. First and foremost, there are genetic risk factors we cannot influence. Furthermore, many risk factors are of an environmental nature. Thus, in many societies, the effects of class on average life expectancies are considerable (Chetty et al., 2016). Therefore, the individual will frequently depend on other actors to exert an influence on relevant processes. This includes technical and financial support for health prevention, such as medical expertise and financial subventions for checkups and effective preventive measures, but also broader socio-political framework conditions, for example, the social living situation or environmental circumstances. Thus, while there may be a justified expectation for people to make reasonable efforts to stay healthy, there can hardly be a moral duty to stay healthy because achieving or maintaining health is simply beyond our causal control.

Of course, the criterion of intention is closely connected to knowledge and control. I may lack motivation to change my lifestyle because I do not know or cannot influence relevant factors. Furthermore, intentionality also points to the evaluative task of the adequate weighing of health prevention compared to other essential things in life. In ethical terms, this points back to the eudemonistic question of conditions and elements of a good life. Thus, most of us would not find it reasonable to subordinate one's whole life to the prevention of diseases and dysfunction in old age—that is, to renounce any pleasures, invest a fortune, or even take serious risks—all the more so if the best outcome is a dubious and remote chance of better health in later life. Health may have significant value, but it is definitely not the only or even most important thing in life. In fact, we may be willing and even justified to put our health at risk for the sake of more significant goods—for example, dangerous activities involving joy and pleasure, a great professional or artistic achievement that promises enduring honor and glory, or simply raising children and taking care of old parents (Pfaller & Schweda, 2019).

5 Conclusion

Medical, gerontological, and social policy developments are changing our views of aging and old age. Later life increasingly appears as a project that has to be carefully planned and systematically executed. Media, aging research, social programs, commercial enterprises, and public health initiatives promote the project view as a positive and empowering perspective on later life.

However, the view of aging as a project has important moral implications. It promotes a moralization of aging, turning a given natural fact, a fundamental structure of the human condition, into an object of moral evaluation, discussion, and justification (or critique). Two questions become especially relevant: How should we want to grow old, that is, what aims are actually desirable with regard to aging and old age? And what should we do to age well, that is, what moral responsibilities do we have with regard to aging and old age?

An ethical analysis can contribute to the clarification of these questions. Thus, eudemonistic ethics provides theories and criteria for the good life that can be applied to the field of aging and old age. We need to include older people's experiences, consider their wishes and desires, and develop a differentiated view of relevant goods. In this regard, a capability approach can provide a starting point (Nussbaum, 2011). With its broad and inclusive scope, it can help detect shortcomings of current ideals of aging and justify the relevance of different aspects of the good life in old age, thus promoting a truly holistic conception and a comprehensive matrix of aspects and dimensions relevant for leading a good life as an older person (Pfaller & Schweda, 2019). Indeed, recent research projects and policy programs in the field of aging and old age are increasingly exploring the theoretical and practical advantages of the capability approach (Falk Erhag et al., 2022).

In a similar way, the debate about the responsibilization of aging can benefit from an ethical analysis of responsibility. Thus, it has to be questioned whether a focus on individual responsibilities is really fair and balanced given the limitations of individual knowledge and control. A more differentiated perspective has to analyze the complex interaction of the various individual, communal, social, economic, political, and scientific actors, factors, and structures influencing health and old age (Schweda & Pfaller, 2020). In a corresponding vein, Minkler proposes the concept of individual "response-ability" to address the role of these manifold actors and (inter-)dependencies that enable individuals to take responsibility. "Response-ability" can be understood as "the capacity of individuals for building on their strengths and meeting the challenges posed by the environment" (Minkler, 1999, 124). As this capacity is embedded in and dependent on a complex web of other individuals, community, research and innovation agendas, and state policies, "response-abilization" would mean a truly empowering approach engaging all these actors to strengthen and support individuals' response-ability and enable them to take control and accept their fair share of responsibility.

Competing Interests The author has no conflicts of interest to declare that are relevant to the content of this chapter.

References

Bearon, L. B. (1996). Successful aging: What does the 'good life' look like. *The Forum, 1*(3).

Blumenthal, H. T. (2003). The aging-disease dichotomy: True or false? *The Journals of Gerontology. Series A: Biological Sciences and Medical Sciences, 58*(2), M138–M145.

Bülow, M. H., & Söderqvist, T. (2014). Successful ageing: A historical overview and critical analysis of a successful concept. *Journal of Aging Studies, 31*, 139–149.

Cardona, B. (2008). 'Healthy ageing' policies and anti-ageing ideologies and practices: On the exercise of responsibility. *Medicine, Health Care and Philosophy, 11*, 475–483.

Chetty, R., Stepner, M., Abraham, S., Lin, S., Scuderi, B., Turner, N., Bergeron, A., & Cutler, D. (2016). The association between income and life expectancy in the United States, 2001–2014. *JAMA, 315*(16), 1750–1766.

Clarfield, A. M. (1990). Dr. Ignatz Nascher and the birth of geriatrics. *Canadian Medical Association Journal, 143*(1990), 944–948.

De Grey, A. (2004, December 3). We Will Be Able to Live to 1000. *BBC News*. http://news.bbc.co.uk/2/hi/uk/4003063.stm

Estes, C. L. (1979). *The Aging Enterprise*. Jossey-Bass.

Falk Erhag, H., Lagerlöf Nilsson, U., Rydberg Sterner, T., & Skoog, I. (Eds.). (2022). A Multidisciplinary Approach to Capability in Age and Ageing. Springer.

Faragher, R. G. (2015). Should we treat aging as a disease? The consequences and dangers of miscategorisation. *Frontiers in Genetics, 6*, 171.

Gilleard, C. (2009). Old age in the dark ages: The status of old age during the early middle ages. *Ageing & Society, 29*(7), 1065–1084.
Gilleard, C. (2015). Ageing and the Galenic tradition: A brief overview. *Ageing & Society, 35*(3), 489–511.
Havighurst, R. (1961). Successful aging. *The Gerontologist, 1*, 8–13.
Katz, S. (1996). *Disciplining Old Age: The Formation of Gerontological Knowledge*. University of Virginia Press.
Klatz, R., & Goldman, R. (2003). *The New Anti-ageing Revolution. Stopping the Clock for a Younger, Sexier, Happier You*. Basic Health Publications.
Kotre, J. N. (1996). *Outliving the Self: How We Live on in Future Generations*. W. W. Norton & Co.
Leibing, A., & Schicktanz, S. (Eds.). (2020). Preventing Dementia? Critical Perspectives on a New Paradigm of Preparing for Old Age. Berghahn Books.
Loffeier, I., Majerus, B., & Moulaert, T. (Eds.). (2017). Framing Age: Contested Knowledge in Science and Politics. Taylor & Francis.
McKee, P., & Barber, C. E. (2001). Plato's theory of aging. *Journal of Aging and Identity, 6*(2), 93–104.
Minkler, M. (1999). Personal responsibility for health? A review of the arguments and the evidence at century's end. *Health Education & Behavior, 26*(1), 121–140.
Moulaert, T., & Biggs, S. (2013). International and European policy on work and retirement: Reinventing critical perspectives on active ageing and mature subjectivity. *Human Relations, 66*(1), 23–43.
Mounk, Y. (2017). *The Age of Responsibility: Luck, Choice, and the Welfare State*. Harvard University Press.
Nussbaum, M. C. (2011). *Creating Capabilities: The Human Development Approach*. Belknap Press.
Parfit, D. (1986). *Reasons and Persons*. Oxford University Press.
Pfaller, L., & Schweda, M. (2019). Excluded from the good life? An ethical approach to conceptions of active ageing. *Social Inclusion, 7*(3), 44–53.
Pfaller, L., Schweda, M. (2024). *"Successful Aging": Leitbilder des Alterns in der Diskussion*. Springer.
Powell, J. (2024). *Unravelling Anti-Aging*. Springer.
Reich, A. J., Claunch, K. D., Verdeja, M. A., Dungan, M. T., Anderson, S., Clayton, C. K., Goates, M. C., & Thacker, E. L. (2020). What does 'successful aging' mean to you? Systematic review and cross-cultural comparison of lay perspectives of older adults in 13 countries, 2010-2020. *Journal of Cross-Cultural Gerontology, 35*(4), 455–478.
Rowe, J. W., & Kahn, R. L. (1997). Successful Aging. *The Gerontologist, 37*(4), 433–440.
Schicktanz, S., & Schweda, M. (2012). The diversity of responsibility: The value of explication and pluralization. *Medicine Studies, 3*, 131–145.

Schweda, M., & Pfaller, L. (2020). Responsibilization of ageing? An ethical analysis of the moral economy of prevention. In A. Leibing & S. Schicktanz (Eds.), *Preventing Dementia? Critical Perspectives on a New Paradigm of Preparing for Old Age* (pp. 192–213). Berghahn Books.

Schweda, M., Pfaller, L., Brauer, K., Adloff, F., Schicktanz, S. (eds.). (2017). Planning Later Life. Bioethics and Public Health in Aging Societies. Routledge.

Sokolon, M. K. (2021). Good old age: Aristotle and the 'virtues' of aging. In E. A. Dolgoy, K. Hurd Hale, & B. Peabody (Eds.), *Political Theory on Death and Dying* (pp. 50–59). Routledge.

Tornstam, L. (1989). Gero-transcendence: A reformulation of the disengagement theory. *Aging Clinical and Experimental Research, 1*(1), 55–63.

United Nations Decade of Healthy Aging. (2022). https://www.decadeofhealthyageing.org/about/about-us/what-is-the-decade

Equitable Fair Innings—Revisiting the Rationing of Health Care on the Basis of Age

Nils Hoppe and Marc Stauch

1 Introduction

Scarcity of resources is an inevitable background condition to health care provision in modern societies; in this context, medicine might be described as a victim of its own success: the greater its ability—through making available innovative treatments—to prolong people's lives, including lives attended by chronic morbid health conditions, the greater the further and future claims upon its resources. This effect has been added to by the development of mass health care services, sponsored by the welfare state. The notion of providing to all on the basis of medical need, which has at least in part been fulfilled, has led to increasing numbers of people surviving to an age that few reached even half-a-century ago.

N. Hoppe
Centre for Ethics and Law in the Life Sciences, Leibniz Universität Hannover, Hannover, Germany
e-mail: nils.hoppe@cells.uni-hannover.de

M. Stauch (✉)
Juristische Fakultät, Leibniz Universität Hannover, Hannover, Germany
e-mail: stauch@iri.uni-hannover.de

J. Hinrichsen et al. (eds.), *Ethics of the Sciences*,
https://doi.org/10.1007/978-981-92-2821-8_19

These developments in themselves are to be welcomed, particularly if it is assumed (as we shall in this paper) that added years of life are invariably of benefit to the individual human subjects who enjoy them.[1] Nonetheless, as the population as a whole grows older, and a greater share of its members become recipients and fewer are financers/producers of services (including health care), health care resources are (and will continue to be) subject to increasing strain, potentially to an unworkable degree. Even under 'ideal' conditions, resource scarcity and allocation pressures are challenges that often prove vexing. The recent COVID-19 pandemic has reminded us that allocation dilemmas—which were hitherto sometimes thought to be esoteric philosophical thought experiments—are a reality, and that it would make sense to debate the societal advantages and disadvantages of different models ahead of time (Albertsen, 2023; Jecker, 2022). It is clear that there will have to be an agreed mechanism for the apportionment of scarce health resources. This is required not only to provide fairness to patients who need access to them, but also to establish reflective mechanisms and normative certainty for healthcare professionals who would otherwise suffer moral distress when forced to make a decision.

An intuitive but controversial response, much debated by ethicists and health economists, is to use age as a basis for health care rationing, such that, given the choice between giving treatment to a younger or older person, then—other things equal—the younger person should receive it. There is empirical evidence to suggest that many people would support some such form of rationing (above an initial age threshold) in the sense described (Kitzhaber, 1993); however, the question both as to the basis of such a preference, and the details of how it should cash out in practice, remain fiercely contested. True life-and-death resource allocation issues are often, in practice, resolved on the basis of a complex bundle of more or less objective clinical parameters, and decisions are only indirectly based on the age of the patient (i.e. because of age-related comorbidities). In health care systems based on principles of solidarity, it may in fact seem alien to adopt a stance that overtly uses age as a criterion. Indeed, a number of ethicists and academic writers have rejected such an approach on the ground that collectively financed health care should be about treating on the basis of medical need, and that using age as

[1] We thus not consider scenarios in which the continued provision of health care might be challenged as no longer in the patient's best interests.

a rationing criterion is a form of unwarranted discrimination, namely 'ageism' (Harris, 1987).

As alluded to, the debate is complicated by the indirect use of age as a factor in such decision-making. Thus, on the one hand, a person's age may serve as an indicator in making a consequentialist judgment as to the value (in terms of outcome) of treating one person, or class of persons, rather than another. Best known in this context is the QALY approach, which focuses on the likely effect of treatment on potential recipients and holds that an efficient use of health care resources mandates prioritising those who will survive longest and in best health: on the whole these are likely to be younger persons.[2] On the other hand, there is a separate set of arguments that focus on age-based rationing grounded in considerations of justice or equity; this is generally referred to in the literature as the 'fair innings' approach.

In this chapter our interest is in this latter form of argument, and accordingly we shall say no more about consequentialist approaches. In particular, our specific intention will be to consider the 'fair innings' approach from a societal, political justice perspective. In so doing, we shall argue that a distinction should be preserved between two related, but separate questions that in the debate are often blurred: the one goes to the size, the other to the division, of the 'health care cake'. In the second place it is important to remain mindful of the underlying reason for having a collective health care scheme at all. Following these steps yields in our view a relatively straightforward justification for health care rationing based on an age-based cut-off point in a given society.

2 The 'Equitable Fair Innings' Approach

The original 'fair innings' argument may be traced back to John Harris' work, *The Value of Life*, in which Harris (1985) wrote:

> Let us say that a fair share of life is the traditional three score and 10, 70 years... The fair innings arguments requires that everyone be given an equal chance to have a fair innings, to reach the appropriate threshold but, having reached it, they have received their entitlement. The rest of their life is the sort of bonus which may be cancelled when this is necessary to help others reach the threshold. (p. 91).

[2] See Harris (1987) who argues against QALYs inter alia for his reason.

Harris himself remained at the time ambivalent about this argument and indeed has since been a prominent critic of ageist results in health care distribution that arise as a side effect of the application of the QALY approach (Harris 1987; Harris & Regmi 2012). However, other moral philosophers working in health ethics, such as Norman Daniels and Daniel Callahan, have offered arguments supportive of the fair innings approach. Thus Daniels argues that fair innings can be justified through an application of the Rawlsian: 'veil of ignorance' theory, namely a health care system capped by age-based rationing, so as to maximize provision of treatment below the age-cut-off, is one a rational agent would favour a priori (in a state of ignorance as to their own empirical propensity for health and longevity) (Daniels, 1988). For his part, Callahan linked fair innings to a conception of life as a patterned narrative, in which old age and death are natural stages to be accepted as giving meaning to what has gone before (Callahan, 1987).

Here it is not our purpose to further investigate those arguments—we thus remain neutral as to whether either (or both) Daniels' and Callahan's justifications may be thought plausible. Instead, as hinted at earlier, we seek to present a further and, to our minds, simpler route to a conclusion in favor of fair-innings-based health care rationing. Thus, in our equitable fair innings approach, we begin not from the perspective of the individual subject with an interest in receiving health care resources to continue living a healthy life, but rather (and obviating the need for recourse to Rawl's 'veil of ignorance') that of a just distributor of public resources.

From this perspective, impartially adjudicating between the claims of the many outweighs the needs of the individual in cases where not all individuals can be provided with care; this is therefore a public health argument, rather than one driven by classic approaches of individualized medical ethics. We do acknowledge that—rather than a solution to the conundrum of whose interests come first—this switch in perspective is nothing other than the all-pervasive medico-legal and ethical conflict between benefiting individuals vs. benefiting communities, and that this is not a trivial context. Indeed, while properly debating and reflecting different approaches to just allocation, we are keen not to pave the way for arguments that may be used to elevate the needs of the many over the *dignity* of the individual, and we are alive to the tricky balancing exercise this represents. Nevertheless, it seems to us that the experiences that the recent pandemic gave rise to provide robust evidence that a prioritization of the group over the individual is the most likely approach driven by sheer necessity, and that we must discuss further how to fairly mitigate the impact on individuals of such a practice.

This is also an opportune moment to make explicit what we take to be the two key purposes of expending public resources on health care: the first is—in line with the first purpose of medicine itself—to secure so far as possible the life and health of individuals within the population, where health is a state permitting normal physiological and social human functioning; secondly (insofar as the first purpose is in no longer achievable), it is to palliate the suffering and preserve the dignity of the individual.

Against this background, we submit that the perspective of the just distributor of resources requires them to be blind to the distinguishing empirical characteristics of each individual. Rather, each individual life may be thought of as an abstract envelope, a sine qua non for the particular life-holder's flourishing (to then be filled according to the holder's values and goals). Here, the focus should in our view be upon guaranteeing to each person a share of the overall health care resources needed to allow them to live a life of equal quality and duration to those of other citizens in the same polity.

In this regard, there are two key questions that face the just distributor. The first concerns the amount of overall resources in that society (expressed, say, as a proportion of GDP) that should be spent on health care rather than on satisfying other claims on public expenditure. Here, let us assume that there is a fairly direct link between the answer (determining the size of the 'health care cake') and the average level of health and life expectancy achieved in that society. Accordingly, we could also term this the 'fair age' question—what is a reasonable age for the society, through collective health care, to try to get its citizens to reach?

It is clear that the question of where to fix the 'fair age' is primarily a political and economic question, and its resolution will vary from one society to another, based on its overall wealth at a given time and the shared sense of priorities (health versus other goods and values), including e.g. whether citizens subscribe to a Callahan-type perspective on life, where ageing is to be accepted as a natural life-process, or support expensive interventions that seek to postpone it for as long as possible. However, there next comes the second key question, which is not about the size of the cake, but how to share it. What would be a 'fair' portion to distribute to each individual citizen?

Unlike the first question, we suggest this second question requires no appeal to contingent social or economic factors; rather it can be resolved by reference to the underlying rationale for having a collective health care scheme, in conjunction with basic principles of distributive justice. The latter starts with the strict formal equality of each individual; nonetheless, its application may in practice entail highly unequal divisions of social resources. In the context of health care, this reflects, as further discussed below, the reality of the 'natural lottery', i.e. the unequal predisposition of different persons for enjoying healthy and extended lives.

To explore these points further, let us consider matters as they present themselves (in a developed country such as the UK) to the health care resources distributor (hereafter, the health department) after it has received its share of the public budget for spending. It is charged with providing resources at a given moment to allow the treatment of millions of citizens, some young, some old; these persons (taking men and women together) reach an average life expectancy of say 80 years; however, there is significant individual variation: many persons die (in some cases much) earlier, and many others live (much) longer. There is also considerable variation in the state of health of subjects (health being understood as a state permitting normal physiological and social human functioning): some persons enjoy good health throughout their lives; others are generally healthy but suffer intermittent acute illness; still others suffer from long-term chronic illness and/or disabilities. The factors conducing to these disparities are a mix of genetic and environmental (including socio-economic) differences, which are in the main beyond the health department's control.

Against this backdrop, a fair distributive approach in our view entails that the department distributes its available resources so that, as far as possible, each member of the population attains (in reasonable health) the same age, i.e. that corresponding to the fair age for that population. This is simply to recognize, that as argued earlier, the necessary starting point for a just distributor is that a unit of healthy life is equally valuable to all, i.e. each individual has the same interest in living in a healthy state to a given age.

As noted earlier, in pursuing such an approach, the actual distribution of health resources between individuals in practice will be highly uneven. Those persons who early develop chronic disease will end up having much more spent upon them over the course of their lives than others who remain healthy into old age and then die quickly. Should this fact trouble us as unfair? It is submitted, the answer is no: as argued earlier, the purpose of establishing a collective health care system is precisely to even out the 'natural lottery' by providing treatment designed to achieve or restore health in accord with individual need. The same goes for arguments that would see unfairness in the differential contributions to the system required of persons on different incomes.

By contrast, on the approach we have outlined, the following *would* count as unfair: namely, for the health department to continue providing treatment to those persons who have already attained the socially agreed fair age. This is because (given the fact of finite health resources) such expenditure represents an opportunity cost in terms of treating other persons who have yet to attain the fair age (and who now in some cases never will). To return to the language in which Harris originally presented the 'fair innings' argument, it is to carry on paying some persons a 'bonus', at the expense of the ordinary wage that is due to others (Harris, 1987).

3 Implications

A key implication of the above is that the just polity/state has no duty to continue providing health care to persons who have attained the fair age—indeed, it is positively required not to do so (in order to avoid injustice to others). Nonetheless, an immediate caveat must be entered here, which is that by health care we mean here treatment designed to secure or restore health in line with the first purpose of medicine. By contrast, the provision of general care, designed to palliate suffering in line with the second purpose (maintaining human dignity), would be unaffected. This is based on an assumed consensus (reflected in numerous human rights norms) that even though there may no longer be an obligation to provide a person with life-prolonging/health-restoring (curative) treatment, there remains a basic duty, while they are alive, to secure minimum conditions for their dignified existence, including hygiene, sustenance, and so far as possible absence of pain and physical discomfort (all of which we subsume under the term 'palliative treatment').

A second implication (or qualification) is that the approach outlined in this paper applies necessarily only to the publicly financed provision of health care. It is without prejudice to the treatment of post-fair-age patients, which is arranged for privately. In this context, persons could be encouraged to take out private health insurance policies to cover for the chance that they will exceed the average life expectancy and, having done so, will require specialist treatment at some point. Admittedly, this might be seen as unfair to those post-fair-age patients who have insufficient means to pay for private insurance and/or treatment: we shall consider this argument further below, under objections.

Thirdly, our approach does not address the question of the just allocation of resources between two or more pre-fair age patients within the health system. We have merely excluded a sizeable, growing, and expensive portion of the patient base from the pool of those who can expect a share of the health care cake. This will result in a greater likelihood of receiving the best treatment within the pre-fair age group, but where the resources are ultra-scarce (such as in the context of organ transplantation (Stahl et al, 2008), or the provision of ventilators during a pandemic (Albertsen, 2023)), the system still faces allocation issues within this smaller group. This paper has nothing to say about such cases, which would continue to be dealt with under established mechanisms for allocation based on need and medical indication.

4 Potential Objections

Here we briefly seek to anticipate certain objections that could be raised to the approach outlined. First, at the level of meta-ethics, a criticism may be that, in focusing on the requirements of justice, we have left out of account other possible ethical bases for health care allocation decisions: thus, for example, in consequentialist terms, it is clear our approach may lead to sub-optimal results (a patient above the 'fair age' who has a good chance of benefitting from treatment may be denied it in favour of a patient yet to reach it with a far lower chance of doing so) (Veatch, 1991). Equally, from a deontological standpoint, our model may be criticized for ignoring the curative treatment needs of those patients above the 'fair age'. However, we do not see either point as a difficulty for our argument: from the perspective of a just health distributor, the equal prior interest of all in health care as a basic good simply trumps contingent matters of individual potential to benefit ex post facto. Nor (*pace* Harris) do we believe it is 'ageist' to take account of a patient's exhaustion of their previous entitlement to curative treatment, on attaining the fair age, to deny future treatment: 'ageism' implies unjustified discrimination, whereas here it is simply that persons who reach the 'fair age' will (at least in well-resourced societies) be elderly. As noted, our approach is in any event without prejudice to private or charitable initiatives that aim to secure continuing health care for certain groups or individuals.

As was hinted at earlier, another challenge to our proposal comes in the form of an economic objection: the system disadvantages the poor, as the rich will be able to purchase life-prolonging treatments beyond the fair age, privately. The poor will die earlier than the rich: how can this be equitable? Here a different comparison is being invoked to that which underlines our central thesis, namely between the post-fair-age rich and the post-fair-age poor. In this regard, our response would be that it is more unjust for a pre-fair age person, rich or poor, to be denied treatment so as to maintain the health and life of the post-fair-age patient. More generally, we would argue that it is not within our proposal to address underlying socio-economic differences within society. That is a question of far broader scope and not a matter for the just distributor of health resources—as suggested earlier, the latter's task is to provide health care as one key necessary condition for individuals to flourish.

Nonetheless, this last problem brings us to a further related objection that could be raised against our model, namely its failure to consider issues of the subjective quality of life experienced by the individual patient (Jecker, 2022); instead, we work on a presumption of a universally applicable unit of reasonable healthy life that people enjoy up until reaching the fair age. This could be regarded as unfair to those, who though they have now attained the fair age, have enjoyed less happy lives (e.g. due to low socio-economic status) compared to patients of that age whose lives have been happier. In this regard, the health economist Alan Williams in fact argued that past life quality should be relevant factor when determining an individual's future health care entitlement (Williams, 1997). The obvious difficulty with such an approach, though, is its subjectivity: how could one judge such an issue? We would also reiterate our view that in the context of health care allocation, the primary concern of the just distributor is not to ensure subjective quality of life, but to provide and allocate the resources that, by establishing the objective necessary conditions for a reasonably healthy and extended life, allow citizens to strive for this goal individually.

A final, and possibly in practice the most significant, objection to our approach we here consider concerns how it would impact on the individual relationship between doctor and patient. The physician's primary objective will be to attempt to cure the patient, and this subjective duty of care will not change significantly merely because the patient reaches the socially agreed fair age. Here, requiring the physician to deny such a patient the intervention that s/he was given just a day beforehand seems too onerous on the treating physician. Here there are two main responses we would make. First, we suggest that—in line with clinical practice—a distinction may be observed between patients at the fair age cut-off who are already receiving a course of interventionist treatment and those who are not. As regards, the first group, where there is an existing doctor-patient relationship, it may indeed be regarded as unconscionable to withdraw treatment 'overnight', so to speak. Instead, modalities for managing such situations would clearly need to be worked out, taking full account of the sensitivities involved, e.g. with a default position that life-prolonging treatment be maintained until a natural point is reached when a significant new active intervention is required on the patient. By contrast, as regards the second group, not yet in medical care and receiving treatment, the particular issue would not arise.

In the second place, as previously argued, the fact the health system (at a meso-level) cuts off reimbursement schemes for scarce resources expended on those over the fair age does not preclude private or charitably financed treatment. Nor does it prevent the physician from providing treatment to individuals on a pro bono basis. Moreover, as noted previously, all palliative treatment options will remain available, and indeed required in line with the second (and unqualified) goal of medicine to secure human dignity. Ultimately, choices about treatment options are made the whole time, many on the basis of what type of treatment the patient's health insurer reimburses.[3] This is merely an extension of that system. Borderline treatments, i.e. those which are either not clearly attributable to a curative or a palliative category, or those where the palliative treatment is costlier than the curative treatment, can and should be the subject of special exception rules.

5 Conclusion

In this paper, we propose a highly abstract approach to addressing finite health resources, based not on an equal distribution of those resources but on their just distribution. What our concept seeks to establish is not an exhaustive solution to the myriad intricate problems of allocating scarce resources in health care system. However, it suggests a plausible upper limit to the duty of a just state to provide health care resources to its

[3] See, e.g. Melichar (2009).

citizens in the light of the inescapable reality of having to make allocation decisions. We argue for a firm moral obligation to ensure that scarce resources are not expended on those who have already enjoyed their 'fair' share of quality life at the cost of others, who have yet to do so. In doing so, we adopt a political justice approach in the tradition of Rawls and Daniel Callahan, while providing a further (and in our view) simpler justification for age-based rationing. This framework deploys hard limits and blanket approaches, which may provoke legitimate questions about the pragmatics of implementing such a system. Nonetheless, the basic intuition underlying the approach is in our view a sound one, based on objective considerations, which would permit a significant freeing up of health resources in a manner acceptable to all.

Competing Interests The authors have no conflicts of interest to declare that are relevant to the content of this chapter.

References

Albertsen, A. (2023). Covid-19 and age discrimination: Benefit maximization, fairness, and justified age-based rationing. *Medicine, Health Care and Philosophy, 26*, 3–11.

Callahan, D. (1987). *Setting limits: Medical goals in an aging society*. Simon and Schuster.

Daniels, N. (1988). *Am I my parents' keeper?: An essay on justice between the young and the old*. Oxford University Press.

Harris, J. (1985). *The value of life*. Routledge.

Harris, J. (1987). QALYfying the value of life. *Journal of Medical Ethics, 13*(3), 117–123.

Harris, J., & Regmi, S. (2012). Ageism and equality. *Journal of Medical Ethics, 38*(5), 263–266.

Jecker, N. S. (2022). Too old to save? COVID-19 and age-based allocation of lifesaving medical care. *Bioethics, 36*, 802–808.

Kitzhaber, J. A. (1993). Prioritising health services in an era of limits: The Oregon experience. *British Medical Journal, 307*(6900), 373–377.

Melichar, L. (2009). The effect of reimbursement on medical decision making: Do physicians alter treatment in response to a managed care incentive? *Journal of Health Economics, 28*(4), 902–907.

Stahl, J. E., Tramontano, A. C., Swan, J. S., & Cohen, B. J. (2008). Balancing urgency, age and quality of life in organ allocation decisions—What would you do?: A survey. *Journal of Medical Ethics, 34*(2), 109–115.

Veatch, R. M. (1991). Equality, justice and rightness in allocating health care: A response to James Childress. In B. S. Kogan (Ed.), *A time to be born and a time to die* (pp. 205–216). De Gruyter.
Williams, A. (1997). Intergenerational equity: An exploration of the 'fair innings' argument. *Health Economics, 6*(2), 117–132.

Index

J. Hinrichsen et al. (eds.), *Ethics of the Sciences*,
https://doi.org/10.1007/978-981-92-2821-8

B

L

M

GPSR Compliance

The European Union's (EU) General Product Safety Regulation (GPSR) is a set of rules that requires consumer products to be safe and our obligations to ensure this.

If you have any concerns about our products, you can contact us on ProductSafety@springernature.com

In case Publisher is established outside the EU, the EU authorized representative is:

Springer Nature Customer Service Center GmbH
Europaplatz 3
69115 Heidelberg, Germany

Batch number: 10809399

Printed by Printforce, the Netherlands